Evolutionary Algorithms in Water Resources

In Focus – Special Book Series

Evolutionary Algorithms in Water Resources

Editors

Dasika Nagesh Kumar and Komaragiri Srinivasa Raju

PUBLISHING

Published by **IWA Publishing**
Unit 104–105, Export Building
1 Clove Crescent
London E14 2BA, UK
Telephone: +44 (0)20 7654 5500
Fax: +44 (0)20 7654 5555
Email: publications@iwap.co.uk
Web: www.iwapublishing.com

First published 2022

British Library Cataloguing in Publication Data
A CIP catalogue record for this book is available from the British Library

ISBN: 9781789063240 (paperback)

Contents

Evolutionary Algorithms in Water Resources
doi: 10.2166/h2oj.2020.001

Editorial: Evolutionary Algorithms in Water Resources

Dasika Nagesh Kumar[a] and Komaragiri Srinivasa Raju[b]

[a] Department of Civil Engineering, Centre for Earth Sciences (CEaS), Interdisciplinary Centre for Water Research (ICWaR), Divecha Centre for Climate Change (DCCC), Indian Institute of Science, Bangalore 560 012 India E-mail: nagesh@iisc.ac.in

[b] Department of Civil Engineering, Birla Institute of Technology and Science-Pilani, Hyderabad Campus, Hyderabad 500 078, India E-mail: ksraju@hyderabad.bits-pilani.ac.in

Evolutionary algorithms and allied fields are getting more visibility as well as familiarity due to their numerous flexibilities such as handling high-dimensional non-linear problems and more. Accordingly, IWA Publishing's *H2Open Journal* played a major role to initiate a special issue on Evolutionary Algorithms in Water Resources, which in our opinion will help budding researchers to formulate their research problems. Ten papers (three on optimization, five on machine learning algorithms, one on Internet of Things, one on remote sensing) are part of the special issue, details of which are as follows.

Janga Reddy and Nagesh Kumar (Reddy & Kumar 2020) reviewed extensively on Evolutionary Algorithms and Swarm Intelligence and their applications in water resources engineering. Their paper also highlighted challenges, improvements needed and opportunities for future applications. Gurav & Regulwar (2020) developed a Multi Objective Fuzzy Linear Programming based irrigation planning model to case study Jayakwadi Project Stage I, India. The proposed model effectively considered the uncertainty in decision parameters/variables. Nandi and Janga Reddy (Nandi & Reddy 2020) evaluated the performance of genetic algorithm (GA), shuffled complex evolution (SCE), differential evolution (DE), and self-adaptive DE (SaDE) algorithms for the parameter calibration of a variable infiltration capacity (VIC) model. The methodology is tested on the upper Tungabhadra River basin in India. It is concluded that SaDE facilitates an effective calibration of the VIC model with higher reliability and faster convergence to optimal solutions as compared to the other methods.

Bhar & Bakshi (2020) proposed a three-layered Feed-Forward Back Propagation (FFBP) based artificial neural network (ANN) for simulation of tidal water level for the Hooghly estuary, India. It is concluded that developed ANN models can be used to generate complete cycles of water level data at the remote station, for both spring and neap cycles. Chaudhary & Dhanya (2020) examined the applicability of Classification and Regression Tree (CART) for bias-correction of Integrated Multi-Satellite Retrievals for Global Precipitation Mission (IMERG) precipitation dataset over India. Better performance of the CART model over Linear Scaling (LS) and Equidistant Cumulative Distribution Matching (EDCDF) was observed over India. Moradi *et al.* (2020) used multivariate linear regression, support vector regression, and Gaussian process regression to predict variations in dissolved organic carbon (DOC) concentrations in an Australian catchment. Experimental results showed that the forecasting capability of the support vector regression model outperforms those of other kernel-based models, thereby generating more accurate results. Naidu *et al.* (2020) studied the hydrological responses under climate change using various machine learning algorithms over a semi-arid river basin in India. They used data driven algorithms and various downscaling algorithms. The annual average precipitation was predicted to increase by about 13.12% with temperature increase of 0.6 °C for the period of 2061–2080 compared to the observed period of 1951–1989 with MIROC Global Climate Model outputs. Samantaray & Ghose (2020) applied integration of support vector machine (SVM) with firefly algorithm (FFA) and phase space reconstruction (PSR)

(SVM-FFA-PSR), SVM, SVM-FFA for runoff prediction for five watersheds of Balangir, Odisha, India and found that SVM-FFA-PSR is satisfactory.

Krishnaveni *et al.* (2020) developed an Internet of Things (IoT) based device by integrating sensors and microcontroller that send the real time tank water level data to the cloud. Methodology is applied to Melnelli big tank in Arcot taluk of Vellore district, TamilNadu, India for remote monitoring of tank water level dynamics.

Surwase *et al.* (2020) developed a mechanism to compare the HEC-RAS flood model with the flood footprints extracted from Synthetic Aperture Radar (SAR) image using multi-segmentation and Otsu's thresholding technique. Test data are flood witnessed in Mahanadi River in Odisha stretched between Tikarpara and Mundali during September 2008 and found that model result is satisfactory.

We thank the authors and reviewers for their support and are thankful to *H2Open Journal*, IWA Publishing, who supported the special issue proposal. Special thanks to Emma Buckingham and Emma Gulseven for excellent co-ordination.

REFERENCES

Bhar, K. K. & Bakshi, S. 2020 Application of artificial neural network for predicting water levels in Hooghly estuary, India. *H2Open Journal* **3** (1), 401–415. DOI: 10.2166/h2oj.2020.041

Chaudhary, S. & Dhanya, C. T. 2020 Decision tree-based reduction of bias in monthly IMERG satellite precipitation dataset over India. *H2Open Journal* **3** (1), 236–255. DOI: 10.2166/h2oj.2020.124

Gurav, J. B. & Regulwar, D. G. 2020 Multi-objective fuzzy optimization for sustainable irrigation planning. *H2Open Journal* **3** (1), 373–389. DOI: 10.2166/h2oj.2020.032

Krishnaveni, M., Kumar, S. K. P., Muthusamy, E. A., Kowshick, J. & Arunya, K. G. 2020 Real-time monitoring of water level and storage dynamics of irrigation tank using IoT. *H2Open Journal* **3** (1), 392–400. DOI: 10.2166/h2oj.2020.123

Moradi, S., Agostino, A., Gandomkar, Z., Kim, S., Hamilton, L., Sharma, A., Henderson, R. & Leslie, G. 2020 Quantifying natural organic matter concentration in water from climatological parameters using different machine learning algorithms. *H2Open Journal* **3** (1), 328–342. DOI: 10.2166/h2oj.2020.035

Naidu, G. S., Pratik, M. & Rehana, S. 2020 Modelling hydrological responses under climate change using machine learning algorithms – semi-arid river basin of peninsular India. *H2Open Journal* **3** (1), 481–498. DOI: 10.2166/h2oj.2020.034

Nandi, S. & Reddy, M. J. 2020 Comparative performance evaluation of self-adaptive differential evolution with GA, SCE and DE algorithms for the automatic calibration of a computationally intensive distributed hydrological model. *H2Open Journal* **3** (1), 306–327. DOI: 10.2166/h2oj.2020.030

Reddy, M. J. & Kumar, D. N. 2020 Evolutionary algorithms, swarm intelligence methods, and their applications in water resources engineering: a state-of-the-art review. *H2Open Journal* **3** (1), 135–188. DOI: 10.2166/h2oj.2020.128

Samantaray, S. & Ghose, D. K. 2020 Modelling runoff in an arid watershed through integrated support vector machine. *H2Open Journal* **3** (1), 275–235. DOI: 10.2166/h2oj.2020.005

Surwase, T., Manjusree, P., Prakash, S. & Kuntla, S. 2020 Development of algorithms for evaluating performance of flood simulation models with satellite-derived flood. *H2Open Journal* **3** (1), 222–235. DOI: 10.2166/h2oj.2020.117

Evolutionary Algorithms in Water Resources
doi: 10.2166/h2oj.2020.128

Invited Review

Evolutionary algorithms, swarm intelligence methods, and their applications in water resources engineering: a state-of-the-art review

M. Janga Reddy[a,*] and D. Nagesh Kumar[b]

[a] Department of Civil Engineering, Indian Institute of Technology Bombay, Mumbai 400076, India

[b] Department of Civil Engineering, Indian Institute of Science, Bangalore 560012, India

*Corresponding author. E-mail: mjreddy@civil.iitb.ac.in

Abstract

During the last three decades, the water resources engineering field has received a tremendous increase in the development and use of meta-heuristic algorithms like evolutionary algorithms (EA) and swarm intelligence (SI) algorithms for solving various kinds of optimization problems. The efficient design and operation of water resource systems is a challenging task and requires solutions through optimization. Further, real-life water resource management problems may involve several complexities like nonconvex, nonlinear and discontinuous functions, discrete variables, a large number of equality and inequality constraints, and often associated with multi-modal solutions. The objective function is not known analytically, and the conventional methods may face difficulties in finding optimal solutions. The issues lead to the development of various types of heuristic and meta-heuristic algorithms, which proved to be flexible and potential tools for solving several complex water resources problems. This paper provides a review of state-of-the-art methods and their use in planning and management of hydrological and water resources systems. It includes a brief overview of EAs (genetic algorithms, differential evolution, evolutionary strategies, etc.) and SI algorithms (particle swarm optimization, ant colony optimization, etc.), and applications in the areas of water distribution networks, water supply, and wastewater systems, reservoir operation and irrigation systems, watershed management, parameter estimation of hydrological models, urban drainage and sewer networks, and groundwater systems monitoring network design and groundwater remediation. This paper also provides insights, challenges, and need for algorithmic improvements and opportunities for future applications in the water resources field, in the face of rising problem complexities and uncertainties.

Key words: evolutionary algorithms, optimization, swarm intelligence, water resources

Highlights

- This paper reviews working principles and applications of evolutionary algorithms.
- This paper reviews working principles and applications of swarm intelligence algorithms.
- Discusses the issues and merits of various meta-heuristic algorithms.
- Provides insights and future research directions.

doi: 10.2166/h2oj.2020.128

Graphical Abstract

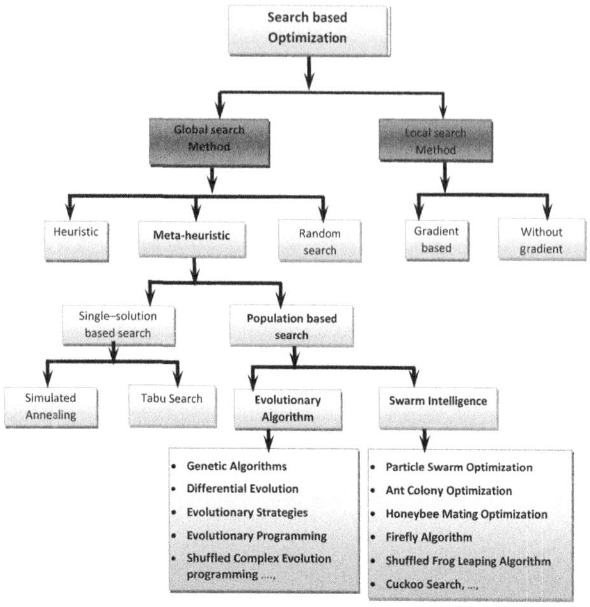

INTRODUCTION

Over the last few decades, there has been rising concern about global warming and associated changes in rainfall, streamflows, and water availability in the river basins. Many regions are often facing a shortage of water, as they receive rainfall in a particular season only, but the water demands have to be satisfied for the entire year. There is an ever-increasing demand for water to meet the diverse needs of society majorly for domestic, industrial, and agricultural purposes. Also, the efficient use of limited water for different users imposes substantial difficulties with conflicting goals (Reddy & Kumar 2006). Therefore, planning, construction, development, and operational activities of water resources projects warrant for solutions using systematic procedures. They can help planners to develop improved designs and operational systems, decide innovative management policies, improve and calibrate simulation models, and resolve conflicts between conflicting stakeholders (Maier *et al.* 2014). The complexity of systems models in water resources engineering has increased tremendously, with several socio-environmental–ecological issues and requires better alternative methods. In general, systems aim to reduce the total system cost or failure risk, maximize net benefits by providing an efficient design or operation policy. One of the important engineering tools that can be used in such events is the optimization tool, which helps to find a set of values of the decision variables subject to the various constraints that will produce the desired optimum response for the chosen objective function. As computers have become more powerful, the size and complexity of problems that can be simulated and solved by optimization techniques have correspondingly expanded.

Today there are a variety of optimization techniques existing to tackle different issues in practical problems of water resources. Some techniques (like exact methods) may provide optimal solutions for smaller problems, and others like meta-heuristic techniques may provide near-optimal solutions while solving large-scale water resources problems. A taxonomy or classification of optimization methods is given in Figure 1. But no single optimization method or algorithm is unanimously declared as the winner that can be applied efficiently to all types of problems. The method chosen for any particular case will depend primarily on (Reddy & Kumar 2012): (1) complexity of the problem, and the character of the objective function whether it is known explicitly, (2) the number and nature of the constraints, equality and inequality constraints, (3) the number of continuous and discrete variables, etc.

doi: 10.2166/h2oj.2020.128

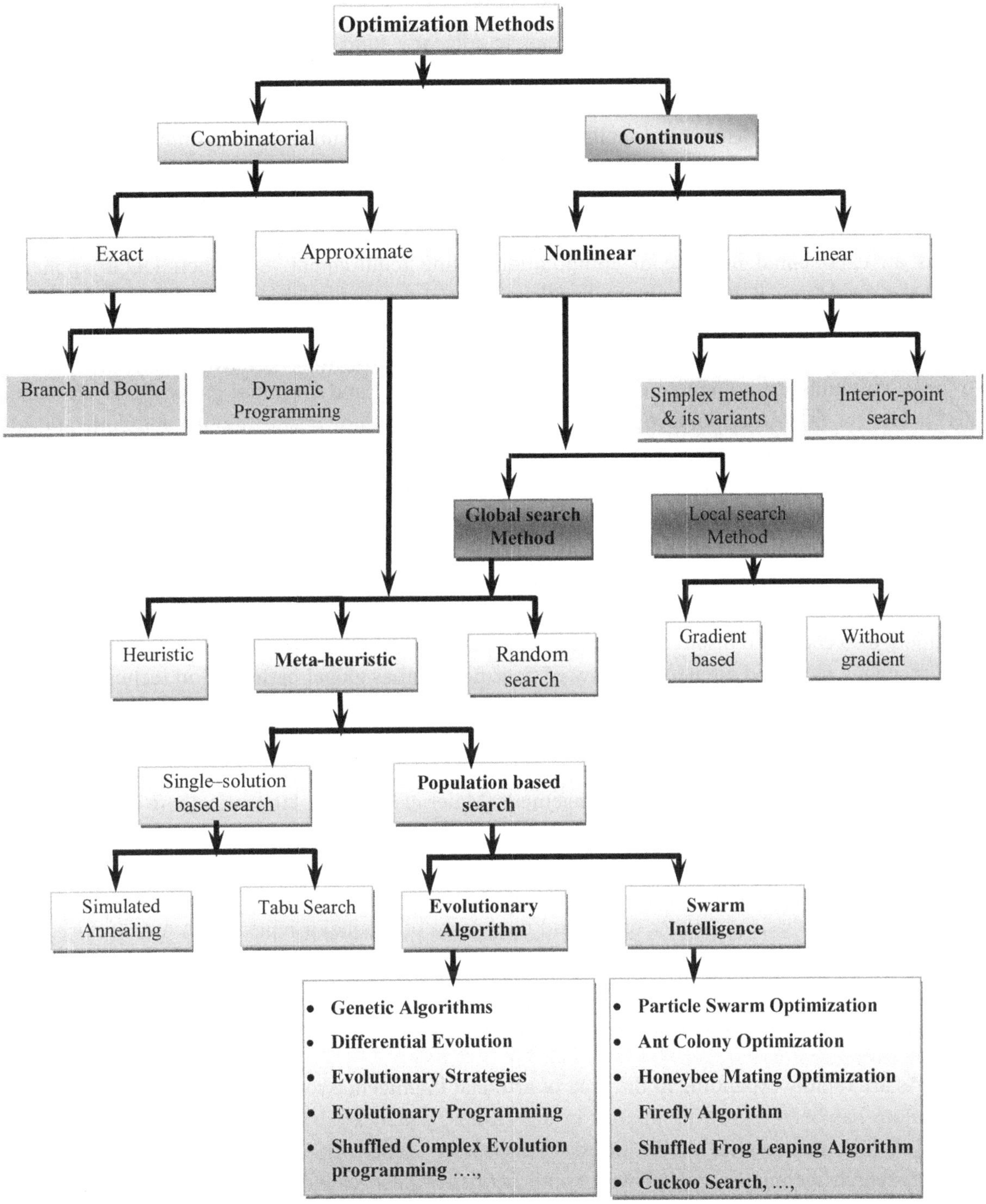

Figure 1 | Taxonomy of optimization methods.

The commonly used methods in water resources include linear programming (LP), dynamic programming (DP), and nonlinear programming (NLP) methods. The LP method can guarantee global-optimal solutions for linear problems and has wider applications in water resources, like for irrigation planning, reservoir operation, conjunctive use planning, crop water allocation, seawater intrusion, command area management, etc. (Yeh 1985). But many practical water resources applications may involve nonlinear functions in optimization modeling for solving the problems. So the popular LP method cannot work in the case of models with nonlinear functions. The DP method

(Bellman 1957) is popularly used for solving sequential decision making or multi-state decision-making problems in water resources. It can handle any kind of functional relationships in the model and can provide optimal solutions based on chosen interval values for the state and decision variables. The main applications in water resources are water allocation, reservoir operation, capacity expansion of water infrastructural facilities, water conveyance/shortest route-finding problems, etc. (Yakowitz 1982). Being a complete enumeration technique, the DP faces computational difficulties while solving large-size problems due to an increase in the number of state variables and the corresponding discrete states, since in the DP method, a linear increase in the number of state variables causes an exponential increase in computational time requirement. So, when DP is applied to larger-size problems, it has the main hurdle of the 'curse of dimensionality'. The gradient-based NLP methods can solve problems with smooth nonlinear objectives and constraints. However, in large and highly nonlinear models, these algorithms may fail to find feasible solutions or converge to local optimum depending upon the degree of nonlinearity and an initial guess (Reddy & Kumar 2012). Hence, these traditional optimization techniques do not ensure global optimum and also have other limitations like requirements of objective functions to be continuous functions and easily differentiable, continuous variables, etc. Lack of ability to obtain a global optimum in the case of traditional nonlinear-optimization techniques and intensity of computational requirements in the case of dynamic programming motivated the search for new approaches, which would conglomerate efficiency and ability to find the global optimum.

In the recent past, nontraditional search and optimization methods based on natural and biological evolution, also called bio-inspired techniques, such as EA and swarm intelligence (SI) algorithms have been receiving increased attention in view of their potential as global optimization techniques for solving complex problems in water resources engineering (Reddy & Kumar 2012). Since, the first applications of genetic algorithms in the water resources area (McKinney & Lin 1994; Ritzel *et al.* 1994) and their acceptance as optimizers have increased tremendously for several practical applications in the water resources planning and management (Maier *et al.* 2014). However, there is a lack of synthesis between common algorithm challenges, common problem behaviors, and needed improvements for different key applications in the field. This paper will help the researchers to comprehend the algorithms and their applications in the planning and management of water resource systems. In the following first, the basic principles of EA and then some of the major types of EA are discussed.

EVOLUTIONARY ALGORITHMS

The EAs are rapidly expanding in the area of artificial intelligence research. During the last two decades, there has been a growing interest in algorithms, which are based on the principle of natural evolution (i.e., the survival of the fittest) (Fogel *et al.* 1966). The EAs are the population-based random search techniques guided with some heuristics (also called as meta-heuristic techniques). The EAs consist of a population of individuals, each representing a search point in the space of feasible solutions and is exposed to a collective learning process which proceeds from generation to generation (Brownlee 2011). The population is randomly initialized and then subjected to the process of selection, recombination, and mutation through several generations, such that the newly created generations evolve towards more favorable regions of the search space. The progress in the search is achieved by evaluating the fitness of all individuals in the population, selecting the individuals with a better fitness value, and combining them to create new individuals with an increased likelihood of improved fitness. After some generations, the program converges, and the best individual represents the optimum (or near-optimum) solution. There exist several EAs, but the basic structure of any evolutionary algorithm is very much the same (Reddy & Kumar 2012). A sample structure is shown in Figure 2.

doi: 10.2166/h2oj.2020.128

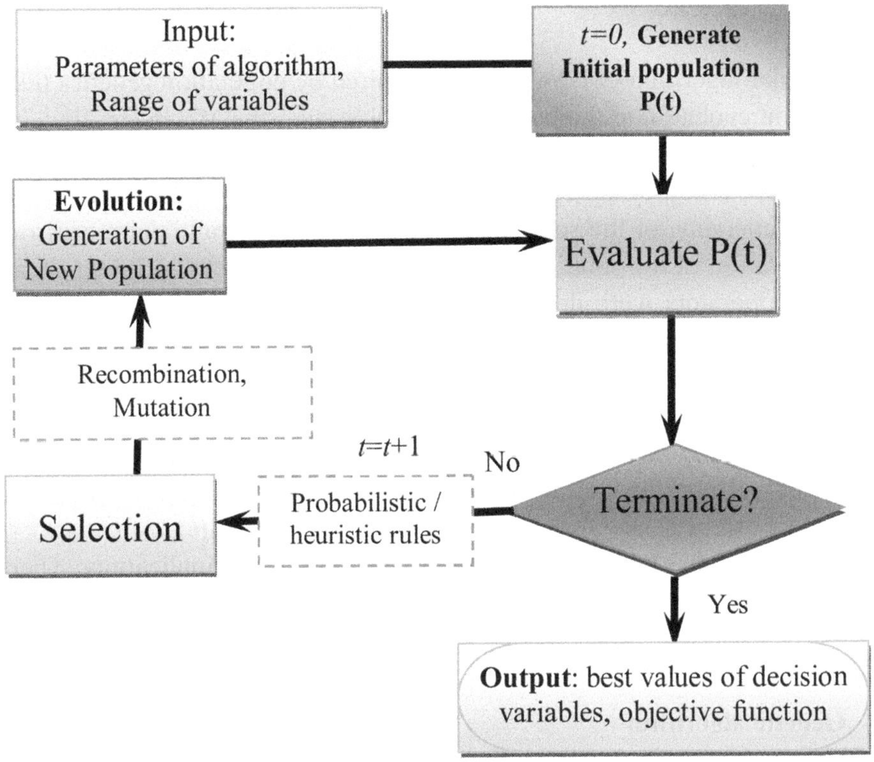

Figure 2 | The basic structure of an evolutionary algorithm (EA).

The key steps involved in EA include:

1. Seeding the population using random generation
2. Evaluate the fitness of each individual in the population
3. Repeat the evolution steps until stopping criterion satisfied:
   (a) Select the individuals for reproduction
   (b) Perform genetic operations to generate the offspring
   (c) Evaluate the individual fitness of the offspring
   (d) Replace the least fit individuals with new best fit individuals
4. Report the best solution of the fittest individual.

The two most important issues in the evolution process are population diversity and selective pressure. These factors are strongly related to each other, i.e., an increase in the selective pressure decreases the diversity of the population and vice versa. In other words, strong selective pressure 'supports' the premature convergence of the search and a weak selective pressure can make the search ineffective. Different evolutionary techniques use different scaling methods and different selection schemes (e.g., probabilistic or proportional selection, ranking, tournament) to strike a balance between these two issues. Furthermore, these algorithms can be easily combined with local search and other exact methods. In addition, it is often straightforward to incorporate domain knowledge in the evolutionary operators and in the seeding of the population. Moreover, EAs can handle problems with any combination of the challenges that may be encountered in real-world applications, such as local optima and multiple objectives. The popular EAs are such as genetic algorithm (GA), evolutionary strategies (ES), evolutionary programming (EP), differential evolution (DE), etc. Some studies also proposed hybrid systems by combining features of more than one EA, and they exhibited significant results in several water resources applications. In the following, a brief overview of popular EAs is given with an emphasis on showing how the various types of algorithms differ, and the stages involved in defining each one.

doi: 10.2166/h2oj.2020.128

Genetic algorithms

GA is the most popular algorithm of EAs. GA was inspired by population genetics (like heredity and gene frequencies), and evolution at the population level besides the Mendelian understanding of the structure (like chromosomes, genes, alleles) and process/mechanisms (like recombination and mutation) (Fogel *et al.* 1966; Goldberg 1989). The basic GA was introduced by Holland (1975) based on the binary encoding of the solution parameters, utilizes multi-point crossover and bit-flip mutation for the evolution of the solutions. Later, several variants of GA (binary/real coding) with a variety of genetic operators were developed and used in various water resources applications. The working of GA involves random initialization of population (i.e., members are randomly generated to cover the entire search space uniformly) to start the process. Then, evaluation of the objective functions, selection of parents, and applying genetic operations, recombination operator for creation of offspring, and mutation operation for perturbing the individuals to produce a new population are conducted. The steps are repeated until a termination condition is satisfied. A sample pseudo-code of GA, describing the key steps in the algorithm, is depicted in Figure 3(a).

Recently, real-coded GA has been receiving more recognition and applications. There exist several variants of crossover operators (arithmetic, simulated binary crossover, Blend crossover, etc.) and

(a) **Genetic algorithm**

begin
define problem dimension (D), generate initial population P_0 of size NP
evaluate the fitness of NP individuals in P_0
initialize generation counter $t \leftarrow 0$
while termination criterion not satisfied **do**
    selection of parents on the basis of their fitness
    create offspring for the new generation P_{t+1} from the parents using crossover operation
    perform mutation on new generation P_{t+1}
    evaluate the fitness of population P_{t+1}
    $t \leftarrow t+1$
    end
end

(b) **DE algorithm**

begin
define problem dimension, generate an initial population with a given size
calculate the fitness of the initial population.
initialize generation counter $g \leftarrow 0$
while termination criterion not satisfied **do**
    for each population in current generation **do**
        select three solutions at random.
        generate one offspring using the DE operators (mutation, crossover).
        if offspring is more fit than parent **then**
            the parent is replaced with offspring
        end
    end
    $g \leftarrow g+1$
end

Figure 3 | The pseudo-codes of evolutionary algorithms: (a) genetic algorithm (GA) and (b) differential evolution (DE).

mutation operators (e.g., Gaussian, polynomial, random mutation, etc.). The basic selection operator for GAs was proportional (or roulette-wheel) selection, but because of its known drawbacks of premature convergence to locally optimal solutions, tournament selection and ranking selection are commonly used nowadays. For numerical optimization, a real-coded GA with Gaussian mutation, arithmetic crossover, and tournament selection is a common choice. Moreover, an operation called elitism is remarkably important for the performance of a GA. The usage of elitism is to leave a certain proportion of the best individuals in every generation untouched by the variation operators. This is to some extent, similar to evolution strategies, where a population of parents generates a new offspring by a mutation in each iteration. The population of the next generation is created by selection from the elite parents and newly created offspring. Nicklow *et al.* (2010) presented an overview of the GA method and its applications in water resources management. By utilizing the strengths of EAs, quick convergence, and yielding efficient solutions for single-objective optimization, researchers also developed multi-objective algorithms by integrating Pareto optimality principles into single-objective genetic algorithms, such as Nondominated Sorting Genetic Algorithms (NSGA-II; Deb *et al.* 2002), Multi-objective Evolutionary Algorithm (MOEA; Reddy & Kumar 2006), etc. Reed *et al.* (2013) discussed the principles of different MOEAs methods and their applications in water resources.

Different variants of GA were developed over the years, like Micro GA (Krishnakumar 1990), Cellular GA (Manderick & Speissens 1989), NSGA (Srinivas & Deb 1994), Contextual GA (Rocha 1995), Grouping GA (Falkenauer 1996), Quantum-inspired GA (Narayanan & Moore 1996), Linkage learning GA (Harik 1997), Island GA (Whitley *et al.* 1998), NSGA-II (Deb *et al.* 2002), Interactive GA (Takagi 2001), Jumping gene GA (Man *et al.* 2004), Dynamic rule-based GA (He & Hui 2006), Hierarchical cellular GA (Janson *et al.* 2006), NSGAIII (Deb & Jain 2014), Tribe competition-based GA (Ma & Xia 2017), Fluid GA (Jafari-Marandi & Smith 2017), Block-based GA (Tseng *et al.* 2018), etc. Historical development of GA variants is also depicted in Figure 4. Although many of these variants use the same basic principles of natural selection and survival of the fittest, they engage different strategies and improved mechanisms in pursuit of better guidance of the search and aiding the enhanced convergence of the method. For example, for multi-objective optimization, the initial version of NSGA (proposed in 1994) was improved over the years and later proposed NSGA-II in 2002 and NSGAIII in 2014 by incorporating additional mechanisms (like nondominated sorting, crowding distance measures, etc.) to handle different issues and complexities of multi-objective optimization problems.

Differential evolution

Storn & Price (1996) proposed DE as a variant of EAs to achieve the goals of robustness in optimization and quick convergence to an optimal solution for numerical optimization. The DE contrasts from other EAs in the evolution process. Here mutation is the main operator, and the crossover is the secondary operator for the generation of new solutions (Reddy & Kumar 2012). After random initialization of the population, the objective functions are evaluated, and the following steps are repeated until a termination condition is satisfied. At each generation, two operators, namely mutation and crossover, are applied to each individual, to produce a new population. In DE, the mutation is the main operator, and each individual is updated using a weighted difference of a selected parent solution and crossover acts as background operator where the crossover is performed on each of the decision variables with a small probability. The offspring replaces the parent only if it improves the fitness value; otherwise, the parent is copied in the new population. The pseudo-code of the DE algorithm is given in Figure 3(b).

There are several variants of DE, depending on the number of weighted differences between solution vectors considered for perturbation, and the type of crossover operator (binary or

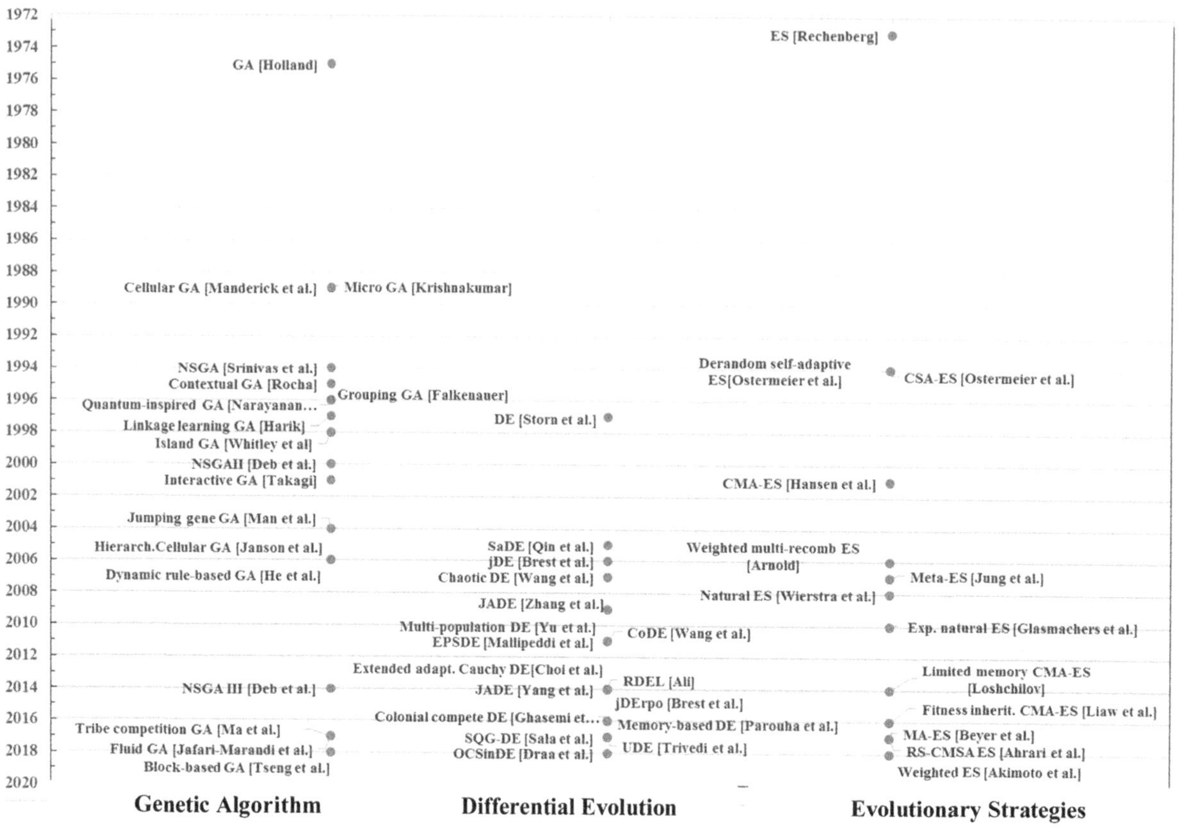

Figure 4 | Evolution of different evolutionary algorithms and their variants developed over the years, showing for genetic algorithms (GA), differential evolution (DE), and evolution strategies (ES).

exponential) used (Storn & Price 1997). For example, in *DE/rand-to-best/1/bin* variant of DE: perturbation is made with the vector difference of best vector of the previous generation (*best*) and current solution vector, plus single vector differences of two randomly chosen vectors (*rand*) among the population. The DE variant uses the binomial (*bin*) variant of crossover operator, where the crossover is performed on each of the decision variables whenever a randomly picked number between 0 and 1 is within the crossover constant (*CR*) value. More details of DE can be found elsewhere (Price *et al.* 2005; Das *et al.* 2016). To generate Pareto-optimal solutions for multi-objective problems, researchers also developed different variants of multi-objective differential evolution algorithms (MODE; Reddy & Kumar 2007c).

Different variants of DE were developed over the years, like DE (Storn & Price 1997), Self-adaptive DE (SaDE; Qin & Suganthan 2005), jDE (Brest *et al.* 2006), Chaotic DE (Wang & Zhang 2007), adaptive DE with optional external archive (JADE; Zhang *et al.* 2009), ensemble of mutation strategies DE (EPSDE; Mallipeddi *et al.* 2011), Composite DE (CoDE; Wang *et al.* 2011b), Multi-population DE (Yu & Zhang 2011), Adaptive Cauchy DE (ACDE; Choi *et al.* 2013), improved JADE (Yang *et al.* 2014), Extended adaptive Cauchy DE (Choi & Ahn 2014), jDErpo (Brest *et al.* 2014), Restart DE algorithm with Local search mutation (RDEL; Ali 2014), Colonial competitive DE (Ghasemi *et al.* 2016), Memory-based DE (Parouha & Das 2016), Stochastic Quasi-Gradient (SQG)-DE (Sala *et al.* 2017), Unified DE (UDE; Trivedi *et al.* 2017), Opposition-based Compound Sinusoidal DE (OCSinDE; Draa *et al.* 2019), etc. The evolution of various DE variants is also depicted in Figure 4. The different variants of DE use different improved strategies and self-adaptive schemes for enhancing the convergence and consistency in solutions for single (or multiple) objective optimization problems. More details can be found in the referred papers.

doi: 10.2166/h2oj.2020.128

Evolutionary strategies

ES model evolution as a process of the adaptive behavior of the individual or in other words ES focus mutational transformations that maintain the behavioral linkage between each parent and its off-spring, respectively, at the level of the individual (Rechenberg 1973; Fogel 1994). ES uses real variables and aims at numerical optimization. Because of that, the individuals incorporated could be a set of strategic parameters. ES rely mainly on the mutation operator (Gaussian noise with zero means). ES evolves by making a series of discrete adjustments (i.e., mutations) to an experimental structure. After each adjustment, the new structure, i.e., the offspring, is evaluated and compared to the previous structure, i.e., the parent. The better of the two is then chosen and used in the next cycle. As selection in this evolutionary cycle is made from one parent and one offspring, the algorithm is known as a '(1 + 1)' ES.

These two-membered ES modify (i.e., mutate) an n-dimensional real-valued vector $x \in \Re^n$ of object variables by adding a normally distributed random variable with expectation zero and standard deviation σ to each of the object variables x_i. The standard deviation is the same for all components of x, i.e., $\forall i \in \{1, 2 \ldots, n\}$: $x'_i = x_i + \sigma N_i(0, 1)$, where x' is the offspring of x and $N_i(0; 1)$ is the realization of a normally distributed random variable with expectation 0 and standard deviation 1. Since the introduction of ES, two additional strategies have been developed: $(\mu + \lambda)$ and (μ, λ). Both of these ES work on populations rather than single individuals and are referred to as multi-membered ES. A $(\mu + \lambda)$ ES creates λ off-springs from μ parents and selects the best μ individuals from the combined set of μ parents plus λ off-springs to make the next population. A (μ, λ) ES, on the other hand, creates λ off-springs and selects the best μ individuals from the off-springs alone (for $1 \leq \mu \leq \lambda$).

Different variants of ES were developed over the years, like Derandomized Self-adaptation ES (Ostermeier *et al.* 1994a), CSA-ES (Ostermeier *et al.* 1994b), CMA-ES (Hansen & Ostermeier 2001), Weighted multi-recombination ES (Arnold 2006), Meta-ES (Jung *et al.* 2007), Natural ES (Wierstra *et al.* 2008), Exponential natural ES (Glasmachers *et al.* 2010), Limited memory CMA-ES (Loshchilov 2014), Fitness inheritance CMA-ES (Liaw & Ting 2016), RS-CMSA ES (Ahrari *et al.* 2017), MA-ES (Beyer & Sendhoff 2017), Weighted ES (Akimoto *et al.* 2018), etc. The historical development of various ES variants is also depicted in Figure 4. The different variants use different strategies/adaptation schemes for better evolution and enhanced performance of the ES algorithm while solving a different kind of optimization problems. Apart from these, other EAs and their hybrid variants were proposed and used in solving water resources problems.

SWARM INTELLIGENCE

The other class of meta-heuristic techniques that are gaining more popularity in recent times for water resources optimization are SI techniques. The SI is based on the claims that intelligent human cognition derives from the interaction of individuals in a social environment. There exist several algorithms that use this socio-cognition, which can be used to solve different optimization tasks (Bonabeau *et al.* 1999). The individual members of a swarm act without supervision, and each of these members has a stochastic behavior due to their perception in the neighborhood. Swarms use their environment and resources effectively by collective group intelligence. The key characteristic of a swarm system is self-organization, which helps in evolving global level response by means of local-level interactions (Reddy 2009). The SI methods are also called behaviorally inspired algorithms.

The main algorithms that fall under SI algorithms include particle swarm optimization (PSO), artificial bee colony (ABC), ant colony optimization (ACO), honey-bee mating optimization (HBMO), firefly algorithms, etc. Similar to EAs, SI models are population-based iterative procedures. The system is randomly initialized with a population of individuals. These individuals are then manipulated and evolved over many iterations by way of mimicking the social behavior of insects or animals in an effort to find the optima. Unlike EAs, SI algorithms do not use evolutionary operators such as recombination and mutation. Basically, a potential solution flies through the search space by modifying itself according to its relationship with other individuals in the population and the environment (Reddy & Kumar 2012). The two algorithms that attracted the interest of many researchers and received wider applications in water resources are PSO and ACO for solving a variety of problems. PSO is based on the social behavior of fish schooling and bird flocking introduced by Eberhart & Kennedy (1995) and has received wider recognition for numerical optimization with continuous variables whereas ACO is basically inspired by the foraging search behavior of real ants and their ability to find shortest paths and was mainly used for discrete combinatorial optimization (Kennedy *et al.* 2001). In the following, a brief description of the basic principles and working of these two SI techniques is presented.

Particle swarm optimization

PSO algorithm proposed by Eberhart & Kennedy (1995) is a population-based meta-heuristic search technique that uses co-operative group intelligence concepts. Here the particle denotes individual in a swarm. Each particle in a swarm behaves in a distributed way using its own or cognitive intelligence and the collective or social (group) intelligence of the swarm. As such, if one particle discovers a good path to food, the rest of the swarm will also be able to follow the good path instantly even if their location is far away in the swarm. PSO shares many similarities with GA (Kumar & Reddy 2007). PSOs are initialized with a population of random solutions and searches for optima by updating iterations. However, in comparison to methods like GA, in PSO, no operators inspired by natural evolution are applied to extract a new generation of candidate solutions. Instead, PSO relies on the exchange of information between individuals (particles) of the population (swarm). In effect, each particle adjusts its trajectory towards its own previous best position and towards the best previous position attained by any other member of its neighborhood (usually the entire swarm) (Kennedy *et al.* 2001).

The PSO algorithm involves the following steps (Kumar & Reddy 2007): initialization of particles with a random position and velocity vectors. Then the fitness of each particle is evaluated by the fitness function. Two 'best' values are defined, the global and the personal bests. The global best is the highest fitness value in an entire iteration (best solution so far), and the personal best is the highest fitness value of a specific particle. Each particle is attracted to the location of the 'best fitness achieved so far' across the whole swarm. In order to achieve this, a particle stores the previously reached 'best' positions in a cognitive memory. The relative 'pull' of the global and the personal best is determined by the acceleration constants called social and cognitive parameters. After this update, each particle is then re-evaluated. If any fitness is greater than the global best, then the new position becomes the new global best. If the particle's fitness value is greater than the personal best, then the current value becomes the new personal best. This procedure is repeated until the termination criteria are satisfied. The pseudo-code of the PSO algorithm is given in Figure 5(a). Further to speed up the convergence and to enhance the reliability in optimal solutions, different studies suggested additional mechanisms, like elitist mutation strategy (Reddy & Kumar 2007a), combining PSO with other local search methods and applied for different kinds of problems in water resources. By utilizing the strengths like faster convergence and efficient optimal solutions for single-objective optimization, researchers also developed multi-objective SI algorithms by integrating nondominance principles into single-objective PSO, for example, elitist-mutated multi-objective PSO (EM-MOPSO; Reddy & Kumar 2007b), etc.

There were several variants of PSO algorithms, and their hybrid algorithms developed over the years, like Constricted PSO (Shi & Eberhart 1998), Adaptive PSO (Clerc1999), Discrete PSO (Clerc 2004), Elitist-mutated PSO (EMPSO) (Reddy 2006), EM-MOPSO (Reddy 2006), Dynamic niching PSO (Nickabadi *et al.* 2008), Adaptive PSO (Zhan *et al.* 2009), Co-evolutionary MOPSO (Goh

doi: 10.2166/h2oj.2020.128

(a) **Particle swarm optimization algorithm**

```
begin
Initialize population: swarm positions X(0) and velocities V(0)
evaluate the fitness of the individuals
initialize iteration counter t ← 0
while termination criterion not satisfied do
    Find personal best of each particle Pbest(t)
    Find global best among all the particles Gbest(t)
    Compute velocity V(t +1)
    Compute position X(t +1)
    Restrict the positions to pre-defined range
    evaluate the fitness of the population
    t ← t+1
End while
end
```

(b) **Ant colony optimization algorithm**

```
begin
Input: size of ants, parameters of algorithms
Structure the given problem as a graph of multi-node/multi-stage problem,
Select heuristic information, pheromone updating rules
initialize iteration counter t ← 0
while termination criterion not satisfied do
    Position each ant in a starting node
    Repeat
        for each ant do
            Choose next node by applying the state transition rule
            Apply step by step pheromone update
        End for
    Until every ant has built a solution
    calculate the fitness and update best solution
    Apply offline pheromone update
    t ← t+1
End while
end
```

Figure 5 | The pseudo-codes of swarm intelligence algorithms: (a) particle swarm optimization (PSO), (b) ant colony optimization (ACO).

et al. 2010), Self-adaptive learning PSO (Wang *et al.* 2011a), Multi-dimensional PSO (Kiranyaz *et al.* 2011), Hybrid niching PSO (NPSO; Li *et al.* 2012), Hybrid PSO-Harmony Search (PSO-HS; Li *et al.* 2012), Hybrid PSO-Firefly Algorithm (PSO-FFA; Bhushan & Pillai 2013), etc. The historical development of the PSO algorithm and its variants are depicted in Figure 6.

Ant colony optimization

The first ACO algorithm was inspired by the foraging behavior exhibited (pheromone trail laying and training behavior) by ant colonies in their search for food (Dorigo *et al.* 1991). ACO was developed as a population-based, heuristic search technique for the solution of difficult combinatorial and complex problems. The main features of the ACO algorithm are pheromone trail and heuristic information (Kumar & Reddy 2006). The working of the ACO algorithm involves the following phases. First, the system is randomly initialized with a population of individuals. These individuals are then manipulated over many iterations by using some guiding principles in their search, such as a probability function based on the relative weighting of pheromone intensity and heuristic information),

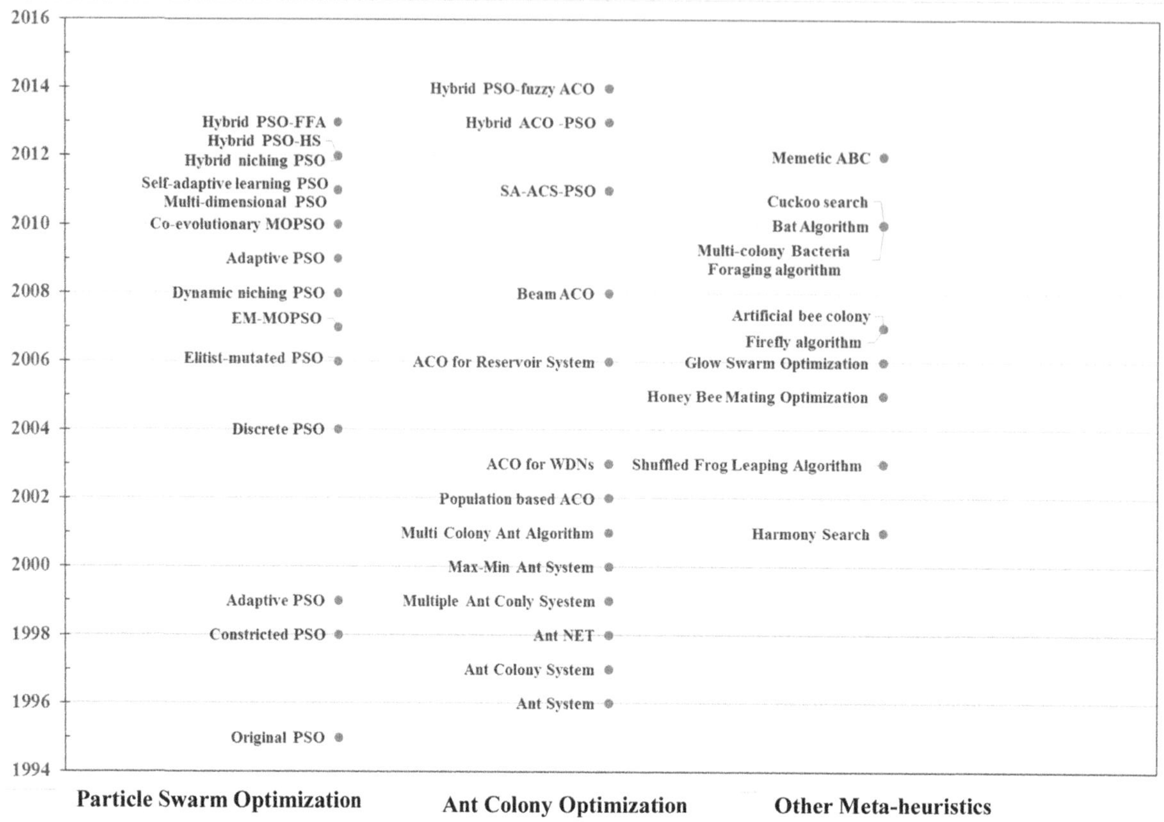

Figure 6 | Evolution of different swarm intelligence algorithms and their different variants developed over the years, showing for particle swarm optimization (PSO), ant colony optimization (ACO), and artificial bee colony (ABC) and other meta-heuristics.

in an effort to find the optima. At the end of each iteration, each of the ants adds pheromone to its path (set of selected options). The amount of pheromone added is proportional to the quality of the solution (for example, in the case of minimization problems, lower-cost solutions are better; hence they receive more pheromone). The pseudo-code of the ACO algorithm, depicting the key steps, is given in Figure 5(b).

An important characteristic of ACO one should be aware of is that it is a problem-dependent application. In order to adopt ACO for application to a particular problem, it requires representation of the problem as a graph or a similar structure easily covered by ants and assigning a heuristic preference to generated solutions at each time step. The ACO has many features, which are similar to that of GA (Dorigo & Stutzle 2004; Kumar & Reddy 2006): (a) both are population-based stochastic search techniques; (b) GA works on the principle of survival of the fittest, whereas ACO works on pheromone trail laying behavior of ant colonies; (c) GA uses crossover and mutation as prime operators in its evolution for next generation, whereas ACO uses pheromone trail and heuristic information; (d) in ACO algorithms, trial solutions are constructed incrementally based on the information contained in the environment and the solutions are improved by modifying the environment through a form of indirect communication called stigmergy, whereas in GA, the trial solutions are in the form of strings of genetic materials and new solutions are obtained through modification of the previous solutions.

There were several variants of ACO algorithms and their hybrid algorithms developed over the years, like Ant System (AS; Dorigo *et al.* 1996), Ant Colony System (ACS; Dorigo & Gambardella 1997), Ant NET (Di Caro & Dorigo1998), Max-Min AS (Stützle & Hoos 2000), Multiple ACS (Gambardella *et al.* 1999), Multi-Colony Ant Algorithms (Iredi *et al.* 2001), Population-based ACO for the dynamic environment (Guntsch & Middendorf 2002), ACO for WDNs (Maier *et al.* 2003), ACO for reservoir system (Reddy 2006), Beam-ACO (Blum 2008), hybrid genetic Simulated Annealing

doi: 10.2166/h2oj.2020.128

(SA) ACS-PSO (Chen & Chien 2011), Hybrid ACO–PSO (Huang *et al.* 2013), Parallel ACO (Delé-Vacq *et al.* 2013), Hybrid PSO-fuzzy ACO (Elloumi *et al.* 2014), etc. The evolution of ACO and its variants are also portrayed in Figure 6.

Other swarm-based meta-heuristics

The other meta-heuristic algorithms that were proposed and applied in various fields for optimization include Harmony Search (HS; Geem *et al.* 2001), Shuffled Frog Leaping Algorithm (SFLA; Eusuff & Lansey 2003), Honey-Bee Mating Optimization (HBMO; Abbass 2001), Glowworm Swarm Optimization (GSO; Krishnanand & Ghose 2006), Firefly algorithm (FFA, Yang 2007), ABC (Karaboga 2005), Cuckoo Search (CS, Yang & Deb 2010), Bat Algorithm (Yang 2010), Multi-colony Bacteria Foraging Optimization (MC-BFO; Chen *et al.* 2010), etc. Table 1 gives brief details of these SI-based meta-heuristic algorithms and their working principles. The evolution of these meta-heuristics over the years and their hybrid variants are also showed in Figure 6.

Table 1 | Swarm intelligence based meta-heuristic algorithms, development and their working principles

Algorithm	Proposed by year	Short description
Harmony Search (HS)	Geem *et al.* (2001)	Search works on the principle of a musician trying to identify a state of pleasing harmony and continuing to play the pitches to seek better harmony
Honey-Bee Mating Optimization (HBMO)	Abbass (2001)	The algorithm inspired by the mating process of bees
Shuffled Frog Leaping Algorithm (SFLA)	Eusuff & Lansey (2003)	The social behavior of frogs inspired SFLA
Artificial Bee Colony (ABC)	Karaboga (2005)	The algorithm simulates the foraging process of the bees
Glowworm Swarm Optimization	Krishnanand & Ghose (2006)	The search imitates the behavior that a glowworm carries a luminescence quantity (called luciferin) along with itself to exchange information with cohorts
Firefly Algorithm (FFA)	Yang (2007)	The algorithm inspired by the fireflies and their ability to emit light through the biochemical process (called bioluminescence)
Bat Algorithm (BA)	Yang (2010)	The algorithm inspired by the echolocation of bats
Cuckoo Search (CS)	Yang & Deb (2010)	The algorithm is inspired by the obligate brood parasitism of some cuckoo species by laying their eggs in the nest of host birds
Multi-colony Bacteria Foraging Optimization (MC-BFO)	Chen *et al.* (2010)	The algorithm integrates the cell-to-cell communication strategies of multi-colony bacterial community with the chemotaxis (optimal foraging search capabilities) behavior of single cell

The performance of meta-heuristic search methods is generally influenced by the parameter of the algorithm. Similar to EAs, these SI algorithms are also quite sensitive to set-up parameters (Reddy 2009). So it is important to fine-tune the parameters for a particular problem of interest before actually applying the same to the problem (Reddy & Kumar 2012).

APPLICATIONS

The EA and SI methods have emerged as a powerful tool for optimization and management of water resources problems. There are numerous applications of EAs for water-related problems, namely, reservoir operation, water distribution systems design, groundwater remediation, parameter estimation in hydrological modeling, watershed management, and fluvial systems, etc. Since there exist several thousands of papers on applications of these algorithms, here, some of the important applications in water resources are reviewed.

Applications in water distribution systems

WDS comprises a system of interconnected nodes, via pipes, supply sources, such as reservoirs, tanks, and a set of hydraulic control elements, such as pumps, valves, regulators, etc. The network of interconnected nodes, pipes, and other hydraulic control elements is collectively termed as a water distribution network (WDN). A typical WDN design is formulated as an optimization problem requiring minimization of cost, satisfying the minimum pressure and flow requirements at different nodes. A variety of EA were applied for design and rehabilitation of WDNs, like GA (Simpson *et al.* 1994; Mackle *et al.* 1995; Savic & Walters 1997; Dandy & Engelhardt 2001; Munavalli & Kumar 2003; van Zyl *et al.* 2004; Keedwell & Khu 2005; Vairavamoorthy & Ali 2005; Wu & Walski 2005; Rao & Salomons 2007; Kadu *et al.* 2008); Ant Colony optimization (ACO; Maier *et al.* 2003), Differential Evolution (DE) (Suribabu 2010; Vasan & Simonovic 2010; Sirsant & Reddy 2018), SFLA (Eusuff & Lansey 2003); Harmony search (HS) algorithm (Geem 2006), Tabu search (TS) algorithm (Cunha & Ribeiro 2004), Honey-bee mating optimization (HBMO; Mohan & Babu 2010) Cross entropy (CE) optimization (Shibu & Reddy 2014); Gravitational search algorithm (GSA, Fallah *et al.* 2019), etc. More details of these applications are given in Table 2.

Further, there were several studies that have used multi-objective EA for multi-objective optimization of WDNs. In order to ensure satisfactory performance of WDNs at different failure conditions, the objectives such as reliability, minimum surplus head, etc., are incorporated into the model in addition to the minimization of cost of the network. The reliability expressed as the performance of the network in terms of demand satisfaction considering these failure conditions. Failures can be hydraulic or mechanical; here, hydraulic failure occurs due to uncertainty in input parameters like nodal demands, and pipe roughness coefficients, whereas mechanical failure occurs due to failure of one or more components such as pipes, pumps, valves, etc. (Sirsant & Reddy 2018). Different reliability indicators are employed in different studies as the objective function to be maximized along with minimization of cost. Different variants of GA techniques were used for the multi-objective design of WDNs such as Multi-objective GA (MOGA; Halhal *et al.* 1997; Savic *et al.* 1997; Vamvakeridou-Lyroudia *et al.* 2005; Dandy & Engelhardt 2006), Strength Pareto Evolutionary Algorithm (SPEA; Cheung *et al.* 2003), Nondominated Sorting GA-II (NSGA-II; Prasad *et al.* 2004; Prasad & Park 2004; Farmani *et al.* 2005; Kapelan *et al.* 2005; Ostfeld & Salomons 2006; Jayaram & Srinivasan 2008; Prasad & Tanyimboh 2008; Creaco *et al.* 2014; Sirsant & Reddy 2020); Multi-objective PSO (Patil *et al.* 2020), etc. More details and discussion of these applications is given in Table 3.

Applications in urban drainage and sewer systems

Urban drainage and sewer systems need to be designed such that the required flow capacity is met at minimum cost. The consideration of networks where both stormwater and sewage are transported through the same channel makes the problem a little more complex. Various studies used different EAs for the design of urban drainage and sewer systems like GA (Walters & Lohbeck 1993; Walters & Smith 1995; Liang *et al.* 2004; Afshar *et al.* 2006; Guo *et al.* 2006) SA (Karovic & Mays 2014), TS (Liang *et al.* 2004), DE method (Yazdi 2018), etc. More details of these applications are explained in Table 4. In addition to carrying the required flow during normal conditions, the high and extreme flow conditions such as flooding overflow should be considered to make the system more robust to such situations. This calls for the need to perform the multi-objective design of these systems considering minimization of the flooding overflow volume or flood damage cost, in addition to the minimization of cost. Different MOEAs were engaged for solving these problems, such as NSGA-II (Barreto *et al.*

doi: 10.2166/h2oj.2020.128

Table 2 | Applications of evolutionary algorithms (EA), swarm intelligence (SI) meta-heuristic methods for single-objective optimization of water distributions systems

Author(s) (year)	Technique used	Work and objectives	Case studies/problems considered	Key findings
Simpson *et al.* (1994)	GA	Application of GA on WDN design problem considering maximization of the reciprocal of total network cost	WDN of Gessler (1985)	GA can provide an optimal and a set of sub-optimal solutions, while other techniques provide only a single solution. Also, complete enumeration and NLP methods can be applied only to small networks
Mackle *et al.* (1995)	GA	Application of GA for optimal pump scheduling for minimization of energy cost subject to reservoir filling and emptying constraints	Example WDN problem	The efficacy of GA proved for solving pump scheduling problems and noted that GA could offer a lot of new possibilities for solving these problems
Savic & Walters (1997)	GA	Development of GANET, and the application of GA for WDN design for minimization of cost of pipes	Benchmark WDN problems: Two-loop, Hanoi and New York Tunnel	The study inferred GA suitable for solving WDN design problems and performed better than LP, NLP, and enumeration methods
Dandy & Engelhardt (2001)	GA	Application of GA for the rehabilitation of WDNs involving the replacement of water supply pipes for the minimization of reciprocal of system cost (capital, repair and damage cost)	EL103N Zone network of Adelaide city	GA performed efficiently for solving rehabilitation problem which included increased complexities such as identifying solutions within budget limits and consideration of the diameter of replaced pipes as decision variables
Maier *et al.* (2003)	ACO, GA	Application of ACO and GA for WDNs and their performance, with the objective of minimization of total cost of WDNs	Two benchmark problems: 14-Pipe network, New York Tunnel WDNs	ACO algorithms outperformed GAs for the two case studies in terms of computational efficiency and their ability to find near global-optimal solutions
van Zyl *et al.* (2004)	GA	Development of a hybrid model by combining GA with a hill climber search method for minimization of cost of energy	Hypothetical WDN problem and a real WDN in the UK	The hybrid method performed significantly better than a GA method for the solution of WDN problems
Vairavamoorthy & Ali (2005)	GA	Use of pipe index vector as a measure of the relative importance of pipes in a network for reducing the search space and improving the efficiency of GA for minimization of capital cost	Alandur and Hanoi WDNs	Different pipes have various degrees of influence on the hydraulic performance of WDN. This information can be employed to reduce the search space and the efficiency of GA

(continued)

Table 2 | continued

Author(s) (year)	Technique used	Work and objectives	Case studies/problems considered	Key findings
Wu & Walski (2005)	GA	Self-adaptive penalty approach for tuning the penalty factor and thus improving the efficacy of the model for minimization of cost and maximization of benefits	Hanoi WDN	Self-adaptive penalty factor approach performed better than other constraint handling methods for GA, such as fixed, dynamic, annealing, niched and self-organizing penalty
Keedwell & Khu (2005)	GA	Proposed cellular automation for network design algorithm–Genetic Algorithm (CANDA-GA) approach for initial seeding of the GA algorithm and applied for WDN design with the objective of minimization of cost	Two-loop, two industry networks	CANDA-GA gave faster convergence of the algorithm and better solutions with the same computational efforts than traditional GA
Rao & Salomons (2007)	GA	The combined use of ANN for predicting the consequences of different pump and valve control settings and GA for selecting the best combination of those settings to minimize the energy cost of meeting the current and future demands	Hypothetical WDN	The GA–ANN model has reduced the run time significantly when compared to the hydraulic simulation model and noted that the meta-modeling could help for large complex networks
Munavalli & Kumar (2003)	GA	Application of niched operator and creep mutation GA for optimal scheduling of chlorine dosage (coded as binary strings) at water quality sources for minimization of the squared difference between the computed chlorine concentration and minimum specified concentration	Three example networks	The improved GA with the niched operator and creep mutation performed better than simple GA, as it produced quick optimal solutions and found well suited for multiple chlorine source problems
Eusuff & Lansey (2003)	Shuffled Frog Leaping Algorithm (SFLA)	Application of SFLA for WDN design problem for minimization of cost of pipes	Two-loop, Hanoi, and New York Tunnel (NYT) WDNs	SLFA obtained the best solutions for Two-loop, and Hanoi, and near-optimal solutions for NYT problem. SLFA found optimal solutions in a lesser number of iterations compared to GA and SA
Geem (2006)	Harmony search (HS) algorithm	Application of HS for WDN design and comparison of results with other meta-heuristic algorithms with the objective of minimization of cost of pipes	New York Tunnel WDNs	HS was able to find the best solution with the least number of iterations when compared with GA, SA, TS, ACO, SFLA methods. The HS showed good convergence and reliability in achieving optimal solutions

(continued)

doi: 10.2166/h2oj.2020.128

Table 2 | continued

Author(s) (year)	Technique used	Work and objectives	Case studies/problems considered	Key findings
Kadu *et al.* (2008)	Modified GA	Application of modified GA for WDN design problem with the objective of minimization of the cost of pipes	A hypothetical network, Hanoi, and TRN	Modified GA improves the efficiency and effectiveness of GA by reducing the search space, random use of various basic and additional GA operators
Cunha & Ribeiro (2004)	Tabu search algorithm	Application of Tabu search algorithms for WDN design problems for minimization of cost of pipes	Two-loop, Hanoi, and NYT	Promising results obtained by using the Tabu search algorithm for WDN design problem
Mohan & Babu (2010)	Honey-bee mating optimization (HBMO)	Application of HBMO for optimal WDN design and its comparison with other algorithms like GA and SA for minimization of cost of pipes	Two-loop and Hanoi WDNs	HMBO was able to arrive at optimal solutions in a lesser number of function evaluations compared to GA and SA. For HMBO, the uniform crossover preferred over the single/ multi-point crossover.
Suribabu (2010)	DE	Application of DE for least-cost design of WDNs for minimization of cost of pipes	Two-loop, Hanoi, New York Tunnel, and 14 pipe network expansion problem	The DE found to be more efficient than GA and SA for the optimal design of WDNs. The role of randomness is less in DE as compared to other meta-heuristic algorithms
Vasan & Simonovic (2010)	DE	Development of DENET, by integrating DE and EPANET for WDN design problem for minimization of cost and maximization of network resilience (two separate cases)	New York Tunnel, Hanoi WDNs	DENET found to be a useful tool, which can be used as an alternative approach for solving WDN design problems
Shibu & Reddy (2014)	Cross entropy (CE) optimization	Application of CE method for optimal WDN design under demand uncertainty for minimization of cost	NYT and a real case study of Tukum-zone, India	The CE method performed quite well for the design of WDNs and noted that the optimal cost goes on increasing with the increase in demand uncertainty
Sirsant & Reddy (2018)	Self-adaptive DE (SADE)	Application and performance evaluation of SADE for the deterministic and reliability-based design of WDNs considering cost minimization subject to satisfying a minimum reliability level	Two-loop, Apulian, BakRyan, Fossolo and a real WDN of Ramnagar zone in India	SADE performed well for both the deterministic and reliability-based designs of WDNs and converged faster than simple DE and other meta-heuristics methods with a higher success rate
Fallah *et al.* (2019)	Gravitational search algorithm (GSA)	Application of GSA for WDN design problem and results compared with those obtained using different EAs for minimization of cost of pipes	Hanoi, Two reservoir network and New York Tunnel and a real WDN situated in Khorramshahr city in Iran	GSA converged faster for Hanoi and Khorramshahr WDNs. For NYT and TR problems, GSA required a larger number of function evaluations compared to previous studies but led to lower-cost solutions. The GSA requires only two parameters

Table 3 | Applications of multi-objective evolutionary algorithms for the multi-objective design of water distributions systems

Author(s) (year)	Technique used	Work and objectives	Case studies/problems considered	Key findings
Halhal *et al.* (1997)	Structured messy GA (SMGA)	Application of SMGA for solving multi-objective rehabilitation of WDNs for minimization of capital cost and maximization of benefits	Small-looped example network, real WDN in Morocco	The SMGA technique proved to be more efficient than traditional GA
Savic *et al.* (1997)	MOGA	Use of pump switching as a surrogate for maintenance cost, improvement of the traditional GA with hybridization involving a local search for minimization of energy and maintenance cost	Example WDN problem	Seeding of the initial population with solutions from previous runs brought improvements in both efficacy and quality of solutions found
Cheung *et al.* (2003)	SPEA	Application of SPEA for the multi-objective design of WDNs, by maintaining an external population for storing the non-dominated solutions from the beginning of the algorithm, for minimization of cost, minimization of the maximum deficit in pressure and maximizing the hydraulic benefits	Hypothetical WDN comprising of 14 pipes and 12 nodes	SPEA performed well for the problem. Uniform crossover was found to be most suitable. The recombination operator has a greater influence than the mutation operator on the results
Prasad *et al.* (2004)	NSGA-II	Investigation of the booster location and scheduling in WDNs for minimization of total disinfectant dosage and maximization of the volumetric demand	Utility located in the eastern USA	The model can determine the optimal locations of booster stations given the number of stations, does not require pruning of the monitoring locations as all the demand nodes modeled as monitoring nodes
Prasad & Park (2004)	NSGA-II	Introduction of the concept of network resilience as a surrogate for WDN reliability and its use for a multi-objective design for minimization of cost and maximization of network resilience	Two-loop and Hanoi WDN	The constraint domination criteria incorporated, thus enabling GA to handle the constraints more efficiently The incorporation of network resilience in multi-objective optimization leads to a considerable reduction in computational time
Kapelan *et al.* (2005)	Robust NSGA-II (RNSGA-II)	Development and application of RNSGA-II (which considers the age of chromosome for calculating its fitness function) for minimization of cost and maximization of robustness	New York tunnel WDN	The RNSGA-II can handle any type of hydraulic uncertainty and model nonlinearity. The drawback is that it needs two additional parameters, minimum chromosome age and several samples, which require some tuning

(continued)

doi: 10.2166/h2oj.2020.128

Table 3 | continued

Author(s) (year)	Technique used	Work and objectives	Case studies/problems considered	Key findings
Farmani *et al.* (2005)	NSGA-II and SPEA-II	Comparison of the performance of two EAs for the multi-objective design of WDNs for minimization of cost and minimization of maximum head deficiency	New York Tunnel and Hanoi	SPEA-II performed slightly better than NSGA-II and noted the need for further improvement to obtain better Pareto fronts
Vamvakeridou-Lyroudia *et al.* (2005)	MOGA	Multi-objective design of WDNs by incorporating fuzzy reasoning for the estimation of benefits considering minimization of cost and maximization of benefits	Anytown WDN	The multiple criteria applied for benefit function are more extensive and strict. Despite this, the model gave solutions that not only satisfy these criteria but also lead to reduced costs
Dandy & Engelhardt (2006)	MOGA	Application of MOGA for replacement scheduling of water pipelines for two cases: single time step and multiple time step considering minimization of economic cost and maximization of reliability	EL103N Zone network of Adelaide city	MOGA performed efficiently for the problem. However, a very simplified framework discussed in this study to solve the problem, such as the replacement of pipes with the same diameter pipe
Ostfeld & Salomons (2006)	NSGA-II	Optimal placement of sensors in WDN using the multi-objective formulation for minimization of the expected time of detection of contamination, projected population affected prior to detection, expected demand of contaminated water prior to detection, and maximization of detection likelihood	Two example WDNs	The presented approach can be applied to real case studies also. The main limitation is how to sample injection events more efficiently
Prasad & Tanyimboh (2008)	NSGA-II	Multi-objective design of WDNs considering tank design procedure (taking tank shape parameters as decision variables) using entropy and resilience as the surrogates for reliability, for minimization of cost and maximization of resilience and entropy (one at a time)	Anytown WDN	Entropy serves as a better surrogate for reliability. The hydraulic performance of the obtained least-cost solutions found to be satisfying all the hydraulic constraints
Jayaram & Srinivasan (2008)	NSGA-II	Introduction of a new index (MRI) as a surrogate for WDN reliability and its application for the multi-objective design of WDNs	A hypothetical network comprising of 14 pipes and 11 nodes	MRI could overcome the drawbacks of the resilience index in terms of its application to WDNs with multiple supply sources
		Three optimization methods are employed: NSGA-II, a heuristic method and a combination of heuristic and NSGA-II for minimization of life cycle cost and maximization of modified resilience index (MRI)		Feeding NSGA-II with initial solutions using heuristic methods reduces the computational burden to a huge extent

(continued)

Table 3 | continued

Author(s) (year)	Technique used	Work and objectives	Case studies/problems considered	Key findings
Raad *et al.* (2010)	AMALGAM	Comparison of four reliability indices as surrogates for reliability for the multi-objective design of WDNs for minimization of cost and maximization of four reliability indices: flow entropy, resilience, network resilience, and a modified reliability surrogate (MRS)	TRP, Hanoi, and NYT WDNs	Although no 'best' RSM was identified, network resilience and MRS showed comparatively better performance than other indices
Creaco *et al.* (2014)	NSGA-II	Comparison of resilience and entropy as reliability surrogate for the multi-objective design of WDNs for minimization of cost and maximization of reliability indices, entropy, and resilience (one at a time)	A hypothetical network consisting of 11 pipes and a real network of Ferrara city, Italy	Resilience index found to be a more suitable surrogate for reliability compared to entropy. Huge computational savings can be obtained by employing these indices in the design procedure
Patil *et al.* (2020)	MOPSO+	MOPSO is augmented with (1) local search, (2) a modified strategy for assigning the leader and (3) a modified mutation scheme termed as MOPSO+ and employed for WDN design for minimization of cost and maximization of network resilience	Hanoi, Blacksburg, New York Tunnel, GoYang WDNs	A significant number of new Pareto-optimal solutions obtained using MOPSO+ compared to previous studies. For medium-sized networks, the number of function evaluations was the same as that of previous studies, but for intermediate-sized problems, more number of function evaluations needed, but the solutions obtained were better The suitability of MOPSO+ still needs to be tested for larger sized networks
Sirsant & Reddy (2020)	NSGA-II	The study applied extended period simulation (EPS) and NSGA-II for the multi-objective optimization of WDNs with maximization of reliability or reliability surrogate measure (RSM) and minimization of cost as two objectives. Investigated alternatives for computationally intensive reliability procedures, and evaluated the efficacy of RSMs like entropy, resiliency, and network resilience, and their combinations to represent the reliability of WDNs	Case studies of Two-loop, GoYang, and Fossolo WDNs, as well as a real case study of Ramnagar zone WDN in India	NSGA-II provides effective Pareto-optimal solutions for the bi-objective WDN model, which helps to find the associations between RSMs and the reliability of WDNs. The study recommended entropy as a surrogate for mechanical reliability (for larger WDNs), while resiliency for hydraulic reliability; and a combined entropy-resilience index (CERI) for accounting both hydraulic and mechanical reliabilities

2006; Muleta & Boulos 2007; Barreto *et al.* 2010; Penn *et al.* 2013; Vojinovic *et al.* 2014; Yazdi *et al.* 2017a; Wang *et al.* 2018), Nondomination Sorting Differential Evolution (NSDE; Yazdi *et al.* 2017b), etc. The specific details of these applications are also given in Table 4.

Applications in reservoir operation and irrigation systems

Reservoirs and irrigation systems need to be operated in a cost-effective manner such that the deficits are minimum as well the benefits achieved in terms of minimum cost or maximum energy production. Thus, the optimization problem is formulated as determination of the optimal release or operating policies such that the deficits are minimum and the benefits are maximum. Several meta-heuristic techniques were applied to solve different problems (Rani & Moreira 2010). To solve single-objective problems, the techniques used include GA (Oliveira & Loucks 1997; Wardlaw & Sharif 1999; Nixon *et al.* 2001; Merabtene *et al.* 2002; Dessalegne *et al.* 2004; Raju & Kumar 2004; Kerachian & Karamouz 2006, 2007; Kumar *et al.* 2006; Kuo *et al.* 2006; Zahraie *et al.* 2008), Shuffled complex evolution (SCE) algorithm (Lerma *et al.* 2015), PSO (Kumar & Reddy 2007; Reddy & Kumar 2007a; Ghimire & Reddy 2013); for solving multi-objective problems, the MOGA (Reddy & Kumar 2006), elitist-mutated MOPSO (Reddy & Kumar 2007b, 2009), multi-objective DE (MODE, Reddy & Kumar 2007c, 2008; Raju *et al.* 2012), etc. More details of these applications are described in Table 5.

Applications in water supply and wastewater system

The water supply and wastewater systems are subjected to many dynamic loadings, such as rain, the release of stormwater from storage tanks, etc. These loadings need to be regulated efficiently such that the required flow quality and quantity levels are maintained at minimum costs. There were several studies that have used EAs for solving these problems, such as GA (Rauch & Harremoes 1999; Tsai & Chang 2001; Wang & Jamieson 2002; Lavric *et al.* 2005; Murthy & Vengal 2006; Montaseri *et al.* 2015; Swan *et al.* 2017; Raseman *et al.* 2020), SFLA (Chung & Lansey 2009), etc. Also, for considering different conflicting issues, such as minimizing the total system cost and satisfactory performance of the systems under different dynamic loadings or contaminant additions, multi-objective optimization models developed and engaged MOEAs to solve the problems like MOGA (Chen *et al.* 2003), NSGA-II (Guria *et al.* 2005; Yandamuri *et al.* 2006; Fu *et al.* 2008; Muschalla 2008), etc. Table 6 gives more details of the applications and findings of the studies.

Applications in watershed management and fluvial systems

Watershed management and planning require modeling the hydrologic and fluvial characteristics properly and efficiently, such that the required water management conditions can be achieved in a cost-effective manner. Thus, the various watershed management techniques, such as the design of detention systems, flood management practices, and other best management practices (BMPs), need to be designed considering cost minimization and system reliability and efficiency maximization. Several studies used EAs for solving the relevant optimization models like GA (Harrell & Ranjithan 2003; Muleta & Nicklow 2005; Artita *et al.* 2013; Aminjavaheri & Nazif 2018), DE (Hang & Chikamori 2017), PSO (Shamsudin *et al.* 2014), MOGA (Yeh & Labadie 1997; Perez-Pedini *et al.* 2005), NSGA-II (Lee *et al.* 2012; Karamouz & Nazif 2013; Aminjavaheri & Nazif 2018), etc. More details of the applications and their findings are given in Table 7.

Table 4 | Applications of EA and SI meta-heuristic methods for optimization of urban drainage and sewer systems

Author(s) (year)	Technique used	Work and objectives	Case studies/problems considered	Key findings
Walters & Lohbeck (1993)	GA	Optimal layout design of the network using GA for minimization of cost	Example network	GA model with integer coding proved to be very efficient compared to binary-coded GA. The computational and memory requirements for GA found to be very less as compared to DP
Walters & Smith (1995)	GA	Optimal layout design using evolutionary and GA principles for minimization of cost. The use of an efficient tree growing algorithm and incorporation of redundant genetic information within the reproduction phase	Example network	The algorithm found to be highly efficient and is suitable for those networks which have one or more identifiable roots from which flow diverges or to which flow converges
Liang *et al.* (2004)	GA and Tabu search	Application of GA and Tabu search techniques for the design of gravity waste water collection systems for minimization of direct cost (material, backfill, excavation, and soil dump cost)	Example network	Both GA and TS produced optimal designs with shallower pipe elevations downstream compared with traditional design. TS technique lead to lower-cost designs than GA
Afshar *et al.* (2006)	GA	Hydrograph-based sewer design using GA as the optimization tool and SWMM as the hydraulic simulator for minimization of total piping and excavation cost	Two example networks	The proposed model proved to be very efficient as it could incorporate the inflow hydrograph as an input to the system and perform flood routing and is suitable for application to large-scale stormwater networks
Barreto *et al.* (2006)	NSGA-II and ε-MOEA	Multi-tier approach for solving the problem of drainage network rehabilitation for minimization of pipe installation cost and minimization of surcharge in a pipe network	Two example networks	NSGA-II performs better than ε-MOEA for small population size, but its performance degrades with an increase in population size. The performance of ε-MOEA showed less sensitivity to population size
Guo *et al.* (2006)	CA-GASiNO (Cellular Automata and GA)	Application of a hybrid CA-GASiNO method which involves seeding the MOGA with local search based on CA principle for Sewer Network Optimization, with minimization of flooding within the sewer system and minimization of the capital cost of the network as two objectives	One small hypothetical network and one large real network	The CA-GASiNO required comparatively much smaller number of function evaluations as compared to NSGA-II

(continued)

doi: 10.2166/h2oj.2020.128

Table 4 | continued

Author(s) (year)	Technique used	Work and objectives	Case studies/problems considered	Key findings
Muleta & Boulos (2007)	NSGA-II	Application of NSGA-II for the multi-objective design of urban drainage systems by coupling it with EPA SWMM5 for minimization of cost and overflow volume	A hypothetical case study	The tool performed well for the example problem and can be useful to any wastewater utility attempting to optimize its capital improvement program
Barreto *et al.* (2010)	NSGA-II and ε-MOEA	Application and performance evaluation of two EAs for the rehabilitation of urban drainage systems for minimization of rehabilitation cost (cost of pipes, excavation, installation, restoration, and reinstatement) and minimization of flood damage cost	An example network consisting of 12 pipes, 13 manholes, and 11 sub-catchments	NSGA-II performed better for smaller population sizes, but its performance deteriorates for large population sizes; ε-MOEA was less sensitive to population size. However, the diversity of solutions not as good as that of NSGA-II for large population sizes. The number of function evaluations and computational time were better for NSGA-II
Penn *et al.* (2013)	NSGA-II	A multi-objective optimization model for estimating the optimal distribution of different types of GWR homes in an existing municipal sewer system for minimization of wastewater flow at the outlet of the neighborhoods sewer system and minimization of the cost of the on-site GWR treatment system	Sewer system located in central Israel, near the coast (flat terrain)	The results obtained lead to maximum water savings in most houses. Further suggested consideration of the water quality aspect in future studies
Karovic & Mays (2014)	SA	Development of an optimization procedure for designing storm and sanitary sewer systems using SA for minimization of cost	A hypothetical storm sewer system	Better results obtained by applying the SA procedure rather than the straight slope method, implying that optimization procedures should be used for designing the sewer systems
Vojinovic *et al.* (2014)	NSGA-II	Multi-objective design of urban drainage systems under uncertainties from climate change, urbanization, population growth, and aging of pipes considering minimization of damage cost and intervention cost	Case study of Dhaka, Bangladesh	The NSGA-II method produced widespread Pareto-optimal solutions for urban drainage systems
Yazdi *et al.* (2017a)	Non-dominated sorting Harmony Search (NSHS)	Development and application of NSHS for assigning optimal rehabilitation plans for sewer pipe networks. Three optimization tools, NSGA-II, MOPSO, and NSHS, were applied and tested for the problem for minimization of pipe replacement cost and flooding overflow volume	A storm sewer pipe network in the south part of Seoul, South Korea	The NHSH gave the best results for all population sizes considered, whereas MOPSO performed the worst for all the cases. NSGA-II gave solutions closer to the NSHS algorithm when the number of function evaluations increased, NSHS still performed better

(continued)

Table 4 | continued

Author(s) (year)	Technique used	Work and objectives	Case studies/problems considered	Key findings
Yazdi *et al.* (2017b)	SPEA-II, NSGA-II, and two extended versions of HS and DE, NSHS and NSDE	A comparative application of different MOEAs for hydraulic rehabilitation of urban drainage networks for minimization of cost of rehabilitation and minimization of the flooding overflow volume	Gasan, Sungsan, Yongdub, and Singil network	The NSHS outperformed other algorithms by generating the best Pareto-optimal solutions for the problem
Yazdi (2018)	DE	Application of DE to find the best monitoring sites based on maximizing the joint entropy obtained by information associated with water quality time series	Drainage network in Tehran, Iran	The entropy-based method leads to a high level of information content with a significantly smaller number of monitoring sites as compared to the other designs
Wang *et al.* (2018)	NSGA-II, MLOT, and GALAXY	A comparison of three MOEAs for multi-objective urban drainage adaptation problems for minimization of cost and system overloading	A portion of the drainage network in the city of Hohhot	GALAXY was found to be the most efficient among the three MOEAs as it can save substantial time and effort to cope with the parameterization issue of MOEAs
Yazdi (2018)	Non-domination sorting DE (NSDE)	Resiliency-based multi-objective design of urban drainage systems using NSDE considering objectives of minimization of total flooding and rehabilitation cost and minimization of the total flooding volume	The western part of Tehran Stormwater Drainage System (TSDS)	Results showed that the optimal design obtained by the NSDE could decrease network flooding from 3.5×10^6 m^3 to near zero with at most 23% lower investment costs relative to the traditional design

Applications in parameter estimation of hydrological models

For simulating the various hydrological processes, there exist several hydrological models (namely, lumped, semi-distributed, distributed models). The model may consist of a large number of components with several parameters, so one has to choose the appropriate model depending on the purpose and accuracy of the model variables of interest. To represent and simulate various hydrological processes accurately, it may require proper calibration and validation of the model parameters using historical data. A typical calibration process thus requires minimization of the error between the model simulated and actual values of the hydrologic variables (e.g., runoff). Sometimes, the model responses vary and may be suitable for a particular event or application. Also, the different performance indices used for the calibration process may give different results. For improving the performance of the simulation models under different scenarios, multi-variate calibration may be needed, which can be done using multi-objective calibration. For calibration of the hydrological models, several studies used different EAs like GA (Wang 1991; Franchini & Galeati 1997; Liong *et al.* 2001; Zou & Lung 2004), SCE algorithm (Zhang *et al.* 2009; Tigkas *et al.* 2016; Krishnan *et al.* 2018; Shin & Choi 2018), MOPSO (Mostafaie *et al.* 2018), NSGA-II (Khu & Madsen 2005; Alamdari *et al.* 2017; Tian *et al.* 2019), AMALGAM (Koppa *et al.* 2019), etc. More details of EAs and MOEA applications for single and multi-objective parameter estimation of hydrological models are presented in Table 8.

Table 5 | Applications of EA and SI meta-heuristic methods for the operation of reservoirs and irrigation systems

Author(s) (year)	Technique used	Work and objectives	Case studies/problems considered	Key findings
Oliveira & Loucks (1997)	GA	Application of GA for multi-reservoir operation considering minimization of the sum of squared deficits	Two hypothetical systems	GA proved to be an efficient tool for solving the multi-reservoir operation problem and can handle complex situations like side demands or constraints on releases, power production as compared to the DP approach
Wardlaw & Sharif (1999)	GA	Different alternative formulations of GA evaluated for the four reservoir problem for maximization of the benefits	Hypothetical four reservoir problem	A real-value representation, incorporating tournament selection, elitism, uniform crossover, and modified uniform mutation will operate most efficiently for the reservoir operation problem. GA found to be more efficient than traditional methods like DP and has the advantage that no initial trial release policy required
Nixon *et al.* (2001)	GA	Examination of GA optimization to identify water delivery schedules for an open channel irrigation system for minimization of the number of orders shifted and minimization of channel flow rate variations, such that particular size of order shifts encouraged, and others discouraged, and channel capacity not exceeded	A problem involving scheduling irrigation water deliveries in a single channel spur consisting of five irrigators with one order each	GA performed well and found to be suitable for efficiently scheduling irrigation orders
Merabtene *et al.* (2002)	GA	Development of a DSS to assess the susceptibility of water supply systems to droughts. GA used to search the optimal operation, such that drought risk index is minimum	Fukuoka city water supply system, Japan	GA performed well for the problem, but the physical attributes of the problem can be further incorporated using engineers' experience through fuzzy inference
Raju & Kumar (2004)	GA	Application of GA to evolve efficient cropping pattern for maximizing benefits for an irrigation project	Sri Ram Sagar Project (SRSP) situated on river Godavari in Telangana, India	GA is found to be a useful tool for irrigation planning and can even be applied for complex problems involving nonlinear optimization

(continued)

Table 5 | continued

Author(s) (year)	Technique used	Work and objectives	Case studies/problems considered	Key findings
Dessalegne *et al.* (2004)	GA	Evaluation and optimization of dam operations in multi-reservoir river systems for minimization of the maximum water surface elevation deviation at any cross-section and over any incremental time step, and minimization of the maximum rate of water level fluctuation at any cross-section over the entire simulation horizon	A hypothetical three-reservoir river system and a portion of the Illinois River Waterway	GA performed efficiently for both the case studies, and the methodology can be further extended to optimize the operation of multiple reservoir systems and to meet environmental requirements as well
Kuo *et al.* (2006)	GA	Development of a hybrid ANN-GA model for reservoir water quality management, with ANN serving as the model for predicting the phosphorus concentrations and GA for determining the proper reduction rates of nutrient loads from the watershed to minimize the phosphorus reduction rate of the inflow	Feitsui Reservoir situated in Taipei Metropolitan Area	The hybrid model performed efficiently by significantly reducing the computational burden as required in case of other simulation models
Kumar *et al.* (2006)	GA	Development of a GA-based model for obtaining an optimal operating policy and optimal crop water allocation from an irrigation reservoir for maximization of the sum of the relative yields from all crops in the irrigated area	Malaprabha single-purpose irrigation reservoir in Karnataka State, India	The performance of the GA model was found to be suitable for application on real case studies
Kumar & Reddy (2006)	ACO, GA	Evaluation of ACO for the derivation of optimal operation policies for multi-purpose reservoir system, considering short-time and long-time horizons	Case study of Hirakud reservoir system, India	The ACO technique has resulted in superior performance for long-term reservoir operation as compared to GA solutions
Reddy & Kumar (2006)	MOGA	Application of MOGA for multi-purpose reservoir operation for minimization of the sum of squared deficits and maximization of annual energy production	Real case study of Bhadra reservoir system in India	MOGA performed well and generated good Pareto-optimal solutions for the reservoir operation problem
Kumar & Reddy (2007)	Elitist-mutated PSO (EMPSO), GA	Development of EMPSO and its application for multi-purpose reservoir operation considering minimization of the sum of squared deficits for irrigation, and maximization of annual energy production	Hypothetical four reservoir system and real case study of the Bhadra reservoir system in India	The EMPSO consistently performed better than standard PSO and GA techniques, requiring less number of function evaluations

(continued)

Table 5 | continued

Author(s) (year)	Technique used	Work and objectives	Case studies/problems considered	Key findings
Reddy & Kumar (2007a)	EMPSO, GA	Elitist-mutated PSO (EMPSO) proposed for integrated reservoir operation for irrigation and multi-crop water allocation in a reservoir command area. The model considers maximizing crop yields with due consideration of water shortages in different periods	Case study of Malaprabha reservoir system in India	The proposed PSO technique gave efficient solutions to obtain reservoir operation policies and crop water allocations for multiple crops. The approach found to be beneficial for integrated water management under water-scarce conditions
Reddy & Kumar (2007b)	EM-MOPSO, NSGA-II	Development of EM-MOPSO for multi-objective multi-purpose reservoir operation for minimization of the sum of squared deficits for irrigation, and maximization of annual hydropower production	Four test problems and a real case study of the Bhadra reservoir system in India	The EM-MOPSO found to be an efficient tool for multi-objective reservoir operation and produced consistently better Pareto-fronts with good convergence and uniformly widespread solutions as compared to NSGA-I
Reddy & Kumar (2007c)	MODE, NSGA-II	Application of MODE for multi-objective multi-purpose reservoir operation for minimization of the sum of squared deviations for irrigation and maximization of annual hydropower production	Hirakud reservoir project in Orissa state, India	The MODE performed better than NSGA-II for reservoir operation problem. Noted that MODE could serve as an efficient alternative tool for solving multi-objective optimization problems in the water resource systems
Kerachian & Karamouz (2006), 2007)	Variable Length Chromosome GA (VLGA)	The SGA and VLGA were used to reduce the computational burden of GA and applied for reservoir operation problem accounting for inherent uncertainties in reservoir inflows considering maximization of the utility function of different water users	15-Khordad Reservoir in the central part of Iran	The SGA reduced the overall time as compared to GA by dynamically updating the chromosome length. VLGA was found to be an efficient tool for solving the problem and can be even applied to problems comprising longer planning horizons
Zahraie *et al.* (2008)	VLGA	Reservoir operation using two adaptive varying chromosome lengths GA (VLGA) for maximization of the utility function of different water users	Zayandeh-Rud Reservoir, Karoon-I Reservoir, and the system of Bakhtiari and Dez reservoirs in series in Iran	The VLGA found to be efficient in solving the reservoir operation problem
Reddy & Kumar (2008)	MODE	Proposed MODE to evolve different strategies for the simultaneous evolution of optimal cropping pattern and operation policies for irrigation reservoir system with the objectives of maximizing the total net benefits from the irrigation system and maximize the total irrigated area in the command area	Real case study of Malaprabha reservoir systems in India	MODE approach provides a wide spectrum of efficient Pareto-optimal solutions and gives sufficient flexibility to select the best irrigation planning and reservoir operation plan, which could be very useful for multi-crop irrigation reservoir systems

(continued)

Table 5 | continued

Author(s) (year)	Technique used	Work and objectives	Case studies/problems considered	Key findings
Reddy & Kumar (2009)	MODE, NSGA-II	Integrated water resource management model for reservoirs considering minimization of squared irrigation deficits and maximization of hydropower for reservoir system, subject to meeting the flood control regulations	Hirakud reservoir system in India	MODE helps in generation of efficient Pareto-optimal solutions and helps to analyze the trade-offs between multiple objectives, thus facilitated easiness in the decision making for the operator
Raju *et al.* (2012)	MODE	Application of MODE for irrigation planning for maximization of net benefits, agricultural production, and labor employment	Mahi Bajaj Sagar Project, Rajasthan, India	MODE performed well for the problem. Further noted that while applying for real case studies, proper sensitivity analysis should be carried out to determine the best-suited parameters
Lerma *et al.* (2015)	Shuffled complex evolution (SCE-UA) and Scatter search algorithm	Assessment of two EAs to design the optimal operating rules for water resource systems for minimization of the cost of water storage, and costs associated with a deficit of demands	Tirso-Flumendosae-Campidano system located on the island of Sardinia (Italy)	SCE-UA found to be a more efficient algorithm for solving the problem in terms of faster convergence and attaining global optimality

Applications in groundwater remediation, groundwater systems monitoring network design

The purpose of the GW monitoring networks is to capture the information about the contamination of GW and its source/location. In order to design a cost-effective system, the number of monitoring wells should be minimum and should not be redundant while being able to capture the desired information about the contamination. Different types of EAs were used for solving the groundwater systems and monitoring network design models formulated with different complexities. The popular techniques used include GA (McKinney & Lin 1994; Ritzel *et al.* 1994; Chadalavada & Datta 2008), SA (Dougherty & Marryott 1991), DE (Alizadeh *et al.* 2018), Probabilistic Pareto GA (PPGA; Luo *et al.* 2016), NSGA-II (Tang *et al.* 2007), ε-multi-objective noisy memetic algorithm (ε-MONMA; Song *et al.* 2019), etc. More details of the applications of EAs for groundwater systems are given in Table 9.

Groundwater systems are redundantly subjected to various pollutants at different times and locations, which are dynamic in nature. In order to ensure that the required GW quality levels are maintained, observation, as well as pumping wells, need to be designed. The problem can thus be formulated as determining the location and number of wells as well as the required pumping rates at these wells such that the cost is minimum and the desired quality levels are maintained. Several studies used EAs for solving the groundwater remediation problems via the simulation–optimization framework such as GA (Huang & Mayer 1997; Wang & Zheng 1997; Sun & Zheng 1999; Smalley *et al.* 2000; Yoon & Shoemaker 2001; Zheng & Wang 2002; Babbar & Minsker 2006; Wu *et al.* 2006; Park *et al.* 2007; Bayer *et al.* 2008; Seyedpour *et al.* 2019), SA (Kobayashi *et al.* 2008); MOGA (Erickson *et al.* 2002; Mantoglou & Kourakos 2007; Singh *et al.* 2008), NSGA-II (Singh & Chakrabarty 2011; Ouyang *et al.* 2017), etc. More details of these EAs applications are also elaborated in Table 9.

doi: 10.2166/h2oj.2020.128

Table 6 | Applications of EA and SI meta-heuristic methods for single and multi-objective optimization of water supply and/or wastewater systems

Author(s) (year)	Technique used	Work and objectives	Case studies/problems considered	Key findings
Rauch & Harremoes (1999)	GA	A novel approach to control the whole system of the sewer system, treatment plant, and receiving water, with the aim to achieve minimum effects of pollution, considering minimization of the overflow volume as objective	A hypothetical example	The application of predictive model control in conjunction with genetic algorithms offers an attractive tool for real-time control of urban wastewater systems
Tsai & Chang (2001)	GA	Development of a mathematical programming model for water usage and treatment network design for minimization of operating cost and freshwater consumption rate	Two example problems	Search space was reduced significantly by using split fractions as decision variables The appropriate ranges of the design variables were efficiently determined by cascading the evolution processes according to the inducing parameters
Wang & Jamieson (2002)	GA	Development of methodology for water quality management on a river basin scale based on GA/ANN approach for minimization of cost of treatment and conveyance of wastewater	Upper River Thames Basin, England	The proposed approach has several advantages in terms of incorporation of both construction and operating costs, rapid convergence, and vast progress in computational efficiency
Chen *et al.* (2003)	MOGA	Development of a hybrid control algorithm integrating the indiscernibility capability of rough set theory and search capability of genetic algorithms with conventional neural-fuzzy controller design and its application for minimization of the operating cost of wastewater treatment	Example problem	The hybrid fuzzy control system provides immediate guidance and control with respect to multi-objective requirements for distributed control system using on-line process data
Lavric *et al.* (2005)	GA	Application of GA for finding the optimum water resource allocation and the wastewater network for minimization of the supply water consumption	An example network	The algorithm guarantees both the optimal topology of the units' operation network and the minimum supply water consumption
Guria *et al.* (2005)	NSGA-II	Application of NSGA-II for optimization of Reverse Osmosis (RO) desalination units using different adaptations of NSGA-II for maximization of the permeate throughput, minimization of the cost of desalination, and minimization of the permeate concentration	Operation of an existing plant and design of a new plant	NSGA-II and other schemes were used to obtain the optimal solutions and noted that the NSGA-II variant found to be producing good solutions

(continued)

Table 6 | continued

Author(s) (year)	Technique used	Work and objectives	Case studies/problems considered	Key findings
Yandamuri *et al.* (2006)	NSGA-II	A multi-objective optimization framework for optimal waste load allocation in rivers for minimization of total treatment cost and maximization of overall performance	Willamette river system in the state of Oregon	Results indicated that the least inequity Pareto-optimal solution demands significantly more treatment effort and cost compared with that of the least-cost Pareto-optimal solution
Murthy & Vengal (2006)	GA	Application of GA for optimization of RO system for maximization of the rejection of the solute while varying the feed flowrate and the overall flux across the membrane	Case study of the RO system	GA converged quickly to the optimal solution and showed its suitability for the RO problem
Fu *et al.* (2008)	NSGA-II	Application of NSGA-II for multi-objective optimal control of urban wastewater systems considering maximization of DO and Ammonia concentration in the river	An integrated case study was taken from literature	NSGA-II was able to generate good Pareto-optimal solutions for the chosen problem successfully
Muschalla (2008)	Adaptive NSGA-II	Development of a new multi-objective EA combining the advantages of NSGA-II and self-adaptive evolution strategies for optimization of the performance of an urban wastewater system considering minimization of investment cost and maximization of river water quality	The sewer system of Heusenstamm, located in the central part of the catchment of the river Bieber	The developed tool found to be robust and efficient in providing Pareto-optimal solutions. Further, it suggested focusing more on improving the computational efficiency of the model
Chung & Lansey (2009)	Shuffled Frog Leaping Algorithm (SFLA)	Application of SFLA for the optimization of a large-scale water supply system for minimization of the total cost (construction and expansion, and operations and maintenance costs for all components of pipes, canals, pumps, and treatment facilities)	A first system was comprised of a single water and wastewater plants, multiple sources (imported water, groundwater aquifer, and surface water) and two demands centers (domestic and agricultural) A second system comprising of six users – three domestic areas, one industrial, one agricultural, and one large outdoor area – and three wastewater treatment plants	The SFLA was able to generate good solutions for both the problems

(continued)

doi: 10.2166/h2oj.2020.128

Table 6 | continued

Author(s) (year)	Technique used	Work and objectives	Case studies/problems considered	Key findings
Montaseri *et al.* (2015)	GA	The study presented a simulation–optimization model for managing the urban stormwater using the MUSIC model as a simulator and genetic algorithm (GA) as an optimizer. The optimization model involves minimizing the total cost of the life cycle of the urban stormwater treatment devices as the objective function	The case study area located in Central Canberra, Australia	The approach helps in the identification of the most appropriate treatment train setting for the post-development scenario, considering cost and water quality aspects, and concluded that the optimized treatment devices were effective means to remove pollutants from urban stormwater runoff
Swan *et al.* (2017)	NSGA-II	NSGA-II and Monte Carlo approaches were used to optimize operational regimes in water treatment works (WTW) considering the objectives of minimizing the operating cost and failure likelihood of a WTW	The hypothetical case study data were used, which was based in a rural location with water abstracted from a lowland reach of a river that impounded in a reservoir prior to treatment	Static models were found to be more suitable for whole WTW optimization modeling and offered the benefit of the reduced computational burden
Raseman *et al.* (2020)	Borg MOEA	The Borg MOEA is used to optimize water treatment problems with five objectives. The first two objectives represent the risk of violating disinfection byproduct regulations (minimize the frequency of elevated total trihalomethane and minimize the frequency of elevated Haloacetic acids), while the remaining objectives represent operating costs (minimize solids produced, minimize lime dose required, minimize carbon dioxide dose required)	The case study located in the Cache la Poudre (CLP) River in northern Colorado, and considers a hypothetical utility that must treat water from the river	The study infers that improving the resilience and affordability of existing water treatment plants is critical for the safety and financial viability of drinking water systems and simulation–optimization framework with Borg MOEA helps to generate various treatment strategies for decision making by water manager

DISCUSSION

Meta-heuristics for decision making?

The wide range of the above applications shows that meta-heuristics is an emerging research area for solving a variety of water resources problems. There is also a growing interest for integrated water resources management frameworks, where the conventional disciplinary boundaries in water resources need to be reconsidered, and future management frameworks have to address different complexities that may arise due to high nonlinearities, a wider range of uncertainties, integration of large system components, etc. These issues pose significant challenges and motivate the need for EAs applications to advance adaptive decision making under uncertainty. Also, it is important to

doi: 10.2166/h2oj.2020.128

Table 7 | Applications of EA and SI meta-heuristic methods for the watershed management and fluvial systems

Author(s) (year)	Technique used	Work and objectives	Case studies/problems considered	Key findings
Yeh & Labadie (1997)	MOGA	Application of MOGA for multi-objective watershed-level planning of stormwater detention systems for minimization of detention system cost, sediment loading, providing supplemental water supply, and maximizing system reliability	Upstream branch of Pazam River in southern Taiwan	Successive Reaching Dynamic Programming (SRDP) can lead to trade-off solutions between detention system cost and detention effect by executing the algorithm for a range of discrete values of maximum flow. However, MOGA can generate trade-off solutions for multiple objectives much more efficiently
Harrell & Ranjithan (2003)	GA	Planning and design of detention ponds by considering land management as well as decisions on pond location and sizes considering minimization of the cost of pond and maximization of removal efficiency	City Lake watershed in High Point, NC	The integrated planning and design of ponds found more cost-effective than setting one performance and site conditions for all the ponds. Incorporating land-use allocation associated with future growth can lead to lower-cost designs. The GA-based method flexible to allow the incorporation of more complex simulation models to estimate pollutant loading and removal
Muleta & Nicklow (2005)	GA and SPEA	Integration of GA and SPEA with SWAT, and replacement of the SWAT simulations with ANN, considering minimization of erosion and sediment yield or maximization of net agricultural profit (for single-objective formulation); and minimizes erosion and sediment yield while simultaneously maximizing individual farm-based income that accrues from growing corresponding crops (for multi-objective formulation)	Big Creek watershed, Southern Illinois	Replacement of SWAT by ANN results in 84% reduction in computational time required. ANN model trained using hybrid EP and BP algorithms was found to be more efficient and effective than EP or BP alone
Perez-Pedini *et al.* (2005)	MOGA	Distributed hydrologic modeling of an urban watershed and combined with GA to determine the optimal location of infiltration-based best management practices (BMPs), for minimization of project cost and watershed flooding	Aberjona River watershed, Massachusetts	The study found 20% reduction in the watershed peak flow by the application of BMPs to fewer than 200 HRUs (1 HRU = 120 × 120 m plot). An incremental approach targeting the more critical areas at initial stages and less critical areas in the future can be practised, as the optimal locations of only a few BMPs can be a subset of the optimal locations of a much larger set of BMPs

(continued)

doi: 10.2166/h2oj.2020.128

Table 7 | continued

Author(s) (year)	Technique used	Work and objectives	Case studies/problems considered	Key findings
Lee *et al.* (2012)	Scatter search and NSGA-II	Development of System for Urban Stormwater Treatment and Analysis Integration (SUSTAIN) and its application for the watershed-scale design of stormwater BMPs, for minimization of the cost	An urban watershed in Kansas City, USA	SUSTAIN was found to be a useful tool which can be applied by various practitioners, municipalities, and watershed groups at the regional and local levels to address diverse management practice planning questions in a wide range of conditions
Karamouz & Nazif (2013)	NSGA-II	Development of a multi-criteria optimization model to select BMPs for flood management in urban watershed systems considering climate change, considering minimization of flood volume and maximization of the drainage system's reliability	An urban watershed located in the north-eastern part of Tehran, the capital of Iran	The optimization results show that the cost of BMP development is a governing factor in system reliability and expected flood damage assessment. A small change in BMP cost and their application scheme may result in significant improvements in system operation
Artita *et al.* (2013)	Species Conserving Genetic Algorithm (SCGA)	Integration of the semi-distributed watershed model, Soil and Water Assessment Tool (SWAT) and SCGA for minimization of the total cost of BMPs	Watershed in southern Illinois	SCGA-SWAT is found to be an efficient tool for solving watershed planning and management problem. Further developments may include maintenance cost and land-use decisions
Shamsudin *et al.* (2014)	PSO	The study concerns detention ponds based on best management practices for the treatment and control of urban stormwater. Analytical Probabilistic Models (APM) and PSO method were used to develop the optimal combination of pond volume and outflow that yield the minimum cost	The case study area of Kuala Lumpur in Malaysia	The comparison of the results of PSO with APM showed that the PSO is more accurate as it does not need discretization of outlet size and found to be more robust, computationally cheaper and faster to implement
Hang & Chikamori (2017)	DE, ES	DE and ES approach applied to the calibration of the long- and short-term runoff model (LST model) to simulate the daily rainfall-runoff process. The study used three objective functions of the Nash–Sutcliffe efficiency coefficient, root mean square error, and mean absolute error, for evaluating the simulation accuracy of the LST model	Case study of the Be River catchment located in southern Vietnam	DE technique found to give slightly better performance and more stable solutions than those found by the ES technique during both calibration and validation periods

(continued)

Table 7 | continued

Author(s) (year)	Technique used	Work and objectives	Case studies/problems considered	Key findings
Aminjavaheri & Nazif (2018)	GA, NSGA-II	The study used GA and NSGA-II, along with SWMM and Management Option Rank Equivalence (M four parameters including the urban percent imperviousness, the depression storage of pervious surfaces, the curve number, and the Manning's roughness coefficient of conduits are the most effective parameters ORE) approach to develop the robust optimal set of BMPs for urban runoff quantity management in data-poor catchments. The three objective optimization model consists of minimizing the outfall runoff volume, minimizing the BMP costs, and minimizing the system vulnerability	The urban basin located in the north of Tehran, Iran	After modeling the study area in the SWMM model, the most effective parameters were determined using the LH-OAT sensitivity analysis and found that four parameters of urban percent imperviousness, the depression storage of pervious surfaces, the curve number, and the Manning's roughness coefficient of conduits were the most effective parameters in calibration. The proposed MORE approach helps to manage the urban surface runoff optimally in a way that the flood damage potential is minimized by the minimum cost

identify the problem properties across the different water resources field domains (such as watersheds, surface water, groundwater, reservoirs, water supply, etc.) that are posing computational barriers for large-scale water resources systems. There is a greater need to engage meta-heuristics-based optimization frameworks for improved decision making in water resources management.

Advantages of EAs as compared with conventional methods of optimization

While applying for practical problems, the EAs offer several benefits that include conceptual simplicity, flexibility, and robust response to changing environments and their ability to self-adapt the search for optimum solutions on the evolution. EAs have broad applicability, since, with the same procedure, they can be applied to any problem that can be expressed as a function optimization task (e.g., discrete combinatorial problems, continuous-valued parameter optimization problems, mixed-integer problems, and others). In contrast, the conventional techniques might be applicable only to continuous values or other restrictions on constrained sets. Many times, the objective or response surfaces modeled in practical problems are often multi-modal, and the conventional gradient-based approaches rapidly converge to local optima, which may return unsatisfactory performance. For simpler problems, where the response surface is strongly convex, the EAs do not perform as well as traditional optimization methods (e.g., gradient-based methods are designed to take advantage of the convex property of surface). EAs offer significant advantages for real-world problems with multi-modal functions. Also, in case of applying LP method to problems with non-linear objectives and/or constraints, which offers an almost certainly incorrect solution because of the simplifications or assumptions required for the technique. In contrast, EAs can directly incorporate any kind of arbitrary objectives and constraints. Also, the conventional methods of optimization are not robust to dynamic changes in the environment and often require a complete restart to provide a solution (such as DP technique). In contrast, EAs can be used to adjust solutions

doi: 10.2166/h2oj.2020.128

Table 8 | Applications of EA and SI meta-heuristic methods for calibration or parameter estimation of hydrological models

Author(s) (year)	Technique used	Work and objectives	Case studies/problems considered	Key findings
Wang (1991)	GA	Application of GA for calibration of a conceptual rainfall-runoff model for minimization of the sum of squares of differences between computed and observed discharge	Xinjiang rainfall-runoff model for the Bird Creek catchment data	GA performed efficiently for the problem, attaining a global peak 8 times out of 10 trial runs
Franchini & Galeati (1997)	GA, Pattern Search (PS)	Analysis of various GA schemes (varying in terms of the mutation and crossover operators) and their robustness and efficiency for calibration of conceptual rainfall-runoff model for minimization of the RMSE between the observed and model-simulated flow values	An 11 parameter distributed model (ADM)	The parameterization of GA does not create any major difficulties, thus making GA a robust algorithm. The PS method performed slightly better than all the GA schemes involved and consumed less computational time compared to GA
Liong *et al.* (2001)	Accelerated Convergence GA (ACGA)	Development of a hybrid GA-NN tool for generating Pareto-optimal solutions of calibration parameters for different model responses for minimization of the residual error for different model responses	UBT catchment in Singapore	ACGA is found to be better than SGA, VEGA, MOGA, and NSGA in terms of generating a wider range of solutions in the initial search space, faster convergence rate, and more optimal Pareto front
Khu & Madsen (2005)	NSGA-II	Multi-objective calibration with Pareto-preference ordering is generated considering different number and combinations of the objective functions which are minimization of the overall volume error, RMSE, RMSE of peak flow events and RMSE of low flow events	NAM rainfall-runoff model of the Tryggevaelde catchment in Denmark	For highly correlated objective functions, a small number of Pareto-optimal solutions were generated. On comparing the solutions obtained using single-objective optimization, all the solutions were within the bounds of those obtained using single-objective calibration and the spread of the Pareto-optimal set was small, indicating good compromise solutions. The NSGA-II can be used for the calibration of model parameters
Zhang *et al.* (2009)	GA, SCE, PSO, DE and artificial immune system (AIS)	Testing of five optimization algorithms for the automatic parameter calibration of the SWAT model considering maximization of NSE	Yellow River headwaters watershed (YRHW), Reynolds Creek Experimental Watershed (RCEW), Little River Experimental Watershed (LREW), and Mahantango Creek Experimental Watershed (MCEW)	No optimization algorithm could perform consistently better than the other algorithms for all four watersheds, implying the complexity of the problem. The study noted that GA could be considered as the first choice for getting global optimum solutions and PSO for getting solutions with lesser computational efforts

(continued)

Table 8 | continued

Author(s) (year)	Technique used	Work and objectives	Case studies/problems considered	Key findings
Zou & Lung (2004)	Alternating Fitness GA (AFGA)	Calibration of water quality models for water quality management using AFGA, which comprises of changing the fitness function evaluations at times throughout the evolution steps for minimization of the sum of squared differences between model generated and observed data	Total phosphorus model of the Triadelphia Reservoir in Maryland	AFGA is capable of generating a higher diversity of solutions compared to the normal GA
Tigkas *et al.* (2016)	SCE, GA, and Evolutionary Annealing-Simplex (EAS)	Comparison of the effectiveness and efficacy of three EAs for automatic calibration of the Medbasin-D Conceptual Hydrological Model for minimization of RMSE, Mean absolute error (MAE), Ratio of mean absolute error to the mean of observed discharge (RMAE), and Maximization of NSE, and Modified NSE (MNSE)	Watershed of the island of Crete in Greece	GA was the most efficient among the three algorithms, while SCE and EAS required a more exhaustive search by reaching the maximum allowed number of functions evaluations in most cases
Alamdari *et al.* (2017)	NSGA-II	Development of an auto-calibration tool to calibrate the Storm Water Management Model (SWMM) and its application for assessing the effects of climate change on water quantity and quality considering maximization of the coefficient of determination (R^2) and Nash Sutcliffe Efficiency (NSE), and minimization of Percent Bias (PBIAS)	Difficult Run watershed, Virginia	The model was calibrated and applied successfully to the considered watershed. However, the sources of uncertainty in GCMs were not considered in the study
Shin & Choi (2018)	SCE algorithm	Application of combined SCE and hydrologic model for calibration of Grid-based rainfall-runoff model (GRM) for maximization of NSE, maximization of correlation coefficient and minimization of normalized RMSE (nRMSE)	The Danseong and Seonsan catchments, which are major tributary catchments of the Nakdong River in South Korea	The performance of the SCE model was found to be appropriate and can be used for the parameter calibration of other hydrologic models

(continued)

doi: 10.2166/h2oj.2020.128

Table 8 | continued

Author(s) (year)	Technique used	Work and objectives	Case studies/problems considered	Key findings
Krishnan *et al.* (2018)	Shuffled Complex Evolutionary metropolis algorithm (SCEM-UA)	Parameter estimation of SWAT using SCEM-UA considering minimization of RMSE	St. Joseph River basin, USA	The major advantage of the SCEM-UA algorithm is that it helps quantify the uncertainty of the model parameters, along with the calibration process and can achieve the calibration process in less number of function evaluations of the model as compared to other optimization algorithms
Mostafaie *et al.* (2018)	NSGA-II, SPEA, MOPSO, and Pareto Envelope-based Selection Algorithm II (PESA-II)	Comparison of the multi-objective optimization techniques for calibration of GR4 J model (a conceptual hydrologic model) using *in situ* runoff and daily GRACE data for maximization of the NSE for flow and daily total water storage	Danube River Basin, located in the Central and Eastern Europe	NSGA-II produced the best results considering the diversity-based metric, while MOPSO produced better results considering the optimality criteria. All the four algorithms found to perform equally well considering the cardinality (no. of Pareto-optimal solutions obtained)
Koppa *et al.* (2019)	AMALGAM	Application of the DREAM simulation model and AMALGAM optimization model for multi-variate calibration of a large-scale distributed hydrologic model (Noah-MP) for minimization of RMSE for a different combination of model responses: Evapotranspiration (ET), surface moisture (SM) and Streamflow (SF)	Mississippi river basin	The multi-variate calibration process found to be much more efficient than a single variate calibration process
Tian *et al.* (2019)	ε-NSGA-II	Development and use of a single-objective function (i.e., minimization of the summation of a power function of the absolute error between observed and simulated streamflow with the exponent of power function) rather than employing multiple objectives for automatic calibration of hydrologic models	Model Parameter Estimation Experiment (MOPEX)	The single-objective optimization can achieve better hydrographs as compared to traditional NSE efficiency for most watersheds

to changing environments; and the available population of evolved solutions offers a basis for further improvement, and in most cases, it is not needed to reinitialize the population at random. Thus, EAs proved to be effective, especially for problems that are intractable by classic methods of optimization, and for cases where heuristic solutions are not accessible or generally lead to unsatisfactory results.

Table 9 | Applications of EA and SI meta-heuristic methods for groundwater remediation, groundwater systems, and monitoring network design

Author(s) (year)	Technique used	Work and objectives	Case studies/problems considered	Key findings
Dougherty & Marryott (1991)	Simulated Annealing (SA)	Application of SA for application of groundwater management problems considering minimization of cost (cost of well installation, operation, and maintenance, costs associated with water treatment and disposal and costs of failing to achieve specific management goals)	Dewatering problem with zooming and contamination problem	Simulated annealing finds good solutions to problems with highly variable objective functions, whereas gradient methods are very good at finding local solutions for the smooth objective function
McKinney & Lin (1994)	GA	Application of GA for maximization of pumping; minimization of cost of water supply; and minimum cost aquifer remediation (solved as three separate problems)	Three hypothetical examples	GA proves to be more efficient than traditional methods like LP, NLP, and DP, especially for complex problems that comprise of discontinuous or highly nonlinear or non-convex problems
Ritzel *et al.* (1994)	GA	Application of GA (and two variants of GA, vector evaluated GA (VEGA) and a Pareto GA) to a multi-objective groundwater pollution contaminant problem considering maximization of reliability (as a fraction of plumes captured) and minimization of system total cost	Hypothetical contamination problem	Pareto GA performs better than VEGA for zero fixed cost solution and obtains a Pareto front similar to that obtained via mixed-integer chance-constrained programming (MICCP)
Huang & Mayer (1997)	GA	Dynamic optimization formulation for groundwater remediation using GA considering well locations and pumping rates as the decision variables considering minimization of cost (installation, pumping and treatment cost)	A hypothetical, homogeneous aquifer system and a spatially correlated heterogeneous aquifer system	The moving well location model produced better results as compared to fixed well location models. The model results are found to be more sensitive to well locations rather than the pumping rates. GA found to be an effective technique for solving the GW remediation problem, but more efforts suggested for improving computational efficiency for a larger problem with more number of decision variables
Wang & Zheng (1997)	GA	Simulation–optimization model for GW remediation by coupling GA with MODFLOW and MT3D, considering dynamic pumping rates for minimization of the net present value of remedial and operational cost	A hypothetical homogeneous test problem and a real case study at a site near Granger, Indiana	GA-based method has specific advantages as compared to gradient-based methods such as ease of handling non-linearity, discrete decision variables, able to find optimal (or at least sub-optimal solutions), by considering proper population size and number of iterations

(continued)

Table 9 | continued

Author(s) (year)	Technique used	Work and objectives	Case studies/problems considered	Key findings
Sun & Zheng (1999)	GA, DP	Dynamic optimization of long-term groundwater management problem (to determine the optimal pumping rates) using DOMODF optimization tool based on differential dynamic programming linked with MODFLOW as the simulation tool for minimization of cost	Two hypothetical problems	The combined use of GA (or DDP) and MODFLOW could form an effective tool for solving GW management problems. The method has certain shortcomings as DDP converges only locally while DP converges globally; DDP is also more error sensitive than GA; DOMODF also subjected to the effects of uncertainties in calibrated flow models
Smalley *et al.* (2000)	Noisy GA	Risk-based in situ bioremediation design (which considers risks related to human health and the environment associated with a contaminated site) for minimization of the total cost (capital and operating costs for remediation wells, site monitoring costs, additional capital and operating costs associated with remedial systems)	An in situ bioremediation case study and a realistic case study from Borden site	Noisy GA is capable of identifying highly reliable designs from a small number of samples. It can serve as an efficient tool for modeling parameter uncertainty and variability
Yoon & Shoemaker (2001)	Real-coded GA (RGA)	Application of RGA to bioremediation problem and compared with that of binary-coded GA (BCGA) for minimization of the total pumping cost	Two hypothetical aquifers	RGA found to be more computationally efficient and accurate than BCGA
Zheng & Wang (2002)	GA	Application of the simulation-optimization (SO) methodology (comprising of MODFLOW as the simulation tool and GA as the optimization tool) on real case study for maximizing Trichloroethylene (TCE) mass removal	Massachusetts Military Reservation (MMR) in the north-eastern United States	The GA-based SO methodology could be successfully applied on desktop PCs. Considerable cost savings can be achieved by adopting dynamic pumping
Erickson *et al.* (2002)	Niched Pareto GA (NPGA)	Application of the multi-objective optimization algorithm NPGA for the optimal design of GW remedial systems for minimization of remediation cost and contaminant mass remaining at the end of the remediation horizon	A hypothetical contaminated GW site for two, three and 15 fixed well locations cases	NPGA performed better than SGA and RS, especially in the case of 15 well scenarios requiring 30% less computational efforts as compared to SGA. NPGA lead to better solutions in terms of 75% less remedial contaminant mass
Shieh & Peralta (2005)	Hybrid SA and GA termed as parallel recombinative SA (PRSA)	Application of PRSA for optimizing in situ bioremediation system design considering minimization of total system cost, and minimization of cost of time-varying pumping strategy	Hypothetical study area	PRSA minimizes the total system cost better than SA and GA. Optimal time-varying pumping strategy requires 31% less pumping cost than optimal steady pumping strategy

(continued)

Table 9 | continued

Author(s) (year)	Technique used	Work and objectives	Case studies/problems considered	Key findings
Babbar & Minsker (2006)	Multi-scale GA	Application of multi-scale GA (different than the normal GA in the way it acquires information about the fitness function and the way it manipulates the new individuals) for GW remediation problem with minimization of the total treatment cost	Hypothetical GW remediation problem and a field-scale GW application at Umatilla Chemical Depot in Oregon	The multi-scale GA leads to computational savings of around 77% for a hypothetical case and 78% for the real case study
Wu *et al.* (2006)	MCSGA and noisy GA	A comparison of Monte Carlo Simulation GA (MCSGA) and Noisy GA (NGA) for the optimal design of sampling network design under uncertainty in hydraulic conductivity for minimization of the total of well installation and sampling cost	A two-dimensional confined aquifer measuring600 m long in the x-direction and 400 m wide in the y-direction	The solution obtained using NGA leads to 45% cost saving as compared to that obtained using MCSGA. Compared to MCSGA, NGA reduces the optimization runtime by a factor of 6.5
Park *et al.* (2007)	GA	Development of an Enhanced natural attenuation (ENA) optimization process for optimal GW remediation design for minimization of the remediation cost	A hypothetical 2D unconfined aquifer	The ENA process led to improved efficiency of the remediation strategy
Mantoglou & Kourakos (2007)	MOGA	Development of a methodology for optimal remediation of GW aquifers under hydraulic conductivity uncertainty for minimization of contaminated groundwater in the aquifer and minimization of remediation cost	A hypothetical orthogonal unconfined aquifer	The MOGA has produced good results. In the example considered, only 11 realizations out of 100 were found to be critical, implying 89% saving in computational time
Tang *et al.* (2007)	ε-NSGA-II	Demonstration and evaluation of the Master-Slave (MS) and Multiple-Population (MP) parallelization schemes for NSGA-II considering minimization of the monitoring cost (for GW monitoring application)	DTLZ optimization function, hydrologic model calibration test case for the Leaf River near Collins, Mississippi, and a GW monitoring application	The ε-NSGA-II method found to be generating good Pareto-optimal solutions. Multi-population was found to be superior to the Master-slave scheme only for the DTLZ optimization problem, while for other problems, MS is found to perform better than MP
Chadalavada & Datta (2008)	GA	Development of a model for optimal GW pollution monitoring network design to determine the optimal and efficient sampling locations for detecting pollution in GW aquifers considering minimization of the summation of unmonitored concentrations at different potential monitoring locations; and minimization of estimation variances of pollutant concentrations at various unmonitored locations	An illustrative study area comprising of a hypothetical aquifer	The designed network of optimal monitoring wells is dynamic in nature, which results in economically efficient designs. However, the stochastic nature of the processes involved has not been explicitly incorporated into the model

(continued)

Table 9 | continued

Author(s) (year)	Technique used	Work and objectives	Case studies/problems considered	Key findings
Kobayashi *et al.* (2008)	SA	Development of a SA based SO model for multiphase/ multicomponent flow simulation, by incorporating a parallelization procedure for the optimization model considering maximization of the amount of methane extracted from the wells	A two-phase (liquid, gas)/ three-component (water, methane, air) conceptual model, and a coal mining site in the Ruhr, Germany	The SA based methodology and parallelization scheme enhanced the computational efficiency of the overall procedure almost linearly by the number of processors in the parallel computer
Bayer *et al.* (2008)	GA	Development of a computationally efficient method for stochastic optimization problems which involves multiple equally probable realisations of the uncertain parameters for maximization of the pumping rates of all wells	A synthetic groundwater model set-up	The computational efficiency is improved manifolds by applying the stack ordering procedure. In the case of more advanced stack ordering strategies, almost 97% of the computational efficiency can be achieved. The presented approach is also independent of the problem to be solved
Singh *et al.* (2008)	Interactive MOGA (IMOGA)	Development of an interactive multi-objective model for groundwater inverse modeling for estimating heterogeneous GW parameters for minimization of the calibration error with respect to field measurements, and a qualitative objective that reflects an expert's qualitative preferences	A hypothetical aquifer case study consisting of six pumping wells and 16 observation wells	The results obtained without considering the quantitative preferences lead to sub-optimal solutions. IMOGA found to be an efficient tool for incorporating expert knowledge into the multi-objective formulation
Singh & Chakrabarty (2011)	NSGA-II	Application of NSGA-II to solve GW remediation problem considering minimization of remediation cost and minimization of remediation time	A hypothetical confined multi-layered aquifer	The NSGA-II generated a good number of Pareto-optimal solutions; and noted that remediation time plays a crucial role in the optimal design of GW remedial systems
Luo *et al.* (2016)	Probabilistic Pareto GA (PPGA)	Optimal design of long-term groundwater monitoring network (LTGM), considering the uncertainty in K-fields for minimization of (i) the total sampling costs for monitoring contaminant plume, (ii) mass estimation error, (iii) the first-moment estimation error, and (iv) the second-moment estimation error of the contaminant plume	A two-dimensional hypothetical example and a three-dimensional field application in Indiana (USA)	The optimization results show the high accuracy of the PPGA. However, the computational burden was extensively large when considering the uncertainties
Ouyang *et al.* (2017)	AMALGAM	Application of AMALGAM for GW remediation design and comparison with NSGA-II for minimization of the total remediation cost, and minimization of the remediation time	A hypothetical perchloroethylene (PCE) contaminated site	The AMALGAM results incurred less remediation cost and required less computational time as compared to NSGA-II

(continued)

Table 9 | continued

Author(s) (year)	Technique used	Work and objectives	Case studies/problems considered	Key findings
Alizadeh *et al.* (2018)	DE	Development and application of an entropy-based method for GW monitoring network design for maximization of the actual (net) value of the information among wells	Eshtehard aquifer monitoring network in the central part of Iran	The DE optimization-entropy yields better performance with a larger amount of information content than the clustering and the estimation error methods
Seyedpour *et al.* (2019)	GA	Development of a coupled groundwater flow and reactive transport of contaminant and oxidant using the Multi-quadratic Radial Basis Function (MRBF) in the Meshfree method, and optimization of the GW remediation problem considering maximization of remediation contaminant concentration and minimization of the cost of the remediation process	A point source of 0.5% w/v potassium permanganate solution was constructed in a sandbox to map the change groundwater plume distribution over time	The optimization methodology employed leads to not only delay in the reaching time of the contaminant to the d/s region but also decrease in the contaminant concentration
Song *et al.* (2019)	ε-multi-objective noisy memetic algorithm (ε-MONMA)	Optimal design of groundwater monitoring network using a surrogate for modeling the uncertainty in k-field considering minimization of the monitoring cost, and error objective functions in terms of the mass of contaminant plume, the estimated center of contaminant plume, and contaminant concentrations	A hypothetical unconfined aquifer 600 m in longitudinal extent, 400 m in transverse extent, and a model thickness of 10 m	The ε-MONMA gave a good performance. The surrogate assisted design outperforms the deterministic and Monte Carlo simulation-based design in terms of huge savings in the computational time

Which evolutionary algorithm to use?

There were several variants of EAs. Each of the meta-heuristic algorithms has its own advantages and disadvantages. Among the nature-inspired EAs, GA is one of the oldest and popular techniques that has several applications in the water resources field. The main advantages include its ability to handle nonlinear, nonconvex, nondifferentiable functions, multi-modal solutions and can provide optimal or near-optimal solutions to a given problem. Real-coded GA has advantages over binary-coded GA for real-valued decision variable problems. However, GA requires the selection of appropriate genetic operators among several versions available and proper tuning of the parameters such as probabilities of crossover and mutation, population size, etc. Also, the optimization process may take higher computational time for complex water resources problems such as groundwater systems monitoring design and remediation, WDN problems, etc., as it may involve a time-consuming simulation–optimization process. The other popular algorithm, differential evolution, also has advantages similar to that of GA; apart from that, it is proved that DE has faster convergence and reliable, optimal solutions for numerical optimization. DE also has limitations in selecting an appropriate version of DE and algorithm parameters. Similarly, other EAs (like SCE, ES) are found to have similar capabilities and difficulties while solving water resources optimization problems. A brief summary of the comparison of basic characteristics and quality performance features of different EA (GA, DE, ES) are given in Table 10. However, the recent developments of self-adaptive EAs (e.g., SADE) are found to be overcoming this issue and helping the tuning of the algorithm parameters as the search progresses.

doi: 10.2166/h2oj.2020.128

Table 10 | Comparison of basic characteristics and quality performance features of different evolutionary algorithms of genetic algorithms (GA), differential evolution (DE), and evolution strategies (ES); and swarm intelligence algorithms of particle swarm optimization (PSO), ant colony optimization (ACO), and artificial bee colony (ABC) algorithms

Feature/algorithm	GA	DE	ES	PSO	ACO	ABC
Representation	Binary or real valued	Real-valued	Real-valued	Real valued	Discrete, graphical representation	Real valued
Fitness	Scaled objective value	Objective function value	Objective function value	Objective function value	Scaled objective value	Scaled objective value
Operations for the evolution of new solutions	Crossover (main operator), Mutation (secondary operator)	Mutation (main operator), Crossover (secondary operator)	Mutation operator (only operator)	Particle velocity and position updating rules	Updating of pheromone intensities, heuristic information	Bee position updating rules and stochastic process
Selection process	Deterministic, extinctive	Probabilistic, preservative	Probabilistic, preservative	Deterministic, extinctive	Probabilistic, preservative	Probabilistic, preservative
Suitable for nonlinear problem? type of variables	Yes, Real/discrete value	Yes, Real/discrete value	Yes, Real/ discrete value	Yes, Real/ discrete value	Yes, Discrete value	Yes, Discrete/ real value
Preferred population size, generations (or iterations)	Small to medium, large	Small to medium, large	Medium, large	Larger, medium	Medium, large	Medium, large
Influence of population size on solution time	High	Low	Low	Low	Medium-to-high	Medium-to-high
Influence of best solution on population	Medium	Less	Less	High	Medium	Medium
Ability-to-parallelization of the search	Yes	Yes	Yes	Yes	Yes	Yes
Tendency for premature convergence	Medium	Low	Low-to-medium	Medium-to-high	Medium	Medium-to-high

Which SI algorithm to use?

Several variants of the SI algorithms (like PSO, ACO, ABC, HBMO, FFA, BA, etc.) were proposed in different studies. The SI methods are basically inspired by co-operative group intelligence principles of the swarm and proved to be other classes of alternative meta-heuristic techniques for solving different kinds of optimization problems in water resources. Similar to EAs, SI algorithms are population-based random search techniques guided with some heuristics, which can also handle different types of problem complexities (like nonlinear, nonconvex, nondifferentiable functions, multi-modal solutions, etc.). But their applicability and convergence characteristics may vary from problem to problem. For example, PSO is a technique applicable for real-valued decision variables, has advantages of easiness in coding the algorithm, provides fast convergence to simple numerical problems, and requires low computational time. However, fine-tuning of algorithm parameters (e.g., inertia weight, social, and cognitive parameters) is required for getting optimal solutions and consistently good performance from the method. But, the basic PSO may face difficulties like premature convergence to locally optimal solutions for higher dimensional, large-scale water resources problems. Several studies proposed different strategies for the improvement of the PSO performance (like EMPSO proposed by Kumar & Reddy 2007). The ACO method is also a random search method guided by probabilistic rules and some heuristics (i.e., pheromone laying behavior of ants) and can produce global-optimal solutions. But ACO is applicable effectively only for a set of problems that involve decision variables with discrete values (for example, the WDNs problem can be represented in graphical form/shortest path that can be easily covered by the ants, and it can be modeled with pheromone trail laying behavior). There are several variants of ACO algorithms. It has similar advantages to other EAs in handling different kinds of complexities. However, the performance of ACO is sensitive to its algorithmic parameters, so it may require tuning of the parameters such as relative pheromone trail, heuristic information, evaporation rate, etc. while solving large-scale problems. Other SI algorithms (like ABC, HBMO, CS, FFA, BA algorithms) also have similar capabilities and limitations like that of PSO. A brief summary of basic characteristics and quality performance features a comparison of different EAs (PSO, ACO, and ABC) that are presented in Table 10.

It should also be noted that different studies have presented different algorithms for solving water problems and noted that some algorithms are relatively better than other methods. But, the inferences made may be applicable for those problems, which may not be generalized to one or not applicable for all types of problems, as each problem may have varying complexity depending on the dimension of the problem and existing relationships for the problem in hand. So, proper selection and usage of the algorithm for a given problem should be based on the problem type and its characteristics.

Hybrid meta-heuristic methods?

Research studies also reported that hybrid meta-heuristic methods that combine global-search with local search algorithms found to be providing improved performance for large-scale water resources optimization problems. Also, several strategies like problem-specific heuristics, chaotic concepts into the initialization of population, elitism concepts, self-adaptive schemes, etc., are incorporated into SI algorithms to improve the performance of the existing methods and to get consistent satisfactory results. Also, to handle the issues related to multi-objective optimization, different studies have suggested different schemes for getting an array of uniform widespread and true Pareto-optimal solutions for different kinds of practical problems.

FUTURE WORK AND RESEARCH DIRECTIONS

There are several issues which require more research efforts to improve the solutions to the practical problems in the water resources engineering area. Some of these issues and future research directions are listed below.

Improvement in algorithm and solution methodology

There is a great deal of research on-going on the development of the new algorithms, a better understanding of the algorithm's performance and search behavior, analyzing the suitability in handling the different complexities of the problems, etc. As the meta-heuristic methods are population-based search algorithms and involve several parameters (like population size, maximum generations/iterations, other algorithm-specific parameters) to start the optimization procedure, and also the model performance is sensitive to the selected parameters. This warrants for proper fine-tuning of parameters before applying to a given problem. To address this issue of how to overcome the effect of

doi: 10.2166/h2oj.2020.128

the sensitivity of the parameters on the performance of the algorithm, recently, several studies explored self-adaptive schemes for overcoming this issue. For improving computational efficiency, studies also explored guided initialization of population, parallelization schemes (parallel processing that uses two more CPUs for computations saves time, which is well suited for population-based EA/SI methods that use simulation–optimization framework for solving complex real-world problems), the multi-algorithm search that combines two or more (global and/or local) search algorithms, etc. But still, there is a scope for improvements and more research is needed in this area. Moreover, it is important to develop knowledge related to the suitability of various optimization methods for a particular type of water resource problem and the incorporation of domain knowledge in the search, which can also help in improved solution methodology.

Search space reduction and improving computational efficiency

The complexity of the real-world problems (especially for problems in the areas of water distribution systems, groundwater systems management, calibration of distributed hydrological models) is increasing day-by-day, and it is expected to continue critical and promising research areas in the future also. Recently several studies are focusing on the issues of how to reduce the search space and guide the algorithms to speed up the convergence and/or improve the reliability in achieving global-optimal solutions. Schemes like fitness approximation, parallelization, and multi-algorithm search frameworks have been explored. But, still much more research is needed to come up with better strategies.

Model inputs and uncertainty

Since the model inputs may involve different kinds of uncertainty, there is a necessity to explore how to best represent various types of uncertainties (such as aleatoric and epistemic uncertainty) in the optimization model and incorporate them into decision making. Also, future uncertainties due to climate change, urbanization, etc., might affect the planning of water resources systems, and exploring new frameworks and evolving flexible design is one of the promising areas for future research.

Solution post-processing and decision making

The solutions obtained by solving an optimization model are sensitive to a given input variables condition. This poses doubts of how reliable are the solutions, considering the changing conditions or input variable values. In the case of multi-objective optimization (with more than two objectives), the MOEAs generate a large number of Pareto-optimal solutions. Then the challenge is how to use them for decision making, as it is difficult to visualize and choose the solutions. So, more efforts are needed to come up with effective decision-making procedures, which can help and equip the users in envisaging the trade-offs and assist in arriving at few practical alternatives and/or facilitate effective decision making under several conflicting goals.

CONCLUDING REMARKS

Real-life water resource planning and management problems may involve several complexities and may pose several challenges for decision-makers. The use of evolutionary algorithms for solving numerical and practical optimization problems has become very attractive and extensive in the last three decades. Several new meta-heuristic algorithms including EA (such as GA, DE, SCE, etc.) and SI algorithms (such as PSO, ACO, ABC, SFL, HBMO, HS, FFA, CS, BA, etc.) are being proposed, which showed improved performance for solving a variety of water resources problems. The main advantage of EAs is the usage of a population of potential solutions that explore the search space simultaneously, exchanging information among them, and uses only objective function values. Also, EAs are stochastic search algorithms, can move to any complicated search apace and locate near-optimal (or optimal) solutions in reasonable computational time. They can provide solutions to any complex optimization problem that is difficult to be solved with the conventional NLP methods due to their nature that may imply discontinuities of the search space, nondifferentiable objective functions, imprecise arguments and function values.

Some of the remarks on the current state-of-the-art in EAs:

- There is no general algorithm that can be applied efficiently to all problems, as the efficiency varies as a function of problem size and complexity. However, the incorporation of problem-specific knowledge and heuristics may help to achieve faster and efficient solutions to a real-world problem.
- EAs may require calibration of the search parameters to ensure efficient convergence. The self-adaptive EA as part of improving the solution methodology can help to arrive quickly at near-optimal (or optimal) solutions, thereby helping water resources engineers in a better decision-making process.
- For complex problems, EAs may require a large number of simulations to find optimal solutions. In such cases, the use of meta-modeling, fitness approximation, parallelization schemes may help to speed up the simulation–optimization process, thereby helping in improving computational efficiency.
- The different types of problem complexities and their associated uncertainties motivate a great need for advances in multi-objective search, interactive optimization, and multi-algorithm search frameworks.

Therefore, EAs continue growing interest among the research community, and there is a tremendous potential for solving many practical water resources problems.

ACKNOWLEDGEMENTS

Authors acknowledge Ms Swati Sirsant (Civil Engineering, IIT Bombay) for assistance in compiling some of the literature used in the study.

REFERENCES

Abbass, H. A. 2001 Marriage in honey bees optimisation: a haplometrosis polygynous swarming approach. In: *Congress on Evolutionary Computation, CEC2001*. Vol. 1. Seoul, Korea, pp. 207–214.

Afshar, M. H., Afshar, A., Marino, M. A. & Darbandi, A. A. 2006 Hydrograph-based storm sewer design optimization by genetic algorithm. *Canadian Journal of Civil Engineering* **33** (3), 319–325.

Ahrari, A., Deb, K. & Preuss, M. 2017 Multimodal optimization by covariance matrix self-adaptation evolution strategy with repelling subpopulations. *Evolutionary Computation* **25** (3), 439–471.

Akimoto, Y., Auger, A. & Hansen, N. 2018 Quality gain analysis of the weighted recombination evolution strategy on general con-vex quadratic functions. *Theoretical Computer Science*. https://doi.org/10.1016/j.tcs.2018.05.015.

Alamdari, N., Sample, D. J., Steinberg, P., Ross, A. C. & Easton, Z. C. 2017 Assessing the effects of climate change on water quantity and quality in an urban watershed using a calibrated stormwater model. *Water* **9** (7), 464.

Ali, W. M. 2014 RDEL: Restart differential evolution algorithm with local search mutation for global numerical optimization. *Egyptian Informatics Journal* **15** (3), 175–188.

Alizadeh, Z., Yazdi, J. & Moridi, A. 2018 Development of an entropy method for groundwater quality monitoring network design. *Environmental Processes* **5** (4), 769–788.

Aminjavaheri, S. M. & Nazif, S. 2018 Determining the robust optimal set of BMPs for urban runoff management in data-poor catchments. *Journal of Environmental Planning and Management* **61** (7), 1180–1203.

Arnold, D. V. 2006 Weighted multi recombination evolution strategies. *Theoretical Computer Science* **361** (1), 18–37.

Artita, K. S., Kaini, P. & Nicklow, J. W. 2013 Examining the possibilities: generating alternative watershed-scale BMP designs with evolutionary algorithms. *Water Resources Management* **27** (11), 3849–3863.

Babbar, M. & Minsker, B. S. 2006 Groundwater remediation design using multiscale genetic algorithms. *Journal of Water Resources Planning and Management* **132** (5), 341–350.

doi: 10.2166/h2oj.2020.128

Barreto, W. J., Vojinovic, Z., Price, R. K. & Solomatine, D. P. 2006 Approaches to multi-objective multi-tier optimization in urban drainage planning. In: *7th International Conference Proceedings on Hydroinformatics*, HIC, Nice, France, pp. 1–8.

Barreto, W., Vojinovic, Z., Price, R. & Solomatine, D. 2010 Multiobjective evolutionary approach to rehabilitation of urban drainage systems. *Journal of Water Resources Planning and Management* **136** (5), 547–554.

Bayer, P., Bürger, C. M. & Finkel, M. 2008 Computationally efficient stochastic optimization using multiple realizations. *Advances in Water Resources* **31** (2), 399–417.

Bellman, R. E. 1957 *Dynamic Programming*. Princeton University Press, Princeton, New Jersey.

Beyer, H. G. & Sendhoff, B. 2017 Simplify your covariance matrix adaptation evolution strategy. *IEEE Transaction on Evolutionary Computation* **21** (5), 746–759.

Bhushan, B. & Pillai, S. S. 2013 Particle swarm optimization and firefly algorithm: performance analysis. In: *3rd IEEE International Advance Computing Conference*. IEEE, Ghaziabad.

Blum, C. 2008 Beam-ACO for simple assembly line balancing. *INFORMS Journal on Computing* **20** (4), 618–627.

Bonabeau, E., Dorigo, M. & Theraulaz, G. 1999 *Swarm Intelligence: From Natural to Artificial Systems*. Oxford University Press, New York, NY.

Brest, J., Greiner, S., Boskovic, B., Mernik, M. & Zumer, V. 2006 Self-adapting control parameters in differential evolution: a comparative study on numerical benchmark problems. *IEEE Transactions on Evolutionary Computation* **10** (6), 646–657.

Brest, J., Zamuda, A., Fister, I. & Boskovic, B. 2014 Some improvements of the self-adaptive jDE algorithm. In: *IEEE Symposium on Differential Evolution (SDE)*. pp. 1–8.

Brownlee, J. 2011 *Clever Algorithms: Nature-Inspired Programming Recipes*. LuLu.com, Australia.

Chadalavada, S. & Datta, B. 2008 Dynamic optimal monitoring network design for transient transport of pollutants in groundwater aquifers. *Water Resources Management* **22** (6), 651–670.

Chen, S.-M. & Chien, C.-Y. 2011 Solving the traveling salesman problem based on the genetic simulated annealing ant colony system with particle swarm optimization techniques. *Expert Systems with Applications* **38** (12), 14439–14450.

Chen, W. C., Chang, N. B. & Chen, J. C. 2003 Rough set-based hybrid fuzzy-neural controller design for industrial wastewater treatment. *Water Research* **37** (1), 95–107.

Chen, H., Zhu, Y. & Hu, K. 2010 Multi-colony bacteria foraging optimization with cell-to-cell communication for RFID network planning. *Applied Soft Computing* **10**, 539–547.

Cheung, P. B., Reis, L. F. R., Formiga, K. T. M., Chaudhry, F. H. & Ticona, W. G. C. 2003 Multiobjective evolutionary algorithms applied to the rehabilitation of a water distribution system: a comparative study. In: *International Conference on Evolutionary Multi-Criterion Optimization*. Springer, Berlin, Heidelberg, pp. 662–676.

Choi, T. J. & Ahn, C. W. 2014 An adaptive cauchy differential evolution algorithm with bias strategy adaptation mechanism for global numerical optimization. *Journal of Computers* **9** (9), 2139–2145.

Choi, T. J., Ahn, C. W. & An, J. 2013 An adaptive Cauchy differential evolution algorithm for global numerical optimization. *Scientific World Journal* **2013**, 12. article ID: 969734.

Chung, G. & Lansey, K. 2009 Application of the shuffled frog leaping algorithm for the optimization of a general large-scale water supply system. *Water Resources Management* **23** (4), 797.

Clerc, M. 1999 The swarm and the queen: towards a deterministic and adaptive particle swarm optimization. In: *IEEE Congress on Evolutionary Computation*. pp. 1951–1957.

Clerc, M. 2004 *Discrete Particle Swarm Optimization, Illustrated by the Traveling Salesman Problem, New Optimization Techniques in Engineering*. Springer, pp. 219–239.

Creaco, E., Fortunato, A., Franchini, M. & Mazzola, M. R. 2014 Comparison between entropy and resilience as indirect measures of reliability in the framework of water distribution network design. *Procedia Engineering* **70**, 379–388.

Cunha, M. & Ribeiro, L. 2004 Tabu search algorithms for water network optimization. *European Journal of Operational Research* **157** (3), 746–758.

Dandy, G. C. & Engelhardt, M. 2001 Optimal scheduling of water pipe replacement using genetic algorithms. *Journal of Water Resources Planning and Management* **127** (4), 214–223.

Dandy, G. C. & Engelhardt, M. O. 2006 Multi-objective trade-offs between cost and reliability in the replacement of water mains. *Journal of Water Resources Planning and Management* **132** (2), 79–88.

Das, S., Mullick, S. S. & Suganthan, P. N. 2016 Recent advances in differential evolution – an updated survey. *Swarm and Evolutionary Computation* **27**, 1–30.

Deb, K. & Jain, H. 2014 An evolutionary many-objective optimization algorithm using reference-point-based nondominated sorting approach, part I: solving problems with box constraints. *IEEE Transactions on Evolutionary Computation* **18** (4), 577 601.

Deb, K., Pratap, A., Agarwal, S. & Meyarivan, T. 2002 A fast and elitist multiobjective genetic algorithm: NSGA-II. *IEEE Transaction on Evolutionary Computation* **6** (2002), 182–197.

DeléVacq, A., Delisle, P., Gravel, M. & Krajecki, M. 2013 Parallel ant colony optimization on graphics processing units. *Journal of Parallel and Distributed Computing* **73** (1), 52–61.

Dessalegne, T., Nicklow, J. & Minder, E. 2004 Evolutionary computation to control unnatural water level fluctuations in multi-reservoir river systems. *River Research and Applications* **20** (6), 619–634.

Di Caro, G. & Dorigo, M. 1998 AntNET: distributed stigmergetic control for communications networks. *Journal of Artificial Intelligence Research* **1998** (9), 317–365.

Dorigo, M. & Gambardella, L. M. 1997 Ant colony system: a cooperative learning approach to the traveling salesman problem. *IEEE Transactions on Evolutionary Computation* **1** (1), 53–66.

Dorigo, M. & Stutzle, T. 2004 *Ant Colony Optimization*. MIT Press, Cambridge, Massachusetts.

Dorigo, M., Maniezzo, V. & Colorni, A. 1991 *Positive Feedback as A Search Strategy*. Technical Report 91-016. Politecnico di Milano, Italy.

Dorigo, M., Maniezzo, V. & Colorni, A. 1996 Ant system: optimization by a colony of cooperating agents. . *IEEE Transactions on Systems, Man, and Cybernetics, Part-B* **26** (1), 29–41.

Dougherty, D. E. & Marryott, R. A. 1991 Optimal groundwater management: 1. Simulated annealing. *Water Resources Research* **27** (10), 2493–2508.

Draa, A., Chettah, K. & Talbi, H. 2019 A compound sinusoidal differential evolution algorithm for continuous optimization. *Swarm and Evolutionary Computation* **50**, 100450.

Eberhart, R. C. & Kennedy, J. 1995 A New optimizer using particle swarm theory. In: *Proceedings of 6th Symposium on Micro Machine & Human Science*. IEEE Service Center, Piscataway, New Jersey, pp. 39–43.

Elloumi, W., Baklouti, N., Abraham, A. & Alimi, A. M. 2014 The multi-objective hybridization of particle swarm optimization and fuzzy ant colony optimization. *Journal of Intelligent and Fuzzy Systems* **27** (1), 515–525.

Erickson, M., Mayer, A. & Horn, J. 2002 Multi-objective optimal design of groundwater remediation systems: application of the niched Pareto genetic algorithm (NPGA). *Advances in Water Resources* **25** (1), 51–65.

Eusuff, M. M. & Lansey, K. E. 2003 Optimization of water distribution network design using the shuffled frog leaping algorithm. *Journal of Water Resources Planning and Management* **129** (3), 210–225.

Falkenauer, E. 1996 The grouping genetic algorithm. In: *State of the Art in Global Optimization*. pp. 249–265.

Fallah, H., Ghazanfari, S., Suribabu, C. R. & Rashedi, E. 2019 Optimal pipe dimensioning in water distribution networks using Gravitational Search Algorithm. *ISH Journal of Hydraulic Engineering* 1–14. https://doi.org/10.1080/09715010.2019.1624630.

Farmani, R., Savic, D. A. & Walters, G. A. 2005 Evolutionary multi-objective optimization in water distribution network design. *Engineering Optimization* **37** (2), 167–183.

Fogel, D. B. 1994 An introduction to simulated evolutionary optimization. *IEEE Transactions on Neural Networks* **5** (1), 3–14.

Fogel, L. J., Owens, A. J. & Walsh, M. J. 1966 *Artificial Intelligence Through Simulated Evolution*. Wiley, New York.

Franchini, M. & Galeati, G. 1997 Comparing several genetic algorithm schemes for the calibration of conceptual rainfall-runoff models. *Hydrological Sciences Journal* **42** (3), 357–379.

Fu, G., Butler, D. & Khu, S. T. 2008 Multiple objective optimal control of integrated urban wastewater systems. *Environmental Modelling and Software* **23** (2), 225–234.

Gambardella, L. M., Taillard, E. & Agazzi, G. 1999 MACS-VRPTW: a multiple ant colony system for vehicle routing problems with time windows. In: *New Ideas in Optimization* (Corne, D., Dorigo, M. & Glover, F., eds). McGraw-Hill, London, UK, pp. 63–76.

Geem, Z. W. 2006 Optimal cost design of water distribution networks using harmony search. *Engineering Optimization* **38** (03), 259–277.

Geem, Z. W., Kim, J. H. & Loganathan, G. V. 2001 A new heuristic optimization algorithm: harmony search. *Simulation* **76** (2), 60–68.

Gessler, J. 1985 Pipe network optimization by enumeration. In: *Computer Applications in Water Resources*. ASCE, pp. 572–581.

Ghasemi, M., Taghizadeh, M., Ghavidel, S. & Abbasian, A. 2016 Colonial competitive differential evolution: an experimental study for optimal economic load dispatch. *Applied Soft Computing* **40**, 342–363.

Ghimire, B. N. S. & Reddy, M. J. 2013 Optimal reservoir operation for hydropower production using particle swarm optimization and sustainable analysis of hydropower. *ISH Journal of Hydraulic Engineering* **19** (3), 196–210.

Glasmachers, T., Schaul, T., Sun, Y., Wierstra, D. & Schmidhuber, J. 2010 Exponential natural evolution strategies. In: *Proceedings of the Genetic and Evolutionary Computation Conference*. GECCO'10, Portland, Oregon, USA, pp. 393–400. https://doi.org/10.1145/1830483.1830557.

Goh, C. K., Tan, K. C., Liu, D. & Chiam, S. C. 2010 A competitive and cooperative co-evolutionary approach to multi-objective particle swarm optimization algorithm design. *European Journal of Operations Research* **202**, 42–54.

Goldberg, D. E. 1989 *Genetic Algorithms in Search, Optimization and Machine Learning*, 1st edn. Addison-Wesley Longman Publishing Co., Toronto, Ontario, Canada.

Guntsch, M. & Middendorf, M. 2002 A population based approach for ACO. In: *Workshops on Applications of Evolutionary Computation*. Springer, pp. 72–81.

Guo, Y., Walters, G. A., Khu, S. T. & Keedwell, E. 2006 Optimal design of sewer networks using hybrid cellular automata and genetic algorithm. In: *Proceedings of the 5th IWA WorldWater Congress, London*. IWA Publishing, London.

Guria, C., Bhattacharya, P. K. & Gupta, S. K. 2005 Multi-objective optimization of reverse osmosis desalination units using different adaptations of the non-dominated sorting genetic algorithm (NSGA). *Computers and Chemical Engineering* **29** (9), 1977–1995.

Halhal, D., Walters, G. A., Ouazar, D. & Savic, D. A. 1997 Multiobjective improvement of water distribution systems using a structured messy genetic algorithm approach. *Journal of Water Resources Planning and Management ASCE* **123** (3), 137–146.

Hang, N. T. T. & Chikamori, H. 2017 Comparison of efficiency between differential evolution and evolution strategy: application of the LST model to the Be River catchment in Vietnam. *Paddy and Water Environment* **15**, 797–808.

doi: 10.2166/h2oj.2020.128

Hansen, N. & Ostermeier, A. 2001 Completely derandomized self-adaptation in evolution strategies. *Evolutionary Computation* **9** (2), 159–195.

Harik, G. R. 1997 *Learning Gene Linkage to Efficiently Solve Problems of Bounded Difficulty Using Genetic Algorithms.* PhD Thesis, University of Michigan, Ann Arbor, MI.

Harrell, L. J. & Ranjithan, S. R. 2003 Detention pond design and land use planning for watershed management. *Journal of Water Resources Planning and Management* **129** (2), 98–106.

He, Y. & Hui, C.-W. 2006 Dynamic rule-based genetic algorithm for large-size single-stage batch scheduling. *Comput-Aided Chemical Engineering* **21**, 1911–1916.

Holland, J. H. 1975 *Adaptation in Natural and Artificial Systems.* University of Michigan Press, Ann Arbor, Michigan.

Huang, C. & Mayer, A. S. 1997 Pump-and-treat optimization using well locations and pumping rates as decision variables. *Water Resources Research* **33** (5), 1001–1012.

Huang, C.-L., Huang, W.-C., Chang, H.-Y., Yeh, Y.-C. & Tsai, C.-Y. 2013 Hybridization strategies for continuous ant colony optimization and particle swarm optimization applied to data clustering. *Applied Soft Computing Journal* **13** (9), 3864–3872.

Iredi, S., Merkle, D. & Middendorf, M. 2001 Bi-criterion optimization with multi colony ant algorithms. In: *Evolutionary Multi-Criterion Optimization, First International Conference (EMO'01).* Springer Verlag, Zurich, pp. 359–372.

Jafari-Marandi, R. & Smith, B. K. 2017 Fluid genetic algorithm (FGA). *Journal of Computational Design and Engineering* **4**, 158–167.

Janson, S., Alba, E., Dorronsoro, B. & Middendorf, M. 2006 Hierarchical cellular genetic algorithm. In: *Proceedings of the European Conference on Evolutionary Computation in Combinatorial Optimization.* EvoCOP. Lecture Notes in Computer Science. Springer, Berlin, Heidelberg, Vol. 3906, pp. 111–122. https://doi.org/10.1007/11730095.

Jayaram, N. & Srinivasan, K. 2008 Performance-based optimal design and rehabilitation of water distribution networks using life cycle costing. *Water Resources Research* **44** (1), W01417, 1–15. https://doi.org/10.1029/2006WR005316.

Jung, J. J., Jo, G.-S. & Yeo, S.-W. 2007 Meta-evolution strategy to focused crawling on semantic web. In: International Conference on Artificial Neural Networks, ICANN 2007. Lecture Notes in Computer Science, Vol. 4669. Springer, Berlin, Heidelberg, pp. 399–407.

Kadu, M. S., Gupta, R. & Bhave, P. R. 2008 Optimal design of water networks using a modified genetic algorithm with reduction in search space. *Journal of Water Resources Planning and Management* **134** (2), 147–160.

Kapelan, Z. S., Savic, D. A. & Walters, G. A. 2005 Multiobjective design of water distribution systems under uncertainty. *Water Resources Research* **41** (11), W11407, 1–15. https://doi.org/10.1029/2004WR003787.

Karaboga, D. 2005 *An Idea Based on Honey Bee Swarm for Numerical Optimization.* Technical Report-TR06. Erciyes University, Turkey.

Karamouz, M. & Nazif, S. 2013 Reliability-based flood management in urban watersheds considering climate change impacts. *Journal of Water Resources Planning and Management* **139** (5), 520–533.

Karovic, O. & Mays, L. W. 2014 Sewer system design using simulated annealing in excel. *Water Resources Management* **28** (13), 4551–4565.

Keedwell, E. & Khu, S. T. 2005 A hybrid genetic algorithm for the design of water distribution networks. *Engineering Applications of Artificial Intelligence* **18** (4), 461–472.

Kennedy, J., Eberhart, R. C. & Shi, Y. 2001 *Swarm Intelligence.* Morgan Kaufmann, San Francisco.

Kerachian, R. & Karamouz, M. 2006 Optimal reservoir operation considering the water quality issues: a stochastic conflict resolution approach. *Water Resources Research* **42** (12), W12401, 1–17. https://doi.org/10.1029/2005WR004575.

Kerachian, R. & Karamouz, M. 2007 A stochastic conflict resolution model for water quality management in reservoir–river systems. *Advances in Water Resources* **30** (4), 866–882.

Khu, S. T. & Madsen, H. 2005 Multiobjective calibration with Pareto preference ordering: an application to rainfall-runoff model calibration. *Water Resources Research* **41** (3), W03004, 1–14. https://doi.org/10.1029/2004WR003041.

Kiranyaz, S., Pulkkinen, J. & Gabbouj, M. 2011 Multi-dimensional particle swarm optimization in dynamic environments. *Expert Systems with Applications* **38** (3), 2212–2223. https://doi.org/10.1016/j.eswa.2010.08.009.

Kobayashi, K., Hinkelmann, R. & Helmig, R. 2008 Development of a simulation-optimization model for multiphase systems in the subsurface: a challenge to real-world simulation–optimization. *Journal of Hydroinformatics* **10** (2), 139–152.

Koppa, A., Gebremichael, M. & Yeh, W. W. 2019 Multivariate calibration of large scale hydrologic models: the necessity and value of a Pareto optimal approach. *Advances in Water Resources* **130**, 129–146.

Krishnakumar, K. 1990 Micro-genetic algorithms for stationary and non-stationary function optimization. In: *Proceedings: Intelligent Control and Adaptive Systems.* SPIE, Vol. 1196, pp. 289–296. https://doi.org/10.1117/12.969927.

Krishnan, N., Raj, C., Chaubey, I. & Sudheer, K. P. 2018 Parameter estimation of SWAT and quantification of consequent confidence bands of model simulations. *Environmental Earth Sciences* **77** (12), 470.

Krishnanand, K. N. & Ghose, D. 2006 Glowworm swarm based optimization algorithm for multimodal functions with collective robotics applications. *Multiagent and Grid Systems* **2** (3), 209–222. DOI: 10.3233/MGS-2006-2301.

Kumar, D. N. & Reddy, M. J. 2006 Ant colony optimization for multipurpose reservoir operation. *Water Resources Management* **20** (6), 879–898.

Kumar, D. N. & Reddy, M. J. 2007 Multipurpose reservoir operation using particle swarm optimization. *Journal of Water Resources Planning and Management* **133** (3), 192–201.

Kumar, D. N., Raju, K. S. & Ashok, B. 2006 Optimal reservoir operation for irrigation of multiple crops using genetic algorithms. *Journal of Irrigation and Drainage Engineering* **132** (2), 123–129.

Kuo, J. T., Wang, Y. Y. & Lung, W. S. 2006 A hybrid neural-genetic algorithm for reservoir water quality management. *Water Research* **40** (7), 1367–1376.

Lavric, V., Iancu, P. & Pleşu, V. 2005 Genetic algorithm optimisation of water consumption and wastewater network topology. *Journal of Cleaner Production* **13** (15), 1405–1415.

Lee, J. G., Selvakumar, A., Alvi, K., Riverson, J., Zhen, J. X., Shoemaker, L. & Lai, F. H. 2012 A watershed-scale design optimization model for stormwater best management practices. *Environmental Modelling and Software* **37**, 6–18.

Lerma, N., Paredes-Arquiola, J., Andreu, J., Solera, A. & Sechi, G. M. 2015 Assessment of evolutionary algorithms for optimal operating rules design in real Water Resource Systems. *Environmental Modelling and Software* **69**, 425–436.

Li, M., Lin, D. & Kou, J. 2012 A hybrid niching PSO enhanced with recombination-replacement crowding strategy for multimodal function optimization. *Applied Soft Computing* **12**, 975–987.

Liang, L. Y., Thompson, R. G. & Young, D. M. 2004 Optimising the design of sewer networks using genetic algorithms and Tabu search. *Engineering, Construction, and Architectural Management* **11** (2), 101–112. https://doi.org/10.1108/09699980410527849.

Liaw, R. T. & Ting, C. K. 2016 Enhancing covariance matrix adaptation evolution strategy through fitness inheritance. In: *Proceedings of the IEEE Congress on Evolutionary Computation (CEC)*. Vancouver, BC, pp. 1956–1963.

Liong, S. Y., Khu, S. T. & Chan, W. T. 2001 Derivation of Pareto front with genetic algorithm and neural network. *Journal of Hydrologic Engineering* **6** (1), 52–61.

Loshchilov, I. 2014 A computationally efficient limited memory CMA-ES for large scale optimization. In: *Proceedings of the Genetic and Evolutionary Computation Conference, GECCO*, pp. 397–404.

Luo, Q., Wu, J., Yang, Y., Qian, J. & Wu, J. 2016 Multi-objective optimization of long-term groundwater monitoring network design using a probabilistic Pareto genetic algorithm under uncertainty. *Journal of Hydrology* **534**, 352–363.

Ma, B. & Xia, Y. 2017 A tribe competition-based genetic algorithm for feature selection in pattern classification. *Applied Soft Computing* **58**, 328–338.

Mackle, G., Savic, G. A. & Walters, G. A. 1995 Application of genetic algorithms to pump scheduling for water supply. In: *First International Conference on Genetic Algorithms in Engineering Systems: Innovations and Applications*. IET, Sheffield, pp. 400–405.

Maier, H. R., Simpson, A. R., Zecchin, A. C., Maier, H. R., Simpson, A. R., Zecchin, A. C., Foong, W. K., Phang, K. W., Seah, H. Y. & Tan, C. L. 2003 Ant colony optimization for the design of water distribution systems. *Journal of Water Resources Planning and Management* **129** (3), 200–209.

Maier, H. R., Kapelan, Z. & Marchi, A. 2014 Evolutionary algorithms and other metaheuristics in water resources: current status, research challenges, and future directions. *Environmental Modelling and Software* **62** (2014), 271–299.

Mallipeddi, R., Suganthan, P. N., Pan, Q. K. & Tasgetiren, M. F. 2011 Differential evolution algorithm with ensemble of parameters and mutation strategies. *Applied Soft Computing* **11** (2), 1679–1696.

Man, K. F., Chan, T. M., Tang, K. S. & Kwong, S. 2004 Jumping-genes in evolutionary computing. In: *Proceedings of the IEEE IECON2004*. IECON2004, Busan, South Korea, pp. 1268–1272.

Manderick, B. & Spiessens, P. 1989 Fine-grained parallel genetic algorithm. In: *Proceedings of the Third International Conference on Genetic Algorithms*. Morgan Kaufmann Publishers Inc., San Francisco, CA, pp. 428–433.

Mantoglou, A. & Kourakos, G. 2007 Optimal groundwater remediation under uncertainty using multi-objective optimization. *Water Resources Management* **21** (5), 835–847.

McKinney, D. C. & Lin, M. D. 1994 Genetic algorithm solution of groundwater management models. *Water Resources Research* **30** (6), 1897–1906.

Merabtene, T., Kawamura, A., Jinno, K. & Olsson, J. 2002 Risk assessment for optimal drought management of an integrated water resources system using a genetic algorithm. *Hydrological Processes* **16** (11), 2189–2208.

Mohan, S. A. & Babu, K. J. 2010 Optimal water distribution network design with honey-bee mating optimization. *Journal of Computing in Civil Engineering* **24** (1), 117–126.

Montaseri, M., Afshar, M. H. & Bozorg-Haddad, O. 2015 Development of simulation-optimization model (MUSIC-GA) for urban stormwater management. *Water Resources Management* **29**, 4649–4665.

Mostafaie, A., Forootan, E., Safari, A. & Schumacher, M. 2018 Comparing multi-objective optimization techniques to calibrate a conceptual hydrological model using in situ runoff and daily GRACE data. *Computational Geosciences* **22** (3), 789–814.

Muleta, M. K. & Boulos, P. F. 2007 Multiobjective optimization for optimal design of urban drainage systems. In: *World Environmental and Water Resources Congress 2007: Restoring Our Natural Habitat*. pp. 1–10.

Muleta, M. K. & Nicklow, J. W. 2005 Decision support for watershed management using evolutionary algorithms. *Journal of Water Resources Planning and Management* **131** (1), 35–44.

Munavalli, G. R. & Kumar, M. M. 2003 Optimal scheduling of multiple chlorine sources in water distribution systems. *Journal of Water Resources Planning and Management* **129** (6), 493–504.

Murthy, Z. V. P. & Vengal, J. C. 2006 Optimization of a reverse osmosis system using genetic algorithm. *Separation Science and Technology* **41** (4), 647–663.

Muschalla, D. 2008 Optimization of integrated urban wastewater systems using multi-objective evolution strategies. *Urban Water Journal* **5** (1), 59–67.

Narayanan, A. & Moore, M. 1996 Quantum-inspired genetic algorithms. In: *Proceedings of the IEEE International Conference on Evolutionary Computation*, Nagoya, Japan. pp. 61–66.
Nickabadi, A., Ebadzadeh, M. M. & Safabakhsh, R. 2008 DNPSO: a dynamic niching particle swarm optimizer for multi-modal optimization. In: *Evolutionary Computation, 2008. CEC 2008*. IEEE, Hong Kong. pp. 26–32.
Nicklow, J., Reed, P., Savic, D., Dessalegne, T., Harrell, L., Chan-Hilton, A. & Zechman, E. 2010 State of the art for genetic algorithms and beyond in water resources planning and management. *Journal of Water Resources Planning and Management* **136** (4), 412–432.
Nixon, J. B., Dandy, G. C. & Simpson, A. R. 2001 A genetic algorithm for optimizing off-farm irrigation scheduling. *Journal of Hydroinformatics* **3** (1), 11–22.
Oliveira, R. & Loucks, D. P. 1997 Operating rules for multireservoir systems. *Water Resources Research* **33** (4), 839–852.
Ostermeier, A., Gawelczyk, A. & Hansen, N. 1994a A derandomized approach to self-adaptation of evolution strategies. *Evolutionary Computation* **2** (4), 369–380.
Ostermeier, A., Gawelczyk, A. & Hansen, N. 1994b Step-size adaption based on non-local use of selection information. In: *Parallel Problem Solving From Nature -PPSN III*, Jerusalem, Israel, pp. 189–198.
Ostfeld, A. & Salomons, E. 2006 Sensor network design proposal for the battle of the water sensor networks (BWSN). In: *8th Annual Water Distribution Systems Analysis Symposium*, Cincinnati, OH, pp. 1–16.
Ouyang, Q., Lu, W., Hou, Z., Zhang, Y., Li, S. & Luo, J. 2017 Chance-constrained multi-objective optimization of groundwater remediation design at DNAPLs-contaminated sites using a multi-algorithm genetically adaptive method. *Journal of Contaminant Hydrology* **200**, 15–23.
Park, D. K., Ko, N. Y. & Lee, K. K. 2007 Optimal groundwater remediation design considering effects of natural attenuation processes: pumping strategy with enhanced-natural-attenuation. *Geosciences Journal* **11** (4), 377–385.
Parouha, R. P. & Das, K. N. 2016 A memory based differential evolution algorithm for unconstrained optimization. *Applied Soft Computing* **38**, 501–517.
Patil, M. B., Naidu, M. N., Vasan, A. & Varma, M. R. 2020 Water distribution system design using multi-objective particle swarm optimisation. *Sādhanā* **45** (1), 21.
Penn, R., Friedler, E. & Ostfeld, A. 2013 Multi-objective evolutionary optimization for greywater reuse in municipal sewer systems. *Water Research* **47** (15), 5911–5920.
Perez-Pedini, C., Limbrunner, J. F. & Vogel, R. M. 2005 Optimal location of infiltration-based best management practices for storm water management. *Journal of Water Resources Planning and Management* **131** (6), 441–448.
Prasad, T. D. & Park, N. S. 2004 Multiobjective genetic algorithms for design of water distribution networks. *Journal of Water Resources Planning and Management* **130** (1), 73–82.
Prasad, T. D. & Tanyimboh, T. T. 2008 Entropy based design of any town water distribution network. In: *Water Distribution Systems Analysis, 2008*. pp. 1–12.
Prasad, T. D., Walters, G. A. & Savic, D. A. 2004 Booster disinfection of water supply networks: multiobjective approach. *Journal of Water Resources Planning and Management* **130** (5), 367–376.
Price, K., Storn, R. M. & Lampinen, J. A. 2005 *Differential Evolution: A Practical Approach to Global Optimization*. Springer Verlag, Berlin. ISBN 978-3-540-20950-8.
Qin, A. K. & Suganthan, P. N. 2005 Self-adaptive differential evolution algorithm for numerical optimization. In: *IEEE Congress on Evolutionary Computation*. IEEE, Edinburgh, UK, Vol. 2, pp. 1785–1791.
Raad, D. N., Sinske, A. N. & Van Vuuren, J. H. 2010 Comparison of four reliability surrogate measures for water distribution systems design. *Water Resources Research* **46** (5), 1–11. https://doi.org/10.1029/2009WR007785.
Raju, K. S. & Kumar, D. N. 2004 Irrigation planning using genetic algorithms. *Water Resources Management* **18** (2), 163–176.
Raju, K. S., Vasan, A., Gupta, P., Ganesan, K. & Mathur, H. 2012 Multi-objective differential evolution application to irrigation planning. *ISH Journal of Hydraulic Engineering* **18** (1), 54–64.
Rani, D. & Moreira, M. M. 2010 Simulation-optimization modeling: a survey and potential application in reservoir systems operation. *Water Resources Management* **24**, 1107–1138.
Rao, Z. & Salomons, E. 2007 Development of a real-time, near-optimal control process for water-distribution networks. *Journal of Hydroinformatics* **9** (1), 25–37.
Raseman, W. J., Kasprzyk, J. R. & Summers, R. S. 2020 Multi-objective optimization of water treatment operations for disinfection byproduct control. *Environmental Science-Water Research and Technology* **6** (3), 702–714.
Rauch, W. & Harremoes, P. 1999 Genetic algorithms in real time control applied to minimize transient pollution from urban wastewater systems. *Water Research* **33** (5), 1265–1277.
Rechenberg, I. 1973 *Evolution strategie: Optimierung technis-cher systeme nach prinzipien der biologischen evolution*. Frommann-Holzboog Verlag, Stuttgart.
Reddy, M. J. 2006 *Swarm Intelligence and Evolutionary Computation for Single and Multiobjective Optimization in Water Resource Systems*. PhD Thesis, Indian Institute of Science, Bangalore.
Reddy, M. J. 2009 Swarm intelligence techniques and its applications in water resources management. *ISH Journal of Hydraulic Engineering* **15** (SP.1), 151–169.
Reddy, M. J. & Kumar, D. N. 2006 Optimal reservoir operation using multi-objective evolutionary algorithm. *Water Resources Management* **20** (6), 861–878.
Reddy, M. J. & Kumar, D. N. 2007a Optimal reservoir operation for irrigation of multiple crops using elitist-mutated particle swarm optimization. *Hydrological Sciences Journal, IAHS* **52** (4), 686–701.

Reddy, M. J. & Kumar, D. N. 2007b Multi-objective particle swarm optimization for generating optimal trade-offs in reservoir operation. *Hydrological Processes: An International Journal* **21** (21), 2897–2909.
Reddy, M. J. & Kumar, D. N. 2007c Multiobjective differential evolution with application to reservoir system optimization. *Journal of Computing in Civil Engineering* **21** (2), 136–146.
Reddy, M. J. & Kumar, D. N. 2008 Evolving strategies for crop planning and operation of irrigation reservoir system using multi-objective differential evolution. *Irrigation Science* **26** (2), 177–190.
Reddy, M. J. & Kumar, D. N. 2009 Performance evaluation of elitist-mutated multi-objective particle swarm optimization for integrated water resources management. *Journal of Hydroinformatics* **11** (1), 79–88.
Reddy, M. J. & Kumar, D. N. 2012 Computational algorithms inspired by biological processes and evolution. *Current Science* **103** (4), 1–11.
Reed, P. M., Hadka, D., Herman, J. D., Kasprzyk, J. R. & Kollat, J. B. 2013 Evolutionary multiobjective optimization in water resources: the past, present, and future. *Advances in Water Resources* **51** (2013), 438–456.
Ritzel, B. J., Eheart, J. W. & Ranjithan, S. 1994 Using genetic algorithms to solve a multiple objective groundwater pollution containment problem. *Water Resources Research* **30** (5), 1589–1603.
Rocha, L. M. 1995 Contextual genetic algorithms: evolving developmental rules. In: *Advances in Artificial Life* (F. Moran, A. Moreno, J. J. Merelo & P. Chacon, eds). Lecture Notes in Artificial Intelligence, Springer-Verlag. pp. 368–382.
Sala, R., Baldanzini, N. & Pierini, M. 2017 SQG-differential evolution for difficult optimization problems under a tight function evaluation budget. In: *Proceedings of the International Workshop on Machine Learning, Optimization, and Big Data*. Springer, Cham, pp. 322–336.
Savic, D. A. & Walters, G. A. 1997 Genetic algorithms for least-cost design of water distribution networks. *Journal of Water Resources Planning and Management* **123** (2), 67–77.
Savic, D. A., Walters, G. A. & Schwab, M. 1997 April Multiobjective genetic algorithms for pump scheduling in water supply. In: *AISB International Workshop on Evolutionary Computing*. Springer, Berlin, Heidelberg, pp. 227–235.
Seyedpour, S. M., Kirmizakis, P., Brennan, P., Doherty, R. & Ricken, T. 2019 Optimal remediation design and simulation of groundwater flow coupled to contaminant transport using genetic algorithm and radial point collocation method (RPCM). *Science of The Total Environment* **669**, 389–399.
Shamsudin, S., Dan'azumi, S. & Yusop, A. A. Z. 2014 Optimum combination of pond volume and outlet capacity of a stormwater detention pond using particle swarm optimization. *Urban Water Journal* **11** (2), 127–136.
Shi, Y. & Eberhart, R. C. 1998 A modified particle swarm optimizer. In: *Proceedings of IEEE International Conference on Evolutionary Computation*. IEEE, Anchorage, AL, pp. 69–73.
Shibu, A. & Reddy, M. J. 2014 Optimal design of water distribution networks considering fuzzy randomness of demands using cross entropy optimization. *Water Resources Management* **28** (12), 4075–4094.
Shieh, H. J. & Peralta, R. C. 2005 Optimal in situ bioremediation design by hybrid genetic algorithm-simulated annealing. *Journal of Water Resources Planning and Management* **131** (1), 67–78.
Shin, M. J. & Choi, Y. S. 2018 Combining an r-based evolutionary algorithm and hydrological model for effective parameter calibration. *Water* **10** (10), 1339.
Simpson, A. R., Dandy, G. C. & Murphy, L. J. 1994 Genetic algorithms compared to other techniques for pipe optimization. *Journal of Water Resources Planning and Management* **120** (4), 423–443.
Singh, T. S. & Chakrabarty, D. 2011 Multiobjective optimization of pump-and-treat-based optimal multilayer aquifer remediation design with flexible remediation time. *Journal of Hydrologic Engineering* **16** (5), 413–420.
Singh, A., Minsker, B. S. & Valocchi, A. J. 2008 An interactive multi-objective optimization framework for groundwater inverse modeling. *Advances in Water Resources* **31** (10), 1269–1283.
Sirsant, S. & Janga Reddy, M. 2018 Reliability-based design of water distribution networks using self-adaptive differential evolution algorithm. *ISH Journal of Hydraulic Engineering* **24** (2), 198–212.
Sirsant, S. & Reddy, M. J. 2020 Assessing the performance of surrogate measures for water distribution network reliability. *Journal of Water Resources Planning and Management* **146** (7), 04020048.
Smalley, J. B., Minsker, B. S. & Goldberg, D. E. 2000 Risk-based in situ bioremediation design using a noisy genetic algorithm. *Water Resources Research* **36** (10), 3043–3052.
Song, J., Yang, Y., Chen, G., Sun, X., Lin, J., Wu, J. & Wu, J. 2019 Surrogate assisted multi-objective robust optimization for groundwater monitoring network design. *Journal of Hydrology* **577**, 123994.
Srinivas, N. & Deb, K. 1994 Multiobjective optimization using nondominated sorting in genetic algorithms. *Evolutionary Computation* **2** (3), 221–248.
Storn, R. & Price, K. 1996 Minimizing the real functions of the ICEC'96 contest by differential evolution. In: *Proceedings of. IEEE International Conference on Evolutionary Computation*. IEEE, Nagoya, Japan, pp. 842–844.
Storn, R. & Price, K. 1997 Differential evolution – a simple and efficient heuristic for global optimization over continuous spaces. *Journal of Global Optimization* **11** (4), 341–359.
Stützle, T. & Hoos, H. H. 2000 MAX MIN ant system. *Future Generation Computer Systems* **16**, 889–914.
Sun, M. & Zheng, C. 1999 Long-term groundwater management by a modflow based dynamic optimization tool 1. *JAWRA Journal of the American Water Resources Association* **35** (1), 99–111.
Suribabu, C. R. 2010 Differential evolution algorithm for optimal design of water distribution networks. *Journal of Hydroinformatics* **12** (1), 66–82.

doi: 10.2166/h2oj.2020.128

Swan, R., Bridgeman, J. & Sterling, M. 2017 Optimisation of water treatment works performance using genetic algorithms. *Journal of Hydroinformatics* **19** (5), 719–733.

Takagi, H. 2001 Interactive evolutionary computation: fusion of the capabilities of EC optimization and human evaluation. *Proc IEEE* **89** (9), 1275–1296.

Tang, Y., Reed, P. M. & Kollat, J. B. 2007 Parallelization strategies for rapid and robust evolutionary multiobjective optimization in water resources applications. *Advances in Water Resources* **30** (3), 335–353.

Tian, F., Hu, H., Sun, Y., Li, H. & Lu, H. 2019 Searching for an optimized single-objective function matching multiple objectives with automatic calibration of hydrological models. *Chinese Geographical Science* **29** (6), 934–948.

Tigkas, D., Christelis, V. & Tsakiris, G. 2016 Comparative study of evolutionary algorithms for the automatic calibration of the Medbasin-D conceptual hydrological model. *Environmental Processes* **3** (3), 629–644.

Trivedi, A., Sanyal, K., Verma, P. & Srinivasan, D. 2017 A unified differential evolution algorithm for constrained optimization problems. In: *Proceedings of the IEEE Congress on Evolutionary Computation (CEC2017)*, IEEE, San Sebastian, Spain, pp. 1231–1238.

Tsai, M. J. & Chang, C. T. 2001 Water usage and treatment network design using genetic algorithms. *Industrial and Engineering Chemistry Research* **40** (22), 4874–4888.

Tseng, H.-E., Chang, C.-C., Lee, S.-C. & Huang, Y.-M. 2018 A block-based genetic algorithm for disassembly sequence planning. *Expert System Applications* **96**, 492–505.

Vairavamoorthy, K. & Ali, M. 2005 Pipe index vector: a method to improve genetic-algorithm-based pipe optimization. *Journal of Hydraulic Engineering* **131** (12), 1117–1125.

Vamvakeridou-Lyroudia, L. S., Walters, G. A. & Savic, D. A. 2005 Fuzzy multiobjective optimization of water distribution networks. *Journal of Water Resources Planning and Management* **131** (6), 467–476.

Van Zyl, J. E., Savic, D. A. & Walters, G. A. 2004 Operational optimization of water distribution systems using a hybrid genetic algorithm. *Journal of Water Resources Planning and Management* **130** (2), 160–170.

Vasan, A. & Simonovic, S. P. 2010 Optimization of water distribution network design using differential evolution. *Journal of Water Resources Planning and Management* **136** (2), 279–287.

Vojinovic, Z., Sahlu, S., Torres, A. S., Seyoum, S. D., Anvarifar, F., Matungulu, H., Barreto, W., Savic, D. & Kapelan, Z. 2014 Multi-objective rehabilitation of urban drainage systems under uncertainties. *Journal of Hydroinformatics* **16** (5), 1044–1061.

Walters, G. A. & Lohbeck, T. 1993 Optimal layout of tree networks using genetic algorithms. *Engineering Optimization* **22** (1), 27–48.

Walters, G. A. & Smith, D. K. 1995 Evolutionary design algorithm for optimal layout of tree networks. *Engineering Optimization* **24** (4), 261–281.

Wang, Q. J. 1991 The genetic algorithm and its application to calibrating conceptual rainfall-runoff models. *Water Resources Research* **27** (9), 2467–2471.

Wang, C. G. & Jamieson, D. G. 2002 An objective approach to regional wastewater treatment planning. *Water Resources Research* **38** (3), 4–1.

Wang, Y.-J. & Zhang, J.-S. 2007 Global optimization by an improved differential evolutionary algorithm. *Applied Mathematics and Computation* **188**, 669–680.

Wang, M. & Zheng, C. 1997 Optimal remediation policy selection under general conditions. *Groundwater* **35** (5), 757–764.

Wang, Y., Li, B., Weise, T., Wang, J., Yuan, B. & Tian, Q. 2011a Self-adaptive learning based particle swarm optimization. *Information Science* **181**, 4515–4538.

Wang, Y., Cai, Z. & Zhang, Q. 2011b Differential evolution with composite trial vector generation strategies and control parameters. *IEEE Transactions on Evolutionary Computation* **15** (1), 55–66.

Wang, Q., Zhou, Q., Lei, X. & Savić, D. A. 2018 Comparison of multiobjective optimization methods applied to urban drainage adaptation problems. *Journal of Water Resources Planning and Management* **144** (11), 04018070.

Wardlaw, R. & Sharif, M. 1999 Evaluation of genetic algorithms for optimal reservoir system operation. *Journal of Water Resources Planning and Management* **125** (1), 25–33.

Whitley, D., Rana, S. & Heckendorn, R. B. 1998 The island model genetic algorithm: on separability, population size and convergence. *Journal of Computing and Information Science in Engineering* **7**, 33–47.

Wierstra, D., Schaul, T., Peters, J. & Schmidhuber, J. 2008 Natural evolution strategies. In: *Proceedings of the IEEE Congress on Evolutionary Computation*. IEEE, Hong Kong, pp. 3381–3387.

Wu, Z. Y. & Walski, T. 2005 Self-adaptive penalty approach compared with other constraint-handling techniques for pipeline optimization. *Journal of Water Resources Planning and Management* **131** (3), 181–192.

Wu, J., Zheng, C., Chien, C. C. & Zheng, L. 2006 A comparative study of Monte Carlo simple genetic algorithm and noisy genetic algorithm for cost-effective sampling network design under uncertainty. *Advances in Water Resources* **29** (6), 899–911.

Yakowitz, S. 1982 Dynamic programming applications in water resources. *Water Resources Research* **18**, 673–696.

Yandamuri, S. R., Srinivasan, K. & Murty Bhallamudi, S. 2006 Multiobjective optimal waste load allocation models for rivers using nondominated sorting genetic algorithm-II. *Journal of Water Resources Planning and Management* **132** (3), 133–143.

Yang, X. S. 2007 *Firefly Algorithm*. Nature-Inspired Metaheuristic Algorithms, pp. 79–90.

Yang, X. S. 2010 A new metaheuristic bat-inspired algorithm. In: *Nature Inspired Cooperative Strategies for Optimization (NISCO 2010) Studies in Computational Intelligence*. Springer, Berlin and Heidelberg, Vol. 284, pp. 65–74.

Yang, X. S. & Deb, S. 2010 Engineering optimisation by Cuckoo search. *International Journal of Mathematical Modelling and Numerical Optimization* **1** (4), 330.

Yang, M., Cai, Z., Li, C. & Guan, J. 2014 An improved JADE algorithm for global optimization. In: *IEEE Congress on Evolutionary Computation (CEC)*, pp. 806–812.

Yazdi, J. 2018 Water quality monitoring network design for urban drainage systems, an entropy method. *Urban Water Journal* **15** (3), 227–233.

Yazdi, J., Sadollah, A., Lee, E. H., Yoo, D. G. & Kim, J. H. 2017a Application of multi-objective evolutionary algorithms for the rehabilitation of storm sewer pipe networks. *Journal of Flood Risk Management* **10** (3), 326–338.

Yazdi, J., Yoo, D. G. & Kim, J. H. 2017b Comparative study of multi-objective evolutionary algorithms for hydraulic rehabilitation of urban drainage networks. *Urban Water Journal* **14** (5), 483–492.

Yeh, W. W.-G. 1985 Reservoir management and operations models: a state-of-the-art review. *Water Resources Research* **21**, 1797–1818.

Yeh, C. H. & Labadie, J. W. 1997 Multiobjective watershed-level planning of storm water detention systems. *Journal of Water Resources Planning and Management* **123** (6), 336–343.

Yoon, J. H. & Shoemaker, C. A. 2001 Improved real-coded GA for groundwater bioremediation. *Journal of Computing in Civil Engineering* **15** (3), 224–231.

Yu, W. & Zhang, J. 2011 Multi-population differential evolution with adaptive parameter control for global optimization. In: *Proceedings of the GECCO 2011*, Dublin, pp. 1093–1098.

Zahraie, B., Kerachian, R. & Malekmohammadi, B. 2008 Reservoir operation optimization using adaptive varying chromosome length genetic algorithm. *Water International* **33** (3), 380–391.

Zhan, Z. H., Zhang, J., Li, Y. & Chung, H. S. 2009 Adaptive particle swarm optimization. *IEEE Transactions on Systems, Man, and Cybernetics Part B: Cybernetics* **39** (6), 1362–1381.

Zhang, X., Srinivasan, R., Zhao, K. & Liew, M. V. 2009 Evaluation of global optimization algorithms for parameter calibration of a computationally intensive hydrologic model. *Hydrological Processes: An International Journal* **23** (3), 430–441.

Zheng, C. & Wang, P. P. 2002 A field demonstration of the simulation-optimization approach for remediation system design. *Groundwater* **40** (3), 258–266.

Zou, R. & Lung, W. S. 2004 Robust water quality model calibration using an alternating fitness genetic algorithm. *Journal of Water Resources Planning and Management* **130** (6), 471–479.

First received 7 April 2020; accepted in revised form 22 May 2020. Available online 11 June 2020

Evolutionary Algorithms in Water Resources
doi: 10.2166/h2oj.2020.032

Multi-objective fuzzy optimization for sustainable irrigation planning

Jyotiba B. Gurav[a,*] and D. G. Regulwar[b]

[a] Department of Civil Engineering, Amrutvahini College of Engineering, Sangamner 422 608, Maharashtra, India

[b] Department of Civil Engineering, Government College of Engineering and Research, Avasari Khurd, Pune 412 405, Maharashtra, India

*Corresponding author. E-mail: jyotiba.gurav@rediffmail.com

Abstract

The objective of the present work is to determine an optimal cropping pattern under uncertainty, which maximizes four objectives simultaneously, including net benefits (NBF), crop production (CPD), employment generation (EGN) and manure utilization (MUT). Except the objective of maximizing the NBF, the other objectives are related to sustainability. To deal with uncertainty, a multi-objective fuzzy linear programming (MOFLP) model has developed along with fuzziness in decision parameters (objective function coefficient, cost coefficients, technological coefficients and resources) and decision variables (area to be irrigated under each crop in each season) and applied the same to Jayakwadi Project Stage-I, Maharashtra, India. The present study is in the form of a successful attempt to deal with irrigation planning associated with sustainability and uncertainty.

Key words: decision variables and decision parameters, fuzzy optimization, sustainable irrigation planning, uncertainty

Highlights

- The present study deals with the development of the MOFLP model under uncertainty that maximizes NBF, CPD, EGN and MUT simultaneously and its successful application for the case study of Jayakwadi Project Stage-I, Maharashtra State, India.
- The proposed model suggests a basis for irrigation planning as an integrated approach under a fuzzy environment considering the decision parameters and decision variables as fuzzy.

doi: 10.2166/h2oj.2020.032

Graphical Abstract

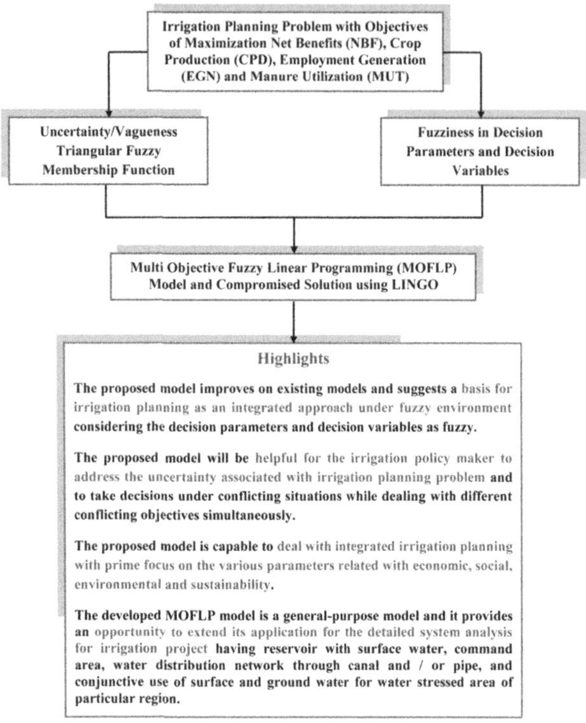

INTRODUCTION

The randomness of hydrologic variables such as stream flow, rainfall and evapotranspiration, and imprecision in management goals/objectives, constraints, crop response, decision makers' interest, which are the most common things closely associated with uncertainty. The term fuzzy introduced and inducted first time by Zadeh (1965), which allows to develop fuzzy set theory in the form of an improvement over classical set theory. The vagueness and uncertainties associated with various kinds of objectives, decision parameters and decision variables can be well described and addressed by a fuzzy set. The concept of decision-making under a fuzzy environment was presented by Bellman & Zadeh (1970).

Gupta *et al.* (2000) have proposed a multi-objective fuzzy linear programming (MOFLP) area allocation model to analyze the conflicting interests of various decision makers such as the irrigation authority (government) and the individual farmers involved in a particular project and applied the developed model to the case study of the Narmada river basin, India. Gasimov & Yenilmez (2002) have discussed the methodology of fuzzy linear programming (FLP) problems. Jimenez *et al.* (2007) have presented an interactive method to solve linear programming with fuzzy numbers, which informs the decision maker in various stages of the decision process. The fuzzy parametric programming based MOFLP model has been proposed with the illustration of numerical example (Arikan & Gungor 2007).

Fuzziness and randomness associated with objective function and constraints can be dealt with the chance-constrained programming model (Nanda *et al.* 2008). The linear programming problem with the triangular fuzzy number (TFN) has been developed and demonstrated with suitable examples (Nasseri 2008).

The FLP problem in which decision parameters and decision variables, fuzzy in nature can be addressed and solved properly by the method proposed by Allahviranloo *et al.* (2008). The application

doi: 10.2166/h2oj.2020.032

of MOFLP in irrigation planning along with fuzziness in objective function only and fuzziness in resources and the objective function is reported in the literature (Raju & Duckstein 2003; Regulwar & Gurav 2010, 2011; Gurav & Regulwar 2012).

Xu *et al.* (2013) have presented a bilevel optimization approach to deal with the regional water resource allocation problem using the technique of interactive fuzzy programming and an entropy-Boltzmann selection-based genetic algorithm under the fuzzy random environment. Guo *et al.* (2014) have presented an inexact fuzzy chance-constrained nonlinear programming (IFCCNP) method for agricultural water resources management to tackle various uncertainties. The model results obtained are helpful for decision makers to deliver insight for tradeoffs related to environmental, economic and system-reliability policy.

Srivastava & Singh (2017) have proposed fuzzy multi-objecive goal programming (FMGP) to find the optimal cropping patterns with objective of maximization of benefit and production along with minimization of investment, fertilizer application and water application to the case study of canal command of Soraon, District Allahabad, Uttar Pradesh (India). Li *et al.* (2016) have proposed the bi-objective programming model with fuzzy inputs such as fuzziness associated with objective functions as well as in constraints, and the developed model has been solved using the fuzzy multi-objective optimization modeling approach for the case study of the Heihe River basin, northwest China. The MOFLP with a two-phase approach can be used for optimal cropping pattern planning, and using this approach, various irrigation planning models have been proposed and presented with its application for different case study of India (Regulwar & Gurav 2013; Mirajkar & Patel 2016).

Zhang *et al.* (2018) have developed an irrigation water allocation model with interval-based fuzzy chance-constrained along with double-sided fuzziness and applied it to the case study of the Heihe River Basin in Northwest China and found the interrelationships to support the decision-making for irrigation water management. Zhang & Guo (2018) have developed a model with the fuzzy linear fractional programming approach with double-sided fuzziness to address and tackle the uncertainty involved for optimal irrigation water allocation in the case study of middle reaches of the Heihe River Basin, in Northwest China. The obtained results help to plan reasonable irrigation water resources management and agricultural production. Banihabib *et al.* (2019) have proposed the fuzzy multi-objective heuristic model to maximize three objective functions along with the uncertainties involved in water resources and economic parameters in a basin and found that optimal operation policies provide better results than the deterministic model. Yue *et al.* (2020) have developed a full fuzzy-interval credibility-constrained nonlinear programming (FFICNP) model to deal with uncertainties in planning irrigation water allocation and applied the same successfully to the case study of the Zhanghe irrigation district in Hubei Province, Central China. The obtained results show lower credibility level with respect to higher net system benefit and system efficiency.

With the above literature review presented, it is found that various models of the LP and MOFLP have been used to tackle the uncertainty and vagueness involved in the irrigation planning scenario. Also, it is found that any model has not been reported in sustainable irrigation planning with fuzzy decision parameters and fuzzy decision variables. Most of the uncertainties associated with irrigation planning are dependent and interrelated with each other in agricultural systems, so it is necessary that in the irrigation planning problem, uncertainties should be considered and addressed properly. Given this, the decision parameters and decision variables are considered fuzzy in nature, which leads in the form of successful attempts in the field of sustainable irrigation planning. The objective of the present study is to develop the MOFLP model and its application to the Jayakwadi Project Stage-I and to find out the optimal cropping pattern for 75% dependable yield for sustainable irrigation planning under the fuzzy environment.

METHODOLOGY AND MODEL DEVELOPMENT

Description of the study area

To represent the applicability of the MOFLP model, a representative agricultural irrigation management problem of the Jayakwadi Project Stage-I is considered. It is an earth type of dam constructed across river Godavari. The reservoir has an active storage capacity of 2,170 Mm^3 and a dead storage capacity of 738 Mm^3. It has an irrigable command area of 141,640 ha (1,416.40 km^2). Also, the project is a hydropower plant, with a pumped storage with a capacity of 12 MW. The main crops grown in the command area are Sugarcane, Banana, Chilies, L S Cotton, Sorghum, Paddy, Wheat, Gram and Groundnut. The location map of the study area is shown in Figure 1.

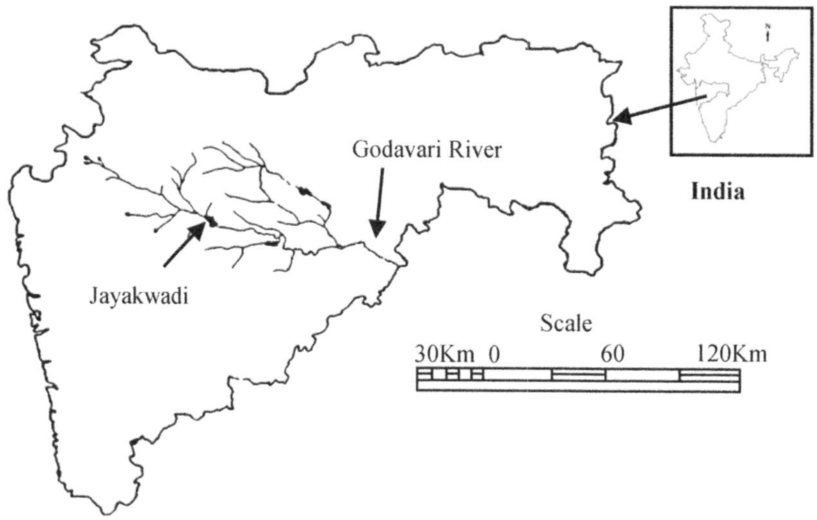

Figure 1 | Jayakwadi Project Stage-I Maharashtra State, India.

Objective function

The various objectives are considered in the formulation of the irrigation planning model depending upon the requirement of the particular region and national importance. In the present study, four objectives of maximization type are considered from an analysis point of view, which are listed as below.

Net benefits

Most of the time, farmers want to maximize the net benefits (NBF) with the cultivation of particular crops for economic prosperity, due to which the decision maker has to incorporate this objective as part of the irrigation planning policy. Gross benefits are calculated by multiplying the average yield of a particular crop per ha and current market price of that crop. In the present study, the net benefits per ha for each crop, which are calculated by subtracting the input cost of the gross benefits. The input cost considered in the present model, which may vary according to the market conditions and different parts of the region, such variations in the input cost can be

doi: 10.2166/h2oj.2020.032

incorporated in the model.

$$\text{Maximize } NBF = \left[\left(\sum_{i=1}^{2} J_i^K F_i^K + \sum_{i=1}^{3} J_i^R F_i^R + \sum_{i=1}^{2} J_i^{TS} F_i^{TS} + \sum_{i=1}^{2} J_i^P F_i^P + \sum_{i=1}^{1} J_i^{HW} F_i^{HW}\right) - \left(\sum_{i=1}^{2} J_i^K G_i^K + \sum_{i=1}^{3} J_i^R G_i^R + \sum_{i=1}^{2} J_i^{TS} G_i^{TS} + \sum_{i=1}^{2} J_i^P G_i^P + \sum_{i=1}^{1} J_i^{HW} G_i^{HW}\right)\right] \quad (1)$$

[I = crop index, which includes various crops in different seasons such as Sugarcane (P), Banana (P), Chilies (TS), L, S Cotton (TS), Sorghum (K), Paddy (K), Sorghum (R), Wheat (R), Gram (R) and Groundnut (HW). P = Perennial, K = Kharif, R = Rabi, TS = Two Seasonal and HW = Hot Weather]

J_i^K – area in ha for ith crop (K);
J_i^R = area in ha for ith crop (R);
J_i^{HW} = area in ha for ith crop (HW);
J_i^P = area in ha for ith crop (P);
J_i^{TS} = area in ha for ith crop (TS);
F_i = benefit coefficient for ith crop;
G_i = input cost for ith crop.

Crop production

To expect maximum production of a particular crop is the natural tendency of every farmer, which needs to be taken into account by the decision maker for the optimal cropping pattern planning with the objective of maximization of crop production (CPD).

The average yield of a crop per ha is taken as CPD coefficient (Commissionerate of Agriculture, Maharashtra State 2006).

$$\text{Maximize } (CPD) = \left[\left(\sum_{i=1}^{2} J_i^K U_i^K + \sum_{i=1}^{3} J_i^R U_i^R + \sum_{i=1}^{2} J_i^{TS} U_i^{TS} + \sum_{i=1}^{2} J_i^P U_i^P + \sum_{i=1}^{1} J_i^{HW} U_i^{HW}\right)\right] \quad (2)$$

U_i = average yield of ith crop (Tons per ha).

The sufficient food availability in the region can be considered with the help of maximization of CPD, which mainly focuses on the survival of people with satisfaction of food demand of the particular region, and given this the second objective can be thought of sustainability-related.

Employment generation

The socio-economic development of a region cannot be possible without providing employment generation (EGN), which requires the decision maker to include the maximization of EGN in irrigation planning.

$$\text{Maximize } (EGN) = \left[\left(\sum_{i=1}^{2} J_i^K M_i^K + \sum_{i=1}^{3} J_i^R M_i^R + \sum_{i=1}^{2} J_i^{TS} M_i^{TS} + \sum_{i=1}^{2} J_i^P M_i^P + \sum_{i=1}^{1} J_i^{HW} M_i^{HW}\right)\right] \quad (3)$$

M_i = requirement of Man Days for ith crop per ha.

doi: 10.2166/h2oj.2020.032

The number of Man Days or labor required for a particular crop per ha has been considered based on discussions with farmers and experts from the agricultural field.

The third objective is associated with sustainability with the view of socio-economic development for a developing country like India, where there is uneven distribution of agricultural land due to which many people are available in the form of labor for the agricultural sector.

Manure utilization

$$\text{Maximize } (MUT) = \left[\left(\sum_{i=1}^{2} J_i^K M_i^K + \sum_{i=1}^{3} J_i^R M_i^R + \sum_{i=1}^{2} J_i^{TS} M_i^{TS} + \sum_{i=1}^{2} J_i^P M_i^P + \sum_{i=1}^{1} J_i^{HW} M_I^{HW}\right)\right] \quad (4)$$

M_i = requirement of manure utilization (MUT) for ith crop (Tons per ha).

The discussion with farmers and experts from the agricultural field, which is considered as a basis to work out the requirement of MUT for a crop per ha. The requirement of MUT, which depends upon various parameters such as market condition and its availability, which may vary from place to place. These changes can be incorporated in the model, which may vary the requirement of MUT.

In India, there is a tradition of preparing the green manure by the farmer with the decomposition of waste from various activities of the farming and livestock, which is the richest, recycled and pure source for various ingredients of manure which does not include any harmful chemicals, fertilizers and pesticides. The manure prepared by this procedure helps to retain nutrient sufficiency of soil for various crops in each crop season. So, more and more utilization of green manure relates to sustainability and to be included in the cropping pattern by the decision maker.

Constraints

Total sowing area

The total sowing area constraint can be represented by the following equation:

$$\left(\sum_{i=1}^{2} J_i^K + \sum_{i=1}^{3} J_i^R + \sum_{i=1}^{2} J_i^{TS} + \sum_{i=1}^{2} J_i^P + \sum_{i=1}^{1} J_i^{HW}\right) \leq TJ \quad (5)$$

TJ = total command area.

Maximum sowing area

The maximum sowing area constraint can be represented by the equation as follows:

Kharif

$$\left(\sum_{i=1}^{2} J_i^P + \sum_{i=1}^{2} K_i^{TS} + \sum_{i=1}^{2} J_i^K\right) \leq TJ_i^P + TJ_i^{TS} + TJ_i^K \quad (6)$$

Rabi

$$\left(\sum_{i=1}^{2} J_i^P + \sum_{i=1}^{2} K_i^{TS} + \sum_{i=1}^{3} K_i^R\right) \leq TJ_i^P + TJ_i^{TS} + TJ_i^R \quad (7)$$

doi: 10.2166/h2oj.2020.032

Hot weather and perennial

$$\left(\sum_{i=1}^{2} J_i^P + \sum_{i=1}^{1} J_i^{HW}\right) \leq TJ_i^P + TJ_i^{HW} \tag{8}$$

TJ_i^K = area for ith crop (K) in ha;
TJ_i^R = area for ith crop (R) in ha;
TJ_i^{HW} = area for ith crop (HW) in ha;
TJ_i^P = area for ith crop (P) in ha.
TJ_i^{TS} = area for ith crop (TS) in ha.

The values for TJ (i.e. for particular crop in particular season) have considered according to the existing cropping pattern of the project.

Affinity constraint

Perennial

$$J_1^P \leq TJ_i^P \tag{9a}$$

J_1^P = area in ha for Sugarcane (P).

$$J_2^P \leq TJ_i^P \tag{9b}$$

J_2^P = area in ha for Banana (P).

Two seasonal

$$J_3^{TS} \leq TJ_i^{TS} \tag{9c}$$

J_3^{TS} = area in ha for Chilies (TS).

$$J_4^{TS} \leq TJ_i^{TS} \tag{9d}$$

J_4^{TS} = area in ha for L S Cotton (TS).

Kharif

$$J_5^K \leq TJ_i^K \tag{9e}$$

J_5^K = area in ha for Sorghum (K).

$$J_6^K \leq TJ_i^K \tag{9f}$$

J_6^K = area in ha for Paddy (K).

Rabi

$$J_7^R \leq TJ_i^R \tag{9g}$$

J_7^R = area in ha for Sorghum (R).

$$J_8^R \leq TJ_i^R \tag{9h}$$

J_8^R = area in ha for Wheat (R).

$$J_9^R \leq TJ_i^R \tag{9i}$$

J_9^R = area in ha for Gram (R).

Hot weather

$$J_{10}^{HW} \leq TJ_i^{HW} \tag{9j}$$

J_{10}^{HW} = area in ha for Groundnut (HW).

Labor availability constraint

With the view of the problem of the unavailability of labor during the farming season, it is suggested that to tackle the problem of uncertainty involved in the availability of labor, the labor requirement should not exceed the total labor availability during that particular crop season.

Kharif

$$\left(\sum_{i=1}^{2} J_i^P M_i^P + \sum_{i=1}^{2} J_i^{TS} M_i^{TS} + \sum_{i=1}^{2} J_i^K M_i^K\right) \leq \sum_{i=1}^{2} AM_i^P + \sum_{i=1}^{2} AM_i^{TS} + \sum_{i=1}^{2} AM_i^K \tag{10}$$

Rabi

$$\left(\sum_{i=1}^{2} J_i^P M_i^P + \sum_{i=1}^{2} J_i^{TS} M_i^{TS} + \sum_{i=1}^{3} J_i^R M_i^R\right) \leq \sum_{i=1}^{2} AM_i^P + \sum_{i=1}^{2} AM_i^{TS} + \sum_{i=1}^{3} AM_i^R \tag{11}$$

Perennial and hot weather

$$\left(\sum_{i=1}^{2} J_i^P M_i^P + \sum_{i=1}^{1} J_i^{HW} M_i^{HW}\right) \leq \sum_{i=1}^{2} AM_i^P + \sum_{i=1}^{1} AM_i^{HW} \tag{12}$$

AM_i = labor availability for ith crop;
M_i = requirement of Man Days for ith crop per ha.

Manure availability constraint

With the view of the scarcity of manure and to ensure the fertility of soil, it is considered that in order to maintain the fertility of the soil, the total manure requirement should not exceed the total availability of the manure in that crop season.

Kharif

$$\left(\sum_{i=1}^{2} J_i^P M_i^P + \sum_{i=1}^{2} J_i^{TS} M_i^{TS} + \sum_{i=1}^{2} J_i^K M_i^K\right) \leq \sum_{i=1}^{2} AM_i^P + \sum_{i=1}^{2} AM_i^{TS} + \sum_{i=1}^{2} AM_i^K \tag{13}$$

Rabi

$$\left(\sum_{i=1}^{2} J_i^P M_i^P + \sum_{i=1}^{2} J_i^{TS} M_i^{TS} + \sum_{i=1}^{3} J_i^R M_i^R\right) \leq \sum_{i=1}^{2} AM_i^P + \sum_{i=1}^{2} AM_i^{TS} + \sum_{i=1}^{3} AM_i^R \tag{14}$$

Perennial and hot weather

$$\left(\sum_{i=1}^{2} J_i^P M_i^P + \sum_{i=1}^{1} J_i^{HW} M_i^{HW}\right) \leq \sum_{i=1}^{2} AM_i^P + \sum_{i=1}^{1} AM_i^{HW} \tag{15}$$

AM_i = manure availability for ith crop;
M_i = requirement of MUT for ith crop per ha.

Water availability constraint

The total water requirement of different crops should not exceed the total water availability in the reservoir,

$$\left(\sum_{i=1}^{2} J_i^K W_i^K + \sum_{i=1}^{3} J_i^R W_i^R + \sum_{i=1}^{2} J_i^{TS} W_i^{TS} + \sum_{i=1}^{2} J_i^P W_i^P + \sum_{i=1}^{1} J_i^{HW} W_i^{HW}\right) \leq AW_i^j \tag{16}$$

$j = 1, 2, 3, 4, 5$ (No of crop seasons);
W_i = irrigation water requirement for ith crop;
AW_i^j = total water availability for ith crop (all crops) for jth interval (all seasons).

Non negativity constraint

$$\begin{gathered} J_i^K, J_i^R, J_i^{TS}, J_i^P, J_i^{HW}, U_i, M_i, TJ, TJ_i^K, TJ_i^R, TJ_i^P, TJ_i^{TS}, TJ_i^{HW}, AM_i^P, AM_i^{TS}, \\ AM_i^K, AM_i^R, AM_i^{HW}, M_i^P, M_i^{TS}, M_i^K, M_i^R, M_i^{HW}, W_i^P, W_i^{TS}, W_i^K, W_i^R, W_i^{HW}, AW_i^j \geq 0 \forall i, j \end{gathered} \tag{17}$$

Triangular fuzzy number

A fuzzy number $\tilde{A}$ is a convex normalized fuzzy set on the real line R such that:
(a) there exists at least one $x_0 \in R$ with $\mu_{\tilde{A}}(x_0) = 1$; (b) $\mu_{\tilde{A}}(x)$ is piecewise continuous.
Let us assume that the membership function of any fuzzy number $\tilde{A}$ is as follows:

$$\mu_{\tilde{A}}(x) = \begin{cases} 1 - \dfrac{m^A - x}{\alpha^A}, & m^A - \alpha^A \leq x < m^A \\ 1 - \dfrac{x - m^A}{\beta^A}, & m^A \leq x \leq m^A + \beta^A \\ 0, & \text{otherwise} \end{cases}$$

where m^A is the mean value of $\tilde{A}$ and α^A and β^A are left and right spreads, respectively, and it is termed as the triangular fuzzy number. The TFN represented as shown in Figure 2.

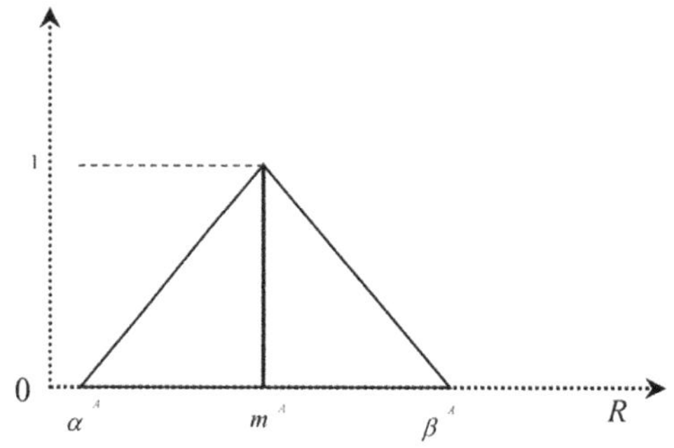

Figure 2 | TFN ($\tilde{A}$).

FLP problem

The FLP model can be represented in the following manner:

$$\begin{aligned} &\max \tilde{z} = \tilde{c}_1 \otimes \tilde{x}_1 \oplus \ldots \tilde{c}_n \otimes \tilde{x}_n \\ &s.t. \\ &\tilde{a}_{11} \otimes \tilde{x}_{11} \oplus \ldots \tilde{a}_{1n} \otimes \tilde{x}_n \leq \tilde{b}_1 \\ &. \\ &. \\ &. \\ &\tilde{a}_{m1} \otimes \tilde{x}_1 \oplus \ldots \tilde{a}_{mn} \otimes \tilde{x}_n \leq \tilde{b}_m \\ &\tilde{x}_1 \geq 0, \tilde{x}_2 \geq 0, \ \ldots, \tilde{x}_n \geq 0. \end{aligned} \tag{I}$$

The matrix form of the above equation is:

$$\begin{aligned} &\max \tilde{z} = \tilde{c} \otimes \tilde{x} \\ &s.t. \\ &\tilde{A} \otimes \tilde{x} \leq \tilde{b} \\ &\tilde{x} \geq 0. \end{aligned} \tag{II}$$

The coefficient matrix $\tilde{A} = [\tilde{a}_{ij}]_{m*n}, \ 1 \leq i, \ j \leq n$ is $m*n$ fuzzy matrix where $\forall i, j, \ \tilde{a}_{ij} \succ 0$ or $\tilde{a}_{ij} \prec 0$ and $\tilde{x}_i, \tilde{b}_j \in F(\aleph)$.

If matrix $\tilde{A}$ denoted by $\tilde{A} = (A, A', A'')$ that $A = [a_{ij}], \ \ A' = [a'_{ij}], \ A'' = [a''_{ij}], \ \tilde{x} = (x, x', x'')$, $\tilde{b} = (b, b', b'')$. Then we have:

$$\begin{aligned} &\max \tilde{z} = (c, c', c'') \otimes (x, x', x'') \\ &s.t. \\ &(A, A', A'') \otimes (x, x', x'') \leq (b, b', b'') \\ &(x, x', x'') \geq 0. \end{aligned} \tag{III}$$

doi: 10.2166/h2oj.2020.032

Algorithm for MOFLP

The detailed solution process for the MOFLP model can be described as follows:

1. Formulate the irrigation planning problem with the help of TFN for decision parameters and decision variables of objective functions and constraints.
2. Solve the model as an FLP problem, considering only one objective individually.
3. Work out the corresponding values of each objective with reference to the procedure adopted in step 2.
4. Pick up the best value (Z_U) and worst value (Z_L) for each objective function comparing the values obtained steps 2 and 3.
5. Formulate the linear membership function with reference to (Z_U) and (Z_L) values for each objective.
6. Introduce a new dummy variable as level of satisfaction (λ), which is subject to the new constraints incorporating vagueness involved in the objective function values and original constraints and maximize it.
7. Identify significant parameters and interactions and formulate the equivalent FLP model as a MOFLP model and solve the formulated model.
8. Analyze the results and find out the compromised solution with the level of satisfaction (λ) under the fuzzy environment.
9. Stop.

The structure of the fuzzy approach for a compromised solution of the MOFLP model is as depicted in Figure 3.

RESULTS AND DISCUSSION

The present study explores the development of the MOFLP model with a TFN of linear membership functions and its applicability for sustainable irrigation planning to the Jayakwadi Project Stage-I, India. Maximization of NBF, CPD, EGN and MUT, these are the objectives for which the irrigation policy maker has to focus with the same priority with available limited resources of the project under uncertainty.

The MOFLP problem considers fuzzy input data by fuzzy membership functions. The objectives (Equations (1)–(4)) and constraints ((5)–(17)), which are imprecise and uncertain, are represented by fuzzy sets in the form of TFN. The fuzziness associated with stipulations (resources) can be well described by considering the tolerance range of membership functions. Various decision parameters (cost coefficients, technological coefficients and resources) and decision variables (area to be irrigated under each crop in each season) are treated as imprecise (fuzzy) in the form of constraints for the existing cropping pattern of the project. Similarly, for objective functions, various decision parameters (objective function coefficients, cost coefficients) and decision variables (area to be irrigated under each crop in each season) considered as fuzzy. The fuzzy objective functions (Equations (1)–(4)) are being maximized individually subject to fuzzy constraints (Equations (5)–(17)) using LINGO 13.0 (Language for INteractive General Optimization). The MOFLP model which gives compromised solution involves total 35 variables and 51 constraints.

Table 1 shows the results of four different objectives for its individual maximization, with the help of which, the linear membership function has constructed for each objective by choosing the best value (z^+) and worst value (z^-), which are indicated in bold figures. The objective functions rise linearly from zero to one for membership functions $\mu_1(X)$, $\mu_2(X)$, $\mu_3(X)$ and $\mu_4(X)$ of the fuzzy sets. The

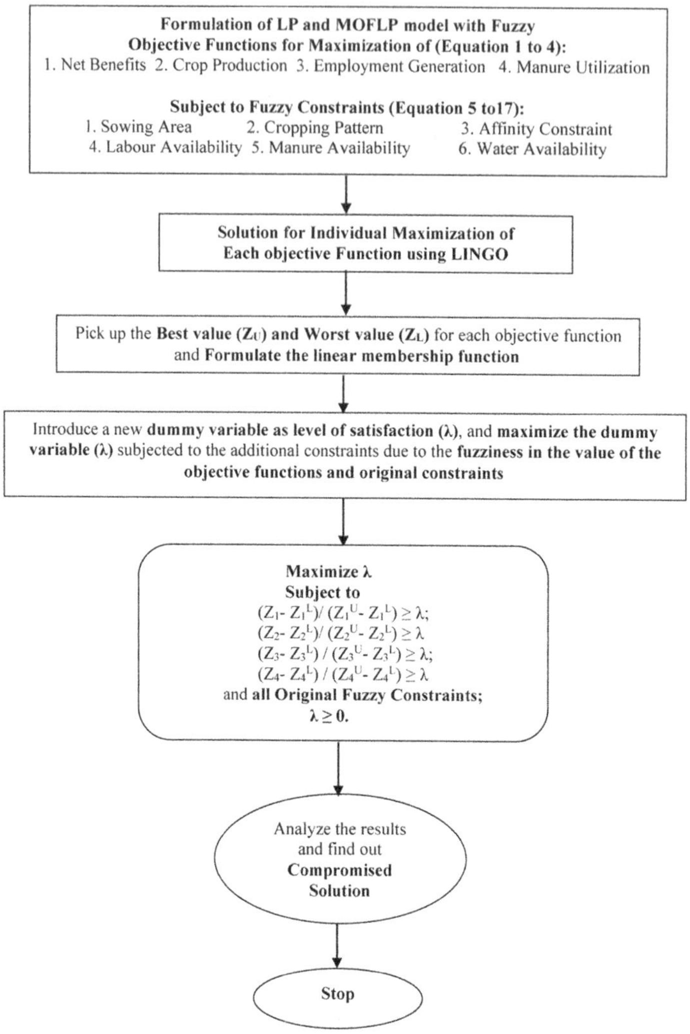

Figure 3 | Structure of the fuzzy approach for the compromised solution of the MOFLP model.

highest values obtained are $Z_1 = 1{,}261.99$ Million Rs, $Z_2 = 355{,}352.60$ Tons, $Z_3 = 25.83$ Million Man Days and $Z_4 = 137{,}038.10$ Tons, respectively.

In case of NBF, $Z_1 = 941.77$ Million Rs or less if level of satisfaction (λ) rises from zero and when it becomes one, $Z_1 = 1{,}261.99$ Million Rs or more. Similarly, for CPD, $Z_2 = 80{,}250.94$ Tons or less if λ rises from zero and when it becomes to one, $Z_2 = 355{,}352.60$ Tons or more. Also, $Z_3 = 17.44$ Million Man Days or less if λ rises from zero and when it becomes to one, $Z_3 = 25.83$ Million Man Days or more for EGN. Similarly, for MUT, $Z_4 = 52{,}932.98$ Tons or less if λ rises from zero and when it becomes to one, $Z_4 = 137{,}038.10$ Tons or more.

The MOFLP model has been formulated for the irrigation planning problem on the basis of FLP formulation and applied successfully to the case study under consideration. The optimal fuzzy solution for the first objective is: $X_1 = (4{,}249.20,\ 4{,}249.20,\ 0)$, $X_2 = (2{,}124.60,\ 2{,}124.60,\ 0)$, $X_3 = (4{,}249.20,\ 4{,}249.20,\ 0)$, $X_4 = (0,\ 0,\ 0)$, $X_5 = (16{,}996.23,\ 16{,}996.23,\ 0)$, $X_6 = (6{,}445.84,\ 6{,}445.84,\ 0)$, $X_7 = (0,\ 0,\ 0)$, $X_8 = (35{,}410,\ 35{,}410,\ 0)$, $X_9 = (7{,}082,\ 7{,}082,\ 0)$, $X_{10} = (0,\ 0,\ 0)$ and the optimal value of the NBF $= 1{,}261.99$ Million Rs.

Similarly, the optimal fuzzy solution for each individual objective and optimal values of the other objectives are represented in Table 1. From the results presented in Table 1, the solution of decision variables for various crops is in the form of the TFN. The TFN has values in the form of left spread, mean value and right spread. When the demand or constraint is satisfied, the membership grade is 1, it

doi: 10.2166/h2oj.2020.032

Table 1 | Solution of the LP model and compromised solution of the MOFLP model

Fuzzy variable	Crop and season	Solution for individual maximization of each objective function NBF (Z_1) (ha)	CPD (Z_2) (ha)	EGN (Z_3) (ha)	MUT (Z_4) (ha)	Compromised solution $\lambda = 0.58$
$\tilde{X}_1$	Sugarcane (*P*)	(4,249.20, 4,249.20, 0)	(4,249.20, 4,249.20, 0)	(0, 0, 0)	(4,247.93, 4,247.93, 0)	(0, 0, 6,498.64)
$\tilde{X}_2$	Banana (*P*)	(2,124.60, 2,124.60, 0)	(2,124.60, 2,124.60, 0)	(0, 0, 0)	(2,124.60, 2,124.60, 0)	(2,124.60, 2,124.60, 0)
$\tilde{X}_3$	Chilies (*TS*)	(4,249.20, 4,249.20, 0)	(4,249.20, 4,249.20, 0)	(4,249.20, 4,249.20, 0)	(4,249.20, 4,249.20, 0)	(4,249.20, 4,249.20, 0)
$\tilde{X}_4$	L S Cotton (*TS*)	(0, 0, 0)	(0, 0, 0)	(35,410, 35,410, 0)	(35,410, 35,410, 0)	(21,403.70, 0, 0)
$\tilde{X}_5$	Sorghum (*K*)	(16,996.23, 16,996.23, 0)	(16,996.23, 16,996.23, 0)	(16,996.23, 16,996.23, 0)	(16,996.23, 16,996.23, 0)	(0, 0, 50,988.68)
$\tilde{X}_6$	Paddy (*K*)	(6,445.84, 6,445.84, 0)	(0, 0, 0)	(14,164, 14,164, 0)	(14,164, 14,164, 0)	(14,164, 14,164, 0)
$\tilde{X}_7$	Sorghum (*R*)	(0, 0, 0)	(14,683.87, 14,683.87, 0)	(0, 0, 0)	(20,276.34, 20,276.34, 0)	(0, 0, 0)
$\tilde{X}_8$	Wheat (*R*)	(35,410, 35,410, 0)	(35,410, 35,410, 0)	(35,410, 35,410, 0)	(0, 0, 0)	(23,846.35, 23,846.35, 0)
$\tilde{X}_9$	Gram (*R*)	(7,082, 7,082, 0)	(0, 0, 0)	(6,437.84, 6,437.84, 0)	(0, 0, 0)	(7,082, 7,082, 0)
$\tilde{X}_{10}$	Groundnut (*HW*)	(0, 0, 0)	(0, 0, 0)	(0, 0, 0)	(4,249.20, 4,249.20, 0)	(0, 0, 0)
NBF (Million Rs)		**1,261.99 (Z_1^+)**	1,240.78	**941.77 (Z_1^-)**	1,094.60	1,127.60
CPD (Tons)		354,319.70	**355,352.60 (Z_2^+)**	**80,250.94 (Z_2^-)**	334,781.70	239,897.50
EGN (Million Man Days)		18.54	**17.44 (Z_3^-)**	**25.83(Z_3^+)**	19.63	22.31
MUT (Tons)		57,480.78	**52,932.98 (Z_4^-)**	119,802.80	**137,038.10 (Z_4^+)**	115,969.60

doi: 10.2166/h2oj.2020.032

means that level of satisfaction is highest (i.e. at apex) or it is having no spread on the left and right of the TFN. When the demand or constraint is not satisfied, the membership grade is lowest, the level of satisfaction is minimum, and corresponding spread on the left and right is observed for a TFN. The situation in which the constraint or demand is not satisfied at all, which can be represented as a horizontal line, i.e. no spread on left and right and zero mean value of the TFN. These features of TFN allow the irrigation policy maker to address and tackle the vagueness and/or uncertainty involved in the various parameters of irrigation planning, and given these, to formulate the irrigation planning problem, the TFN is used for the present study.

From Table 1, when the NBF is maximized, it is observed that the benefit coefficient is low for various crops such as Groundnut (*HW*), Sorghum (*R*) and LS Cotton (*TS*); because of this, the area to be allocated for irrigation under these crops is zero. Similarly, when the CPD is maximized, as the CPD coefficients are low for various crops such as Groundnut (*HW*), Gram (*R*), Paddy (*K*) and LS Cotton (*TS*); because of this, the area to be allocated for irrigation under these crops is zero. Also, in case of maximization of EGN, the area under irrigation is zero for Groundnut (*HW*), Sorghum (*R*), Banana (*P*) and Sugarcane (*P*), which is because the labor requirement per ha for these crops is low and also, the limited area allocation of these crops under the existing cropping pattern of the project. If MUT is maximized, then for crops such as Gram (*R*) and Wheat (*R*), the area under irrigation is zero because the manure requirement per ha is low and also due to the limited area under the existing cropping pattern of the project. If four conflicting objectives are considered simultaneously under the fuzzy environment, then crops such as Groundnut (*HW*) and Sorghum (*R*), the area under irrigation is zero.

The results are shown graphically for Sugarcane (*P*) for the fuzzy optimization of individual objectives in the form of TFN in Figure 4. The maximum level of satisfaction (λ) from the membership functions of four participating/conflicting objectives has been designated as the 'best' achieved/compromised solution. The final modified form of the MOFLP model with adding a new dummy variable $\lambda = \min\, [\mu_1(X), \mu_2(X), \mu_3(X), \mu_4(X)]$ such that:

Maximize λ Subject to:

$$(Z_1 - 941.77 \times 10^6)/(1261.99 \times 10^6 - 941.77 \times 10^6) \geq \lambda;\ (Z_2 - 80250.94)/(355352.60 - 80250.94) \geq \lambda$$

$$(Z_3 - 17.44 \times 10^6)/(25.83 \times 10^6 - 17.44 \times 10^6) \geq \lambda;\quad (Z_4 - 52932.98)/(137038.10 - 52932.98) \geq \lambda$$

And all other original fuzzy constraints given in the model;

$\lambda \geq 0.$

The obtained value of the level of satisfaction (λ) = 0.58 is in the form of compromised solution, and it provides NBF = 1,127.60 Million Rs, CPD = 239,897.50 Tons, EGN = 22.31 Million Man Days and MUT = 115,969.60 Tons, respectively.

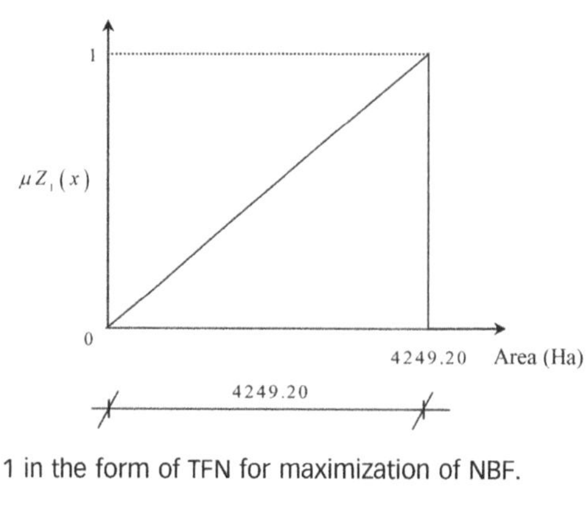

Figure 4 | Optimal solution of crop 1 in the form of TFN for maximization of NBF.

Evolutionary Algorithms in Water Resources
doi: 10.2166/h2oj.2020.032

The problem of irrigation planning can be analyzed to deal with uncertainty in various strategies such as (i) fuzzy objective functions ($\tilde{Z}_k$), (ii) fuzzy cost coefficients ($\tilde{C}_j$), (iii) fuzzy resources ($\tilde{b}_i$), (iv) fuzzy technological coefficients ($\tilde{A}_{ij}$) and fuzzy resources ($\tilde{b}_i$) simultaneously and (v) fuzzy resources ($\tilde{b}_i$) and fuzzy decision variable ($\tilde{X}_j$) simultaneously (Regulwar & Gurav 2010, 2011, 2012; Gurav & Regulwar 2012). The uncertainty/vagueness arises in the values of the objective function because of sharing the same resources for more than one objective(s) ($\tilde{Z}_k$); which can be tackled successfully by trading off between the values of objective functions with fuzzy sets in the first strategy.

Also, the cost coefficients including the fluctuations in market price, input cost and monetary benefits associated with uncertainty/vagueness can be dealt with the second strategy. When the resources (b_i) are limited but with variation to certain extent due the practical condition/difficulties (e.g. migration of labor from one region to the other) which leads to uncertainty/vagueness in resources, that can be tackled by implementing the third strategy. Sometimes, the uncertainty associated with technological coefficients (A_{ij}), and resources (b_i) can be dealt with the fourth strategy.

Similarly, depending upon the field situation the strategy sixth (developed and proposed model for the present study) can be used by irrigation policy makers for irrigation planning. In the present study, the triangular membership function has been used to represent the fuzzy sets. Along with the triangular membership function, the other, such as linear membership functions-monotonously increasing and/or decreasing; linear membership functions; convex and/or concave, exponential membership functions and trapezoidal membership functions, can also be feasible for such problems.

The results which are represented in the form of Table 1 (present study) are improved qualitatively and quantitatively over the results presented in Table 2 (Regulwar & Gurav 2011). The uniqueness of the present manuscript is to deal the uncertainty/vagueness involved in the problem of sustainable irrigation planning issues in developing countries like India. The present paper deals with the uncertainty included in the objective functions ($\tilde{Z}_k$), cost coefficients (C_j), technological coefficients (A_{ij}), resources (b_i) and decision variable (X_j) simultaneously, by treating these as fuzzy. This seems to be probably closer to the real-world problem of irrigation planning associated with uncertainty.

Table 2 | Salient parameters of the optimal cropping pattern planning for $\tilde{Z}$, $\tilde{b}_i$, $\tilde{a}_{ij}$ and $\tilde{a}_{ij}$ and $\tilde{b}_i$ (Regulwar & Gurav 2011)

Sr no.	Crop and season	Compromised solution for $\tilde{Z}$ ($\lambda = 0.580$) Area (ha)	Compromised solution for $\tilde{b}_i$ ($\lambda = 0.503$) Area (ha)	Compromised solution for $\tilde{a}_{ij}$ ($\lambda = 0.501$) Area (ha)	Compromised solution for $\tilde{a}_{ij}$ and $\tilde{b}_i$ ($\lambda = 0.287$) Area (ha)
1	Sugarcane (*P*)	2,166.18	1,871.80	1,872.20	1,839.55
2	Banana (*P*)	2,124.60	1,857.26	2,146.01	2,000.64
3	Chilies (*TS*)	4,249.20	3,714.52	4,357.91	4,090.02
4	LS Cotton (*TS*)	28,567.80	24,725.47	30,124.79	19,102.49
5	Sorghum (*K*)	16,996.80	14,858.08	17,889.34	16,988.02
6	Paddy (*K*)	14,164.00	12,381.73	14,907.79	14,156.69
7	Sorghum (*R*)	0	0	0	0
8	Wheat (*R*)	23,832.78	21,017.54	32,603.18	38,326.00
9	Gram (*R*)	7,082.00	6,190.86	7,263.18	6,816.70
10	Groundnut (*HW*)	0	0	0	0
Net Cropped Area (ha)		99,183.36	86,617.26	111,164.42	103,320.11
NBF (Million Rs)		1,503.73	1,314.87	1,617.66	1,602.43
CPD (Tons)		319,563.50	278,042.5	312,941.30	308,066.20
EGN (Million Man Days)		29.74	25.99	34.03	32.60
MUT (Tons)		154,506.50	134,365.3	163,647.90	131.783.00
Irrigation Intensity (%)		70.02	61.15	78.48	72.94

doi: 10.2166/h2oj.2020.032

CONCLUSION

The present study deals with the development of the MOFLP model under uncertainty that maximizes NBF, CPD, EGN and MUT simultaneously and its successful application for the case study of Jayakwadi Project Stage-I, Maharashtra State, India.

The MOFLP model compromised solution has worked out with $\lambda = 0.58$, which suggests NBF = 1,127.60 Million Rs, CPD = 239,897.50 Tons, EGN = 22.31 Million Man Days and MUT = 115,969.60 Tons respectively. These obtained results of the present model are promising as it considers the uncertainty involved in irrigation planning problem which is closer to the scenario of the real world.

The proposed model improves on existing models and suggests a basis for irrigation planning as an integrated approach under the fuzzy environment considering the decision parameters and decision variables as fuzzy.

The proposed model will be helpful for the irrigation policy maker to address the uncertainty associated with irrigation planning problems and to take decisions under conflicting situations while dealing with different conflicting objectives simultaneously.

The proposed model is capable of dealing with integrated irrigation planning with prime focus on the various parameters related to economic, social, environmental and sustainability.

The MOFLP model developed in this paper is a general-purpose model, and it provides an opportunity to extend its application for the detailed system analysis for irrigation project having reservoir with surface water, command area, water distribution network through canal and/or pipe, and conjunctive use of surface and ground water for water-stressed area of the particular region.

ACKNOWLEDGEMENTS

The authors thank the reviewers for their valuable suggestions and useful comments. The authors convey grateful acknowledgement and thanks to MPKV Rahuri and CADA Aurangabad for providing necessary data for analysis.

DATA AVAILABILITY STATEMENT

Data cannot be made publicly available; readers should contact the corresponding author for details.

REFERENCES

Allahviranloo, T., Lotfi, F. H., Kiasary, M. K., Kiani, N. A. & Alizadeh, L. 2008 Solving fully fuzzy linear programming problem by the ranking function. *Appl. Math. Sci.* **2** (1), 19–32.

Arikan, F. & Gungor, Z. 2007 A two phase approach for multi-objective programming problems with fuzzy coefficients. *Inf. Sci.* **177**, 5191–5202.

Banihabib, M. E., Tabari, M. M. R. & Tabari, M. M. R. 2019 Development of a fuzzy multi-objective heuristic model for optimum water allocation. *Water Resour. Manag.* **33**, 3673–3689.

Bellman, R. & Zadeh, L. A. 1970 Decision making in a fuzzy environment. *Manag. Sci.* **17** (4), 141–164.

Commissionerate of Agriculture, Maharashtra State 2006 Agricultural Statistical Information Maharashtra State, India. The Report of Commissionerate of Agriculture, Maharashtra State part-II (2006).

Gasimov, R. N. & Yenilmez, K. 2002 Solving fuzzy linear programming problems with linear membership functions. *Turk. J. Math.* **26**, 375–396.

Guo, P., Wang, X., Zhu, H. & Li, M. 2014 Inexact fuzzy chance-constrained nonlinear programming approach for crop water allocation under precipitation variation and sustainable development. *J. Water Resour. Plan. Manag.* **140** (9), 05014003.

Gupta, A. P., Harboe, R. M. T. & Tabucanon, T. 2000 Fuzzy multiple-criteria decision making for crop area planning in Narmada river basin. *Agric. Syst.* **63** (1), 1–18.

Gurav, J. B. & Regulwar, D. G. 2012 Multi objective sustainable irrigation planning with decision parameters and decision variables fuzzy in nature. *Water Resour. Manag.* **26** (10), 3005–3021.

Jimenez, M., Arenas, M., Bilbao, A. & Rodriguez, V. M. 2007 Linear programming with fuzzy parameters: an interactive method resolution. *Eur. J. Oper. Res.* **177** (3), 1599–1609.

Li, M., Guo, P. & Singh, V. P. 2016 Biobjective optimization for efficient irrigation under fuzzy uncertainty. *J. Irrig. Drain. Eng.* **142** (8), 05016003.

Mirajkar, A. B. & Patel, P. L. 2016 Multiobjective two-phase fuzzy optimization approaches in management of water resources. *J. Water Resour. Plan. Manag.* **142** (11), 04016046.

Nanda, S., Panda, G. & Dash, J. K. 2008 A new methodology for crisp equivalent of fuzzy chance constrained programming problem. *Fuzzy Optim. Decis. Making* **7**, 59–74.

Nasseri, S. H. 2008 A new method for solving fuzzy linear programming by solving linear programming. *Appl. Math. Sci.* **2** (50), 2473–2480.

Raju, K. S. & Duckstein, L. 2003 Multiobjective fuzzy linear programming for sustainable irrigation planning: an Indian case study. *Soft Comput.* **7**, 412–418.

Regulwar, D. G. & Gurav, J. B. 2010 Fuzzy approach based management model for irrigation planning. *J. Water Resour. Protec.* **2**, 545–554.

Regulwar, D. G. & Gurav, J. B. 2011 Irrigation planning under uncertainty – a multi objective fuzzy linear programming approach. *Water Resour. Manag.* **25** (5), 1387–1416.

Regulwar, D. G. & Gurav, J. B. 2012 Sustainable irrigation planning with imprecise parameters under fuzzy environment. *Water Resour. Manag.* **26** (13), 3871–3892.

Regulwar, D. G. & Gurav, J. B. 2013 Two-phase multi objective fuzzy linear programming approach for sustainable irrigation planning. *J. Water Resour. Protec.* **5** (6), 642–651.

Srivastava, P. & Singh, R. M. 2017 Agricultural land allocation for crop planning in a canal command area using fuzzy multiobjective goal programming. *J. Irrig. Drain Eng.* **143** (6), 04017007.

Xu, J., Tu, Y. & Zeng, Z. 2013 Bilevel optimization of regional water resources allocation problem under fuzzy random environment. *J. Water Resour. Plan. Manag.* **139** (3), 246–264.

Yue, Q., Zhang, F., Zhang, C., Zhu, H., Tang, Y. & Guo, P. 2020 A full fuzzy-interval credibility-constrained nonlinear programming approach for irrigation water allocation under uncertainty. *Agric. Water Manag.* **230** (1), 105961.

Zadeh, L. A. 1965 Fuzzy sets. *Inf. Control* **8**, 338–353.

Zhang, C. & Guo, P. 2018 FLFP: a fuzzy linear fractional programming approach with double-sided fuzziness for optimal irrigation water allocation. *Agric. Water Manag.* **199**, 105–119.

Zhang, C., Engel, B. A. & Guo, P. 2018 An interval-based fuzzy chance-constrained irrigation water allocation model with double-sided fuzziness. *Agric. Water Manag.* **210**, 22–31.

First received 18 April 2020; accepted in revised form 4 August 2020. Available online 15 September 2020

Evolutionary Algorithms in Water Resources
doi: 10.2166/h2oj.2020.041

Application of artificial neural network for predicting water levels in Hooghly estuary, India

Kalyan Kumar Bhar[a,*] and Susmita Bakshi[b]

[a] Department of Civil Engineering, Indian Institute of Engineering Science and Technology, Shibpur, Howrah 711103, West Bengal, India

[b] Department of Civil Engineering, Meghnad Saha Institute of Technology, Kolkata 700150, West Bengal, India

*Corresponding author. E-mail: kalyan@civil.iiests.ac.in

Abstract

Hydrodynamic models for morphodynamic studies in estuaries require continuous tidal water level data as boundary conditions. However, for the Hooghly estuary in India, measurement of continuous tidal water elevation data at the most downstream point is a very difficult task because of the remote location and the confluence with the deep sea. The tidal water level data at this station are measured for a half tidal cycle which is not useful for hydrodynamic modeling. However, at other upstream stations, tide water level data are measured continuously. Accordingly, in this study, an attempt is made to generate continuous tidal water level data at the remote station, using the data of the neighboring stations as input to an artificial neural network (ANN) model. A three-layered feed-forward backpropagation (FFBP) network with two hidden layers is selected and five different combinations of input vectors are used. Simulated water level data obtained from each model are compared with the observed data graphically as well as by estimating the standard error parameters. The best model suitable for prediction of continuous tidal elevation during any time of the tidal cycle and applicable throughout the year is then identified. It is found that tidal data from the nearest neighboring station are more suitable for training.

Key words: artificial neural network, Hooghly estuary, tidal water level prediction

Highlights

- Application of ANN to predict tidal water levels at remote stations in an estuary where only 12-h levels can be recorded.
- Predicted water levels can be used in hydrodynamic modeling as a boundary condition.
- This is the first time this technique has been attempted for Hooghly estuary in India.
- The predicted water levels can also be used for forecasting tidal water levels.

INTRODUCTION

In an estuary, studies on characteristics of sediment transport processes and the resulting changes in the bed morphology are essential tasks for operation and management of ports, construction of jetties and other structures, maintenance dredging works to provide adequate navigational depths for ships, etc. Usually, for these purposes, an appropriate two-dimensional or three-dimensional hydrodynamic

doi: 10.2166/h2oj.2020.041

model is used to assess the sediment transport phenomena and the resulting morphological changes. However, along with other data like bathymetry, sediment characteristics, etc., such a hydrodynamic model requires continuous tidal water level data over the period of simulation, to be provided at the key sections as boundary conditions. Unfortunately, in the large Hooghly estuary in India, it is extremely difficult to measure the tidal water level at the most downstream station near the sea continuously for 24 h because of the limited accessibility and proximity to the turbulent sea. Accordingly, tidal water level data at this station, which is an important boundary section for hydrodynamic modeling, are available only for daytime (12-h period). However, for neighboring upstream stations, tidal water level data are measured continuously. So, since a continuous tidal water level record at the remote station is essential for hydrodynamic modeling, an appropriate model needs to be developed which can simulate the continuous tidal water level at that station. This may preferably be done using the continuous water level data of the neighboring upstream stations, so that the characteristics of the tidal water wave during spring tides and neap tides, peaks, troughs, lags, etc., are adequately reflected in the simulated tidal water levels. Accordingly, in this study, an attempt is made to simulate the continuous tidal water level data at the remote station, by developing artificial neural network (ANN) models and training them with measured continuous water level data of the neighboring upstream stations as input.

Conventional tidal forecasting using least-squares method for the prediction of long-term tidal levels requires a large number of parameters for harmonic analysis. Least Square Estimation (LSE) plays a major role in harmonic analysis. Cai *et al.* (2018) and Li *et al.* (2019) used the Inaction Method (IM), to predict short-term and long-term tidal levels, respectively. A large number of tide gauge data comprising long and continuous records were required for the long-term prediction. ANN models can be used as an alternative to conventional harmonic analysis for forecasting tidal elevation where measured data are limited (Lee & Jeng 2002; Meena & Agrawal 2015; Salim *et al.* 2015).

ANN models have been successfully used in many studies related to hydraulics of flow. Examples include modeling stage discharge relationships (Jain & Chalisgaonkar 2000; Bhattacharya & Solomatine 2005; Habib & Meselhe 2006) and flood forecasting (Campolo *et al.* 1999; Shrestha *et al.* 2004; Peters *et al.* 2006). Important applications in sediment hydraulics include development of sediment rating curves (Jain 2008), sediment yield prediction (Sarangi *et al.* 2005), determination of scour (Akib *et al.* 2014), and sediment transport (Yitian & Gu 2003; Yang *et al.* 2009; Maanen 2010; Haddadchi *et al.* 2013); modeling of salinity in estuaries (Rath *et al.* 2017) and groundwater level predictions (Mohanasundaram *et al.* 2019).

ANNs have also been attempted for simulation of tidal water level, discharge, and wave height in coastal areas, as substitutes of numerical models. Tsai & Lee (1999) used backpropagation ANN to forecast tidal water level variations, using two advance consecutive tidal data and two corresponding shocks as inputs to obtain one prediction value. Lee & Jeng (2002) used ANN models to predict hourly tidal levels over a long duration using a short-term hourly tidal record. Ustoorikar & Deo (2008) used ANN and genetic programming for filling up gaps in wave height data. Panda *et al.* (2010) compared the river stage data obtained from an ANN model with a one-dimensional hydrodynamic model. Feed-forward neural network architecture with Levenberg–Marquardt (LM) backpropagation training algorithm was used to train the neural network model using hourly water level data. Chen *et al.* (2012) applied ANN as an alternative modeling approach to simulate the water stage time-series of the Danshui River estuary in northern Taiwan and compared the results with vertical (laterally averaged) 2D and 3D hydrodynamic models. Hidayat *et al.* (2014a) developed a tidal discharge forecast model for a site on the Mahakam River. Hidayat *et al.* (2014b) used an M5 model tree machine-learning technique to reconstruct a disrupted water level record of the Mahakam Delta in Indonesia. Meena & Agrawal (2015) used ANN models to predict the tidal level variations at stations situated on the southwest coast of India and fill the gap for the missing data. A sequential learning radial basis function (RBF) network was used by Yin *et al.* (2013) for accurate real-time prediction of tidal level. Lee & Resdi (2014) presented a feed-forward backpropagation (FFBP) network and generalized regression neural network (GRNN) model for prediction of coastal high and low water, which provided effective prediction but was not suitable for one-year simulation. Salim *et al.* (2015) used FFBP and Nonlinear Auto Regressive with exogenous input (NARX) network to predict hourly tidal level variations at Mangalore, Karnataka, using a week's hourly tidal levels as input and found the NARX network performance better than the FFBP network.

An adaptive variable-structure online sequential extreme learning machine (OS-ELM) was used by Yin *et al.* (2016) for online tidal level prediction purpose. Tidal prediction simulations were conducted based on the actual measured tidal data and meteorological data and OS-ELM was found effective for short-term tidal predictions in terms of accuracy and rapidness.

In this study, simulation of the continuous tidal water level data at the most remote station in the Hooghly estuary is done with an ANN (FFBP) model using observed continuous tidal water level data of the neighboring upstream stations. This simulated continuous tidal water level data at the remote station is proposed to be used with a hydrodynamic model as one of the important boundary conditions. The simulated water level data may also be used for navigation purposes. The ANN (FFBP) model is selected because of its simplicity of structure, minimum data requirement, good convergence, and acceptable performances.

The following aspects are investigated while developing the models: (i) which of the upstream neighboring stations are suitable for training and whether the inclusion of more stations improves the model's performance; (ii) whether the model's performance is the same for any part of the tidal cycle (i.e., during spring tide and neap tide); (iii) whether 1-day ahead prediction is possible; and (iv) whether increasing the number of lagged inputs actually improves the model's performance.

STUDY AREA

The Hooghly estuary (Figure 1) is the lowest portion of the Ganga–Bhagirathi–Hooghly river system which joins with the Bay of Bengal and lies approximately between 21°31′ N–23°20′ N and 87°45′ E–88°45′ E. Hooghly estuary is a funnel-shaped estuary. Width of this estuary varies from about 25 km at the mouth to about 6 km at the head. The estuary is very shallow with an average depth of 6 m and the maximum depth is about 20 m. It is a well-mixed estuary, and mixing zones of the estuary extend up to Diamond Harbour. The estuary is characterized by the presence of a large number of tidal bars and tidal islands. The large tidal variations, irregular coastlines, the presence of islands, and dredged navigational channels separated by shallow areas make the flow patterns quite complex.

The Hooghly estuary is the gateway to the two major riverine ports in India: the Haldia port and the Kolkata port. The Haldia port is shown in Figure 1, and the Kolkata port is situated further upstream of Diamond Harbour (at the top). The red lines shown in Figure 1 are the shipping channels. On the left side, the Eden Channel and the Jelligham channel take the ships to the Haldia port and the Rangafalla Channel on the right navigates the ships towards Kolkata port.

Operation and management of these two ports are totally dependent on the navigability of these shipping channels. Availability of adequate draft for movement of ships along these channels is largely affected by frequent sediment deposition. This poses a perennial problem to the Kolkata Port Trust (KoPT) Authority, and regular maintenance dredging requires to be employed, which is highly expensive. Disposing the dredged material at appropriate locations in the estuary is also difficult as tidal action brings the disposed dredged material back to the shipping channels. All these tasks require a proper understanding of the hydrodynamics of flow and characteristics of the morphological changes.

doi: 10.2166/h2oj.2020.041

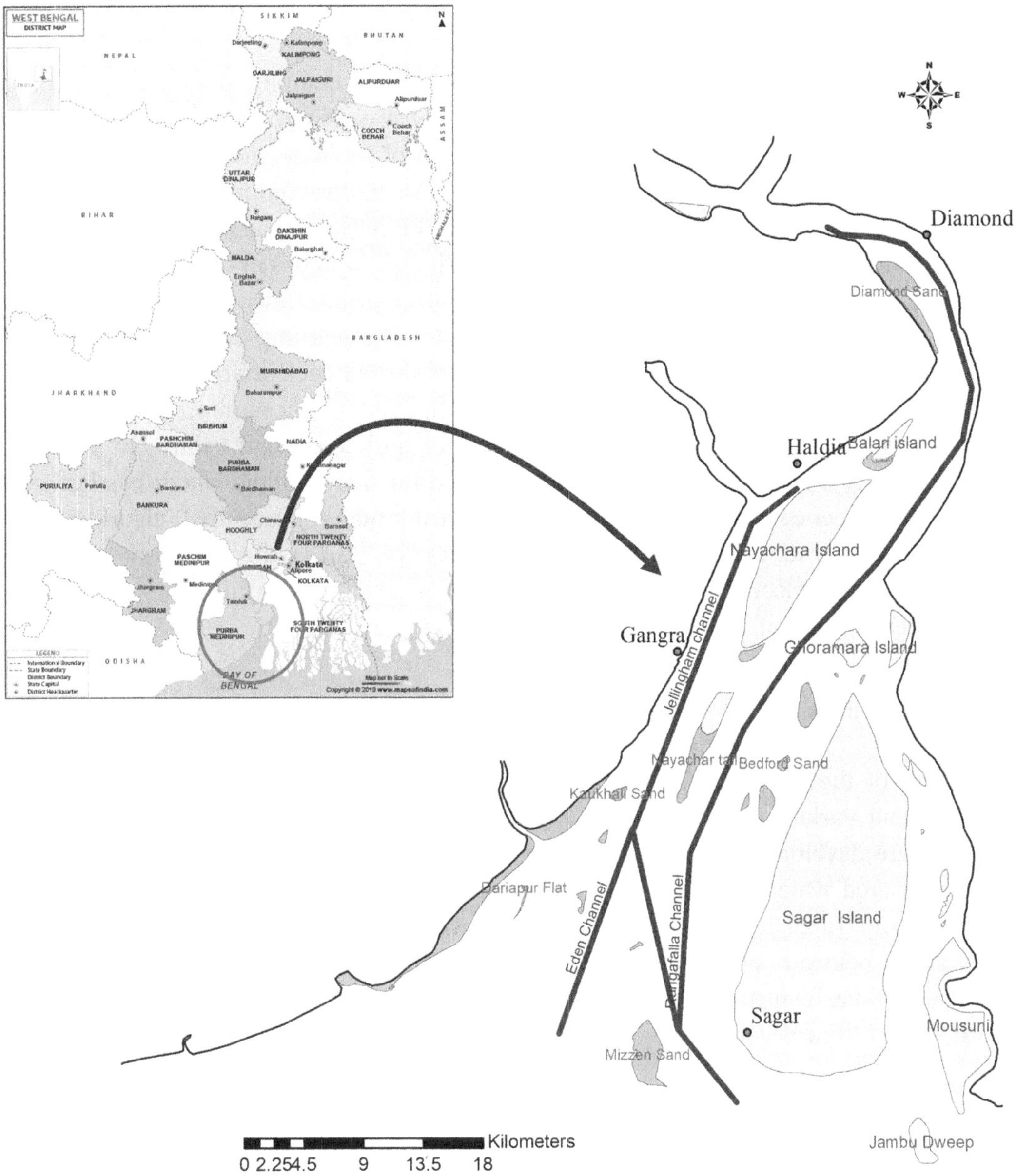

Figure 1 | Study area.

The morphology of the estuary is extremely dynamic with high rates of erosion and accretion, which makes it essential to develop appropriate hydrodynamic models for assessing the morphological behavior of the estuary. These hydrodynamic models require continuous tidal water level data at the key stations as boundary conditions. However, continuous record of tidal water level data is not available at some of the remote stations in the estuary, like Sagar (Figure 1), Dadanpatra and Fraserganj (not shown in Figure 1). The tidal elevation data measured at these stations are usually for a 12-h stretch (daytime). However, at other upstream stations like Gangra, Haldia, Diamond Harbour (Figure 1), tidal water level data are measured continuously for 24 h. Accordingly, in this-study, it is attempted to simulate the continuous tidal water level data of the remote gauging station Sagar, situated at the mouth of the estuary, using continuous tidal water level data of the neighboring upstream stations, Gangra, Haldia, and Diamond Harbour, by developing a number of ANN models.

METHODS

ANN models consist of interconnected nodes which can extract complex nonlinear relationships from a set of input and output data. A typical network consists of one input layer, one or more hidden layers, and one output layer. Each layer is made up of interconnected nodes with a set of associated weights. Multilayer feed-forward network (MLFF) is the commonly used ANN model (Maier *et al.* 2010) in water resources applications. The MLFF network has a unidirectional flow of information. The output from nodes in one layer is used as inputs to nodes in the next layer. Different learning rules are available to train the network, e.g., Delta Learning Rule, Memory-based learning rule, Hebbian Learning Rule, and Boltzman Learning Rule. The learning algorithm is the specific mathematical method, used to update the synaptic weights during each training iteration. A set of optimal values of connection weights and thresholds are thus found that minimize a predetermined error function between the ANN model simulated output and the target values. In the present study, the nonlinear least squares LM algorithm is adopted for this training, which has been found to have a faster convergence rate, compared to the standard gradient descent BP algorithm (Habib & Meselhe 2006; Chen *et al.* 2012). As mentioned by Wu *et al.* (2014), the steps followed in ANN model development are a selection of input variables, division of data, determination of best network architecture, model calibration, and finally model validation.

Selection of input variables

Tidal water levels of the three upstream stations at Gangra, Haldia, and Diamond Harbour are selected as the input variables, and the tidal water at Sagar is taken as the output variable. Five different models are developed and tested, which have the following input variables as shown in Table 1. The measured water level data at all four stations are at half-hourly intervals. Hence, in the second column of Table 1, level at $(t-1)$ denotes level at half an hour before t; level at $(t-2)$ denotes level at 1 h before t, and so on. Similarly, level at $(t-48)$ indicates level at 24 h before t, i.e., 1-day before t. It may be mentioned here that Gangra is the nearest station and Diamond Harbour is the farthest station from Sagar.

Table 1 | Input variables used in different models

Model	Input variables	Model architecture
I	Water level at Gangra at current time, t	1-3-3-1
II	Water levels at Gangra, Haldia, and Diamond Harbour at current time, t	3-5-5-1
III	Water levels at Gangra at time t, $(t-1)$, $(t-2)$, and $(t-3)$	4-6-6-1
IV	Water levels at Gangra at time $(t-48)$, $(t-49)$, $(t-50)$, and $(t-51)$	4-7-7-1
V	Water levels at Gangra at time, t, $(t-1)$, $(t-2)$, $\ldots$, $(t-12)$	13-9-9-1

Performance evaluation of the models

Performances of the developed ANN models are measured with respect to the following statistical parameters, which are calculated from the predicted values of ANN and the target values.

Root mean square error (RMSE)

It is one of the most widely used functions for assessing the performance of a predictive model and is defined as the square root of the average sum of squares of the difference between the target and

doi: 10.2166/h2oj.2020.041

predicted values:

$$\text{RMSE} = \sqrt{\frac{1}{n}\sum_{i=1}^{n}(T_i - P_i)^2} \tag{1}$$

where T_i is the target data, P_i is the corresponding value predicted by the ANN model, and n is the number of data.

Nash–Sutcliffe model efficiency coefficient (E)

This coefficient, E, measures the differences between the observations and predictions relative to the variability in the observed data itself.

$$E = 1 - \frac{\sum_{i=1}^{n}(T_i - P_i)^2}{\sum_{i=1}^{n}(T_i - \bar{T})^2} \tag{2}$$

where $\bar{T}$ denotes the mean of T_i. As per Equation (2), E may range from $-\infty$ to 1.0. A value of $E = 1.0$ indicates a perfect model, whereas $E = 0.0$ indicates that the observed mean is as good as the model predicted, and $E < 0$ indicates that the model is worse than using mean as a predictor.

Correlation coefficient (R)

This is a measure of the linear correlation between the target values and predicted values:

$$R = \frac{n\left(\sum_{i=1}^{n} T_i P\right)_i - \left(\sum_{i=1}^{n} T_i\right)\left(\sum_{i=1}^{n} P_i\right)}{\sqrt{\left[n\sum_{i=1}^{n} T_i^2 - \left(\sum_{i=1}^{n} T_i\right)^2\right]\left[n\sum_{i=1}^{n} P_i^2 - \left(\sum_{i=1}^{n} P_i\right)^2\right]}} \tag{3}$$

Value of R ranges between 0 and 1, with a value of 1 indicating a perfect linear relation.

Mean absolute percentage error (MAPE)

This is an unbiased statistical measure for the predictive capability of a model and is a measure of the relative error in prediction with respect to the actual value of the variable. MAPE calculates the error as a percentage of the actual value.

$$\text{MAPE} = \frac{1}{n}\sum_{i=1}^{n}\frac{|T_i - P_i|}{T_i} \times 100 \tag{4}$$

The combined use of R, RMSE, MAPE, and E helps to assess the performance of a model and compare the accuracy of any two modeling approaches. Statistically, ANN performance can be considered good when the values of R, MAPE, MSE, and RMSE are close to 1, 0, 0, and 0, respectively (Habib & Meselhe 2006).

Network architecture

A typical MLP has an input layer, an output layer, and one or two 'hidden' layers of neurons to connect the input and output layers. The neurons are connected by excitation functions whose weights and biases are estimated during training. The best network architecture for an ANN model is

doi: 10.2166/h2oj.2020.041

determined considering the number of input variables used. Appropriate training parameters are selected and then the network is trained using pre-processed training data. Identification of the best network architecture involves determining the number of hidden nodes, the number of iterations, the learning rate, and the momentum coefficient. This is done by trial and error, based on the mean square error (MSE) value. Parameters like learning rate, epoch, show, goal, and momentum coefficient are used to control the rate of convergence and are obtained by trial and error, based on the MSE values. In this study, a learning rate of 0.1 is used in all models. A maximum epoch of 1,500 is assigned during training and testing of the ANN models. The value of show is taken as 500, that is, the training result is shown after every 500th iteration. In all models, a momentum coefficient of 0.3 is used. For convergence, the acceptable value of MSE is taken as 10^{-6} which is termed as the goal.

Most of the reported works have used one or two hidden layers. The number of neurons in the hidden layer of an MLP determines the potential complexity of the model estuaries (Hidayat *et al.* 2014a; Rath *et al.* 2017). In this study, it is found that using two hidden layers yielded much better results than using a single hidden layer, keeping the total number of neurons in the hidden layers the same. So, two hidden layers with an equal number of nodes in each hidden layer are adopted for further use. A three-layer fully connected feed-forward network (Figure 2) is thus used in all the models with two hidden layers and a single output layer, and different activation function for different layers. Tan-sigmoid function is used in the first hidden layer, log-sigmoid function is used in the second hidden layer, and pure-line function is used in the output layer.

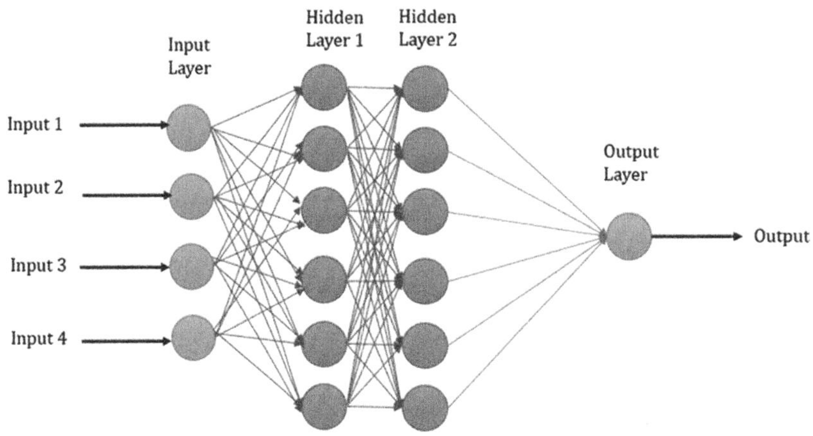

Figure 2 | Network architecture (Model III).

The optimum number of nodes in the hidden layer is selected by varying them and checking the improvement in the *R* value (Figure 3).

In Table 1, the column 'Model architecture' indicates the type of layers and nodes used in the particular model. For example, in Model III, the term '4-6-6-1' indicates four inputs in the input layer, two hidden layers with six nodes, and one output layer. In the present study, design of the network, its training and simulation are performed using the Neural Network Toolbox MATLAB 13.0 developed by MathWorks®, Inc.

The backpropagation algorithm requires initial values of layer weights and biases to start the training process. In the Neural Network Toolbox, the layer weights and biases are initialized using the Nguyen-Widrow method. Once a network is properly trained with acceptable levels of accuracy using a known dataset of input and target, it can be used for simulation. This implies that the trained network can be used to predict possible outputs for any other dataset of input, not used in the training

doi: 10.2166/h2oj.2020.041

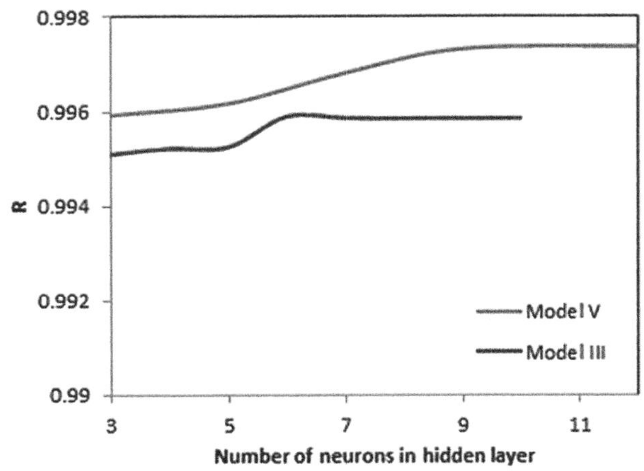

Figure 3 | Effects of variation in the number of neurons in the hidden layer.

process. Levels of acceptance of the predicted output can be verified from the parameters *R*, RMSE, MAPE, and *E*, as defined in Equations (1)–(4).

Dataset

Three different datasets are used in the study. Dataset 1 and Dataset 2 contain 14 days' tidal water level data measured continuously at an interval of 0.5 h at all the four stations: Gangra, Haldia, Diamond Harbour, and Sagar. Dataset 1 was for spring tide measured during April 2014 and Dataset 2 was for neap tide measure during January 2015. On the other hand, Dataset 3 (spring tide measured during November 2016) contains 14 days' tidal water level data measured continuously at an interval of 0.5 h at stations Gangra, Haldia, and Diamond Harbour, but only 12-h data (during daytime) at an interval of 0.5 h over a period of 14 days measured at station Sagar. So, Dataset 3 contains half-cycle data at Sagar, whereas Dataset 1 and Dataset 2 contain full-cycle data at Sagar. In fact, Dataset 1 and Dataset 2 are the only two available datasets with full-cycle data at Sagar.

In this study, Dataset 1 is used for training the networks, Dataset 2 is used for validation of the network outputs, and Dataset 3 is used for simulation (prediction). To improve the forecasting capability of ANN outside the range of training datasets, the input data are generally normalized in the range of 0 to $+1$ or -1 to $+1$ (Kisi 2004; Panda *et al.* 2010). Standardization of the input and output vector is, therefore, done to keep the data within the interval of $[-1, 1]$.

RESULTS AND DISCUSSION

During training with Dataset 1, following *R* values are obtained from the five models: (i) $R = 0.8732$ for Model I; (ii) $R = 0.8667$ for Model II; (iii) $R = 0.9993$ for Model III; (iv) $R = 0.9907$ for Model IV; and (v) $R = 0.9994$ for Model V. Since Model I and Model II yielded low values of *R*, these two models were not considered for further analysis. Since Model I contains only one data as input, it might not be able to represent the variabilities in the output water level. Model II has three input variables which include data of stations which are located further upstream and away from the station Sagar and in these stations, there are appreciable lags in the tidal data and the tidal ranges are also larger compared to that of Sagar. These might have resulted in a lower *R* value.

Accordingly, it may be concluded that not all of the upstream stations are suitable for training and inclusion of more stations does not necessarily improve the performance of the model.

Next, validation is done for Models III, IV, and V, with Dataset 2. The statistical parameters *R*, MAPE, RMSE, and *E* estimated from the results of this validation are shown in Table 2.

doi: 10.2166/h2oj.2020.041

Table 2 | Statistical parameters estimated from validation and prediction results

	Validation with Dataset 2				Prediction with Dataset 3			
Model	***R***	**MAPE**	**RMSE**	***E***	***R***	**MAPE**	**RMSE**	***E***
III	0.9336	15.91	0.5184	0.8212	0.9495	20.29	0.5994	0.8373
IV	0.9256	15.57	0.5053	0.8246	0.9297	14.94	0.7914	0.7527
V	0.9349	15.96	0.6096	0.7793	0.9551	11.75	0.3654	0.9103

It may be seen from Table 2 that all three models yielded very close values of parameters *R*, MAPE, RMSE, and *E*, with Model IV resulting in marginally better values.

The water levels obtained from these three models with Dataset 2 are plotted with the corresponding observed levels at Sagar, and these are shown in Figures 4(a), 5(a), and 6(a), for Models III, IV, and V, respectively. It may be observed from these figures that all the three models resulted in water levels very close to the observed levels. However, the peak values of the simulated levels are a little lower than the observed levels. One probable reason for this could be that the Dataset 1 which is used in training, contains the water level data of spring tides, whereas Dataset 2 used in validation has the water level data of neap tides (lower peaks). The purpose behind the use of these two datasets of different tides (spring and neap) was to examine whether the trained model could be used to predict any part of the tidal cycle (spring or neap), which was one of the objectives of this study. Based on the results, it may be stated that the trained network can be used to predict any part of the tidal cycle satisfactorily.

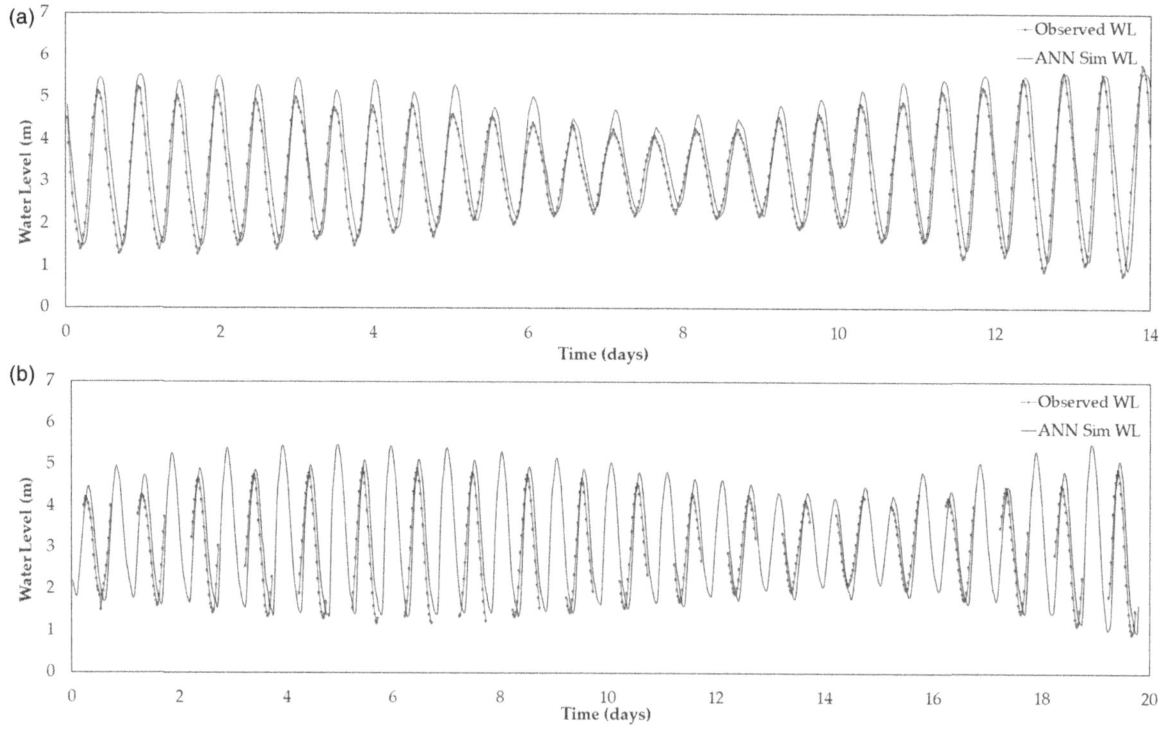

Figure 4 | Comparison of observed and ANN predicted level by Model III: (a) validation with Dataset 2 and (b) simulation with Dataset 3.

All the three models are considered for simulation of the discontinuous Dataset 3. The simulated water levels are plotted with the observed water levels at Sagar for assessing the performances of the three models in terms of predicting full-cycle data. These plots are shown in Figures 3(b), 4(b), and 5(b) for Models III, IV, and V, respectively.

doi: 10.2166/h2oj.2020.041

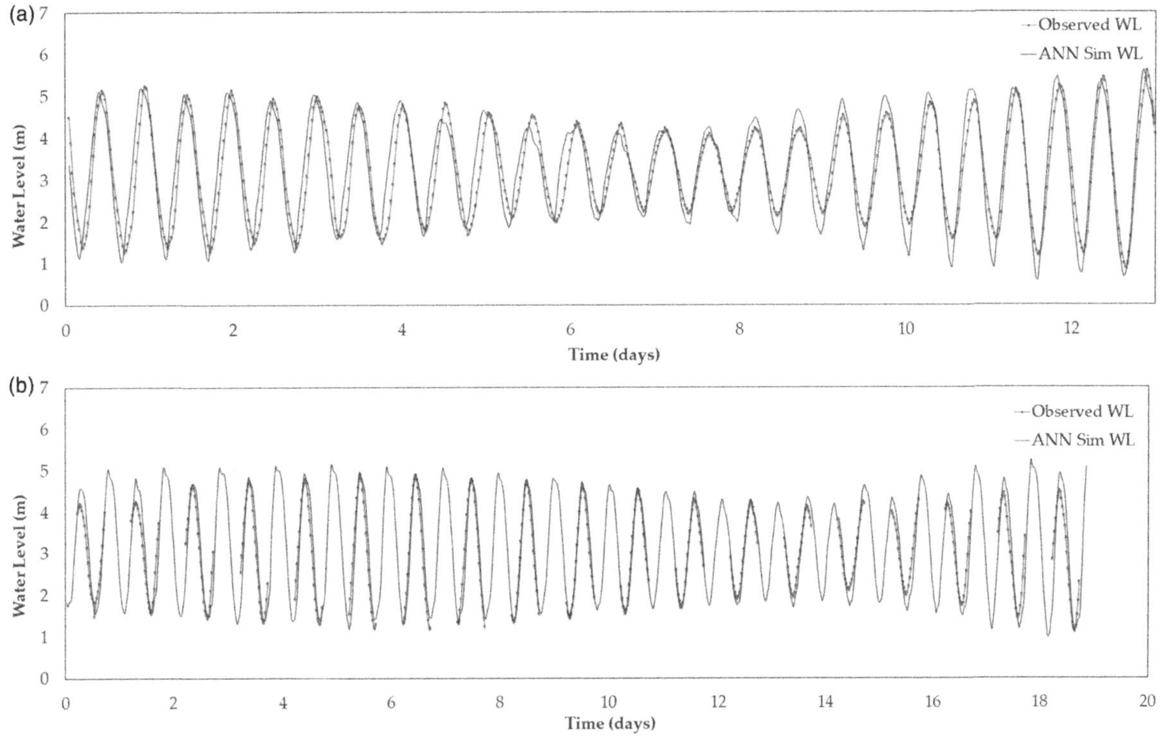

Figure 5 | Comparison of observed and ANN predicted Level by Model IV: (a) validation with Dataset 2 and (b) simulation with Dataset 3.

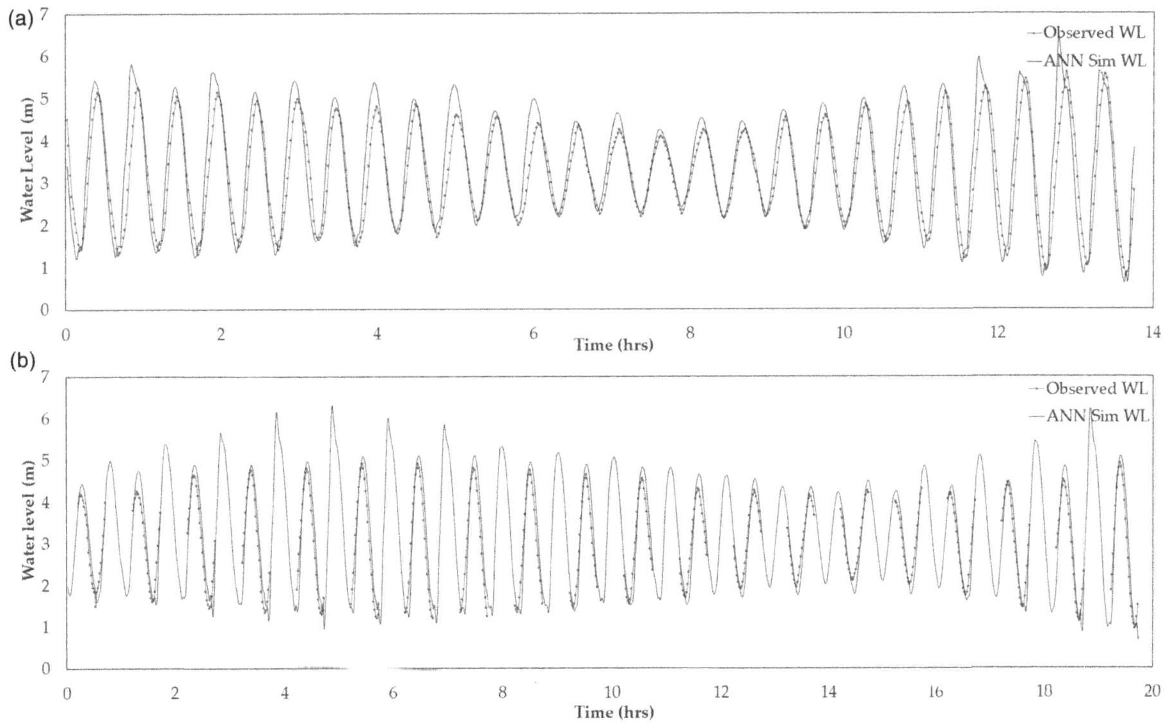

Figure 6 | Comparison of observed and ANN predicted level by Model V: (a) validation with Dataset 2 and (b) simulation with Dataset 3.

Scatterplots of observed vs. ANN predicted levels are developed for all the three models, and these are shown in Figure 7 for Dataset 2 and in Figure 8 for Dataset 3. It may be observed that in case of Model III, the scatter and dispersion around the regression line is less, and the points are more

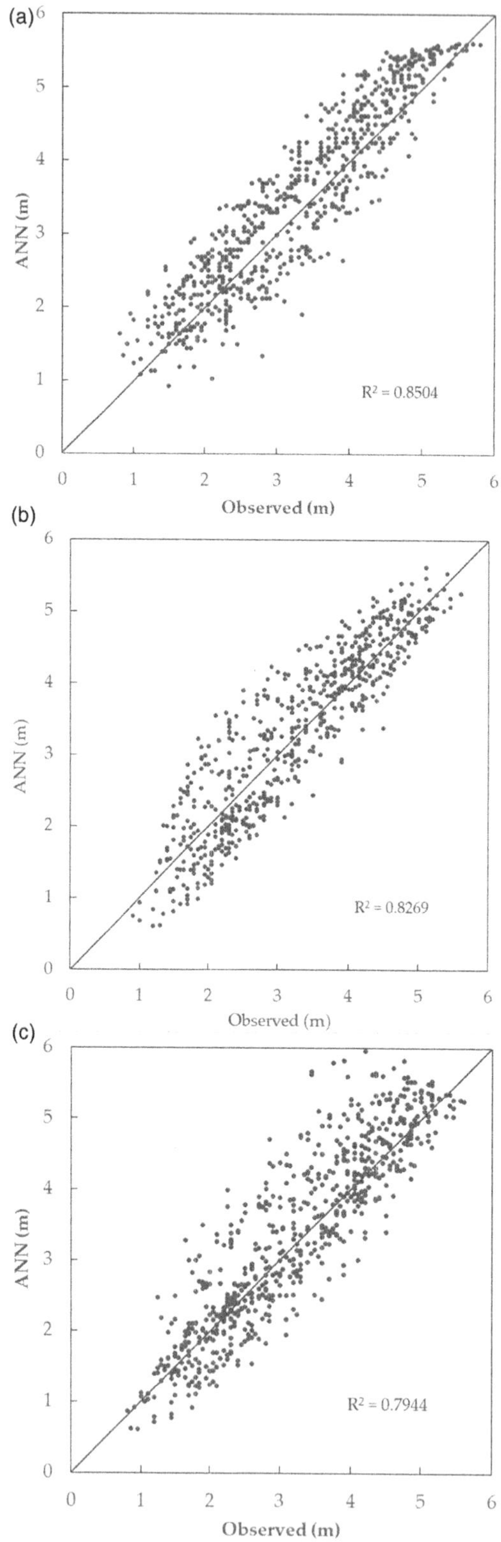

Figure 7 | Scattered plots with Dataset 2 and ANN results: (a) Model III, (b) Model IV, and (c) Model V.

cohesive compared to those of Model IV and Model V. Scatter is more prominent in the case of Model V. The situation is reversed in case of simulation with Dataset 3 (Figure 8), with Model V having more cohesive scatter, closer to the regression line. Consistent and low scatter is observed for Model IV, in both Figures 7 and 8.

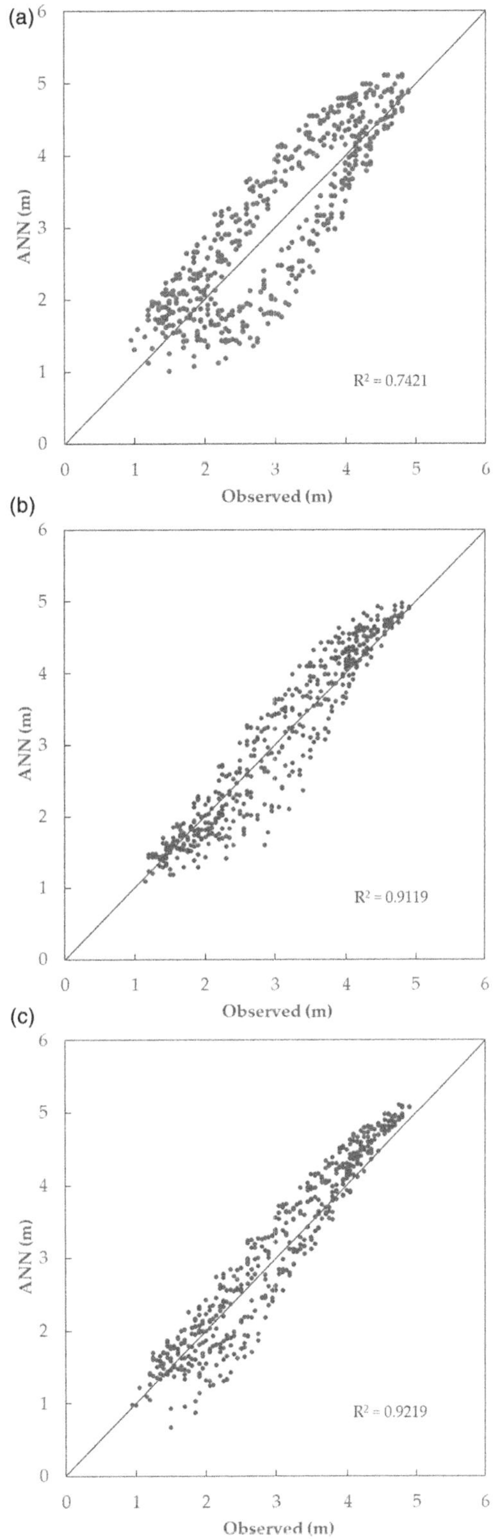

Figure 8 | Scattered plots with Dataset 3 and ANN results: (a) Model III, (b) Model IV, and (c) Model V.

It may be noted that, both Model III and Model IV have four input variables, and in terms of statistical parameters mentioned in Table 2, Model IV apparently performed better. Considering the variations in the tidal range, as well as high water levels and low water levels (peaks and troughs), it can be seen from Figure 5 that Model IV performed better. Model III marginally overpredicted the high water levels compared to Model IV.

doi: 10.2166/h2oj.2020.041

Model V overpredicted the high water levels (Figure 6(b)), more than that of Model III (Figure 4(b)). It may be noted that Model V has input variables similar to those of Model III but has nine additional variables. Inclusion of these additional variables was expected to improve the performance, but it did not happen. This indicates that the increasing number of variables does not necessarily improve the performance of an ANN model.

Based on the validation and simulation results, Model IV may be considered as the preferred one, with the advantage of having 1 day before data (i.e., already observed data) of the neighboring station as input. Hence, it may also be used for 1-day ahead forecast.

CONCLUSION

In this paper, an ANN-based approach is presented for prediction of complete cycles of tidal water levels at a remote station in an estuary where only half-cycle data can be physically measured. Based on the objectives of this study, the following conclusions made from the results obtained.

- Comparisons of the graphical plots of observed and predicted tidal water levels, and the comparisons of statistical errors parameters (*R*, MAPE, RMSE, and *E*) estimated from the predicted and observed tidal water levels, indicated that the developed ANN models may be accepted for tidal water level prediction.
- Water level data of the nearest upstream station are more suitable for training. Inclusion of data of additional upstream stations further away from the station concerned not necessarily improve the performance of the model.
- The developed ANN models can be used to generate complete cycles of water level data at the remote station, for both spring and neap cycles.
- Performance of Model IV indicated that 1-day ahead prediction is possible.
- Performance of Model V indicated that increasing the number of lagged input variables does not improve the performance of the model.
- The generated tidal data can be used as a boundary condition in the hydrodynamic models.

However, it should be noted that, because ANN is a data-driven model, the above conclusions are primarily related to the system under study. Also, it should be mentioned here that availability of adequate and synchronous data at different stations is a concern. In this study, only two sets of continuous tidal water level data were available at Sagar, which were used in training and validation. With more continuous dataset available, additional training or validation could have been done. Data during the monsoon period were not available because the sea becomes very rough and there are administrative restrictions in movements at that time. Nevertheless, this study successfully identified the applicability of ANN in such field conditions.

ACKNOWLEDGEMENTS

The authors gratefully acknowledge the support and contributions of the Hydraulic Study Department, Kolkata Port Trust (KoPT), West Bengal, India, in carrying out this research work. All data for this work have been obtained from KoPT.

DATA AVAILABILITY STATEMENT

All relevant data are included in the paper or its Supplementary Information.

doi: 10.2166/h2oj.2020.041

REFERENCES

Akib, S., Mohammadhassani, M. & Jahangirzadeh, A. 2014 Application of ANFIS and LR in prediction of scour depth in bridges. *Computers and Fluids* **91**, 77–86.

Bhattacharya, B. & Solomatine, D. P. 2005 Neural networks and M5 model trees in modelling water level-discharge relationship. *Neurocomputing* **63**, 381–396.

Cai, S., Liu, L. & Wang, G. 2018 Short-term tidal level prediction using normal time-frequency transform. *Ocean Engineering* **156**, 489–499.

Campolo, M., Andreussi, P. & Soldati, A. 1999 River flood forecasting with a neural network model. *Water Resources Research* **35** (4), 1191–1197.

Chen, W. B., Liu, W. C. & Hsu, M. H. 2012 Comparison of ANN approach with 2D and 3D hydrodynamic models for simulating estuary water stage. *Advances in Engineering Software* **45**, 69–79.

Habib, E. H. & Meselhe, E. A. 2006 Stage-discharge relations for low-gradient tidal streams using data-driven models. *Journal of Hydraulic Engineering* **132** (5), 482–492.

Haddadchi, A., Mohavedi, N., Vahidi, E., Omid, M. H. & Dehghani, A. A. 2013 Evaluation of suspended load transport rate using transport formulas and artificial neural network models (Case study: Chelchay catchment). *Journal of Hydrodynamics, Series B* **25** (3), 459–470.

Hidayat, H., Hoitink, A. J. F., Sassi, M. G. & Torfs, P. J. J. F. 2014a Prediction of discharge in a tidal river using artificial neural networks. *Journal of Hydrologic Engineering* **19** (8), 04014006-1–8.

Hidayat, H., Setiawan, F. & Handoko, U. 2014b Reconstructing disrupted water level records in a tide dominated region using data mining technique. *Indonesian Journal of Geospatial* **3** (2), 9–16.

Jain, S. K. 2008 Development of integrated discharge and sediment rating relation using a compound neural network. *Journal of Hydrologic Engineering* **13** (3), 124–131.

Jain, S. K. & Chalisgaonkar, D. 2000 Setting up stage-discharge relations using ANN. *Journal of Hydrologic Engineering* **5** (4), 428–433.

Kisi, O. 2004 River flow modeling using artificial neural networks. *Journal of Hydrologic Engineering* **9** (1), 60–63.

Lee, T. L. & Jeng, D. S. 2002 Application of artificial neural networks in tide forecasting. *Ocean Engineering* **29**, 1003–1022.

Lee, W.-K. & Resdi, T. A. B. T. 2014 Neural network approach to coastal high and low water level prediction. R. Hassan, M. Yusoff, Z. Ismail, N. Amin & M. Fadzil (eds). Proceeding of the International Civil and Infrastructure Engineering Conference (InCIEC 2013), Kuching, Malaysia, pp. 121–125.

Li, S., Liu, L., Cai, S. & Wang, G. 2019 Tidal harmonic analysis and prediction with least-squares estimation and inaction method. *Estuarine, Coastal and Shelf Science* **220**, 196–208.

Maanen, B. 2010 The use of artificial neural networks to analyze and predict alongshore sediment transport. *Nonlinear Processes in Geophysics* **17**, 395–404.

Maier, H. R., Jain, A., Dandy, G. C. & Sudheer, K. P. 2010 Methods used for the development of neural networks for the prediction of water resource variables in river systems: current status and future directions. *Environmental Modelling Software* **25**, 891–909.

Meena, B. L. & Agrawal, J. D. 2015 Tidal level forecasting using ANN. *Procedia Engineering* **116**, 607–614.

Mohanasundaram, S., Suresh Kumar, G. & Narasimhan, B. 2019 A novel deseasonalized time series model with an improved seasonal estimate for groundwater level predictions. *H2Open Journal* **2** (1), 25–44.

Panda, R. K., Pramanik, N. & Bala, B. 2010 Simulation of river stage using artificial neural network and MIKE11 hydrodynamic model. *Computers and Geosciences* **36**, 735–745.

Peters, R., Schmitz, G. H. & Cullmann, J. 2006 Flood routing modelling with artificial neural networks. *Advances in Geosciences* **9**, 131–136.

Rath, J. S., Hutton, P. H., Chen, L. & Roy, S. B. 2017 A hybrid empirical-Bayesian artificial neural network model of salinity in the San Francisco Bay-Delta estuary. *Environmental Modelling & Software* **93**, 193–208.

Salim, A. M., Dwarakisha, G. S., Liju, K. V., Thomas, J., Devi, G. & Rajeesh, R. 2015 Weekly prediction of tides using neural networks. *Procedia Engineering* **116**, 678–682.

Sarangi, A., Madramootoo, C. A., Enright, P. & Prasher, S. 2005 Performance evaluation of ANN and geomorphology-based models for runoff and sediment yield prediction for a Canadian watershed. *Current Science* **89** (12), 2022–2033.

Shrestha, R. R., Theobald, S. & Nestmann, F. 2004 Combined hydrodynamic-neural network model for flood flow simulation. In *Proceedings of the 6th International Conference on Hydroinformatics*, Vol. 2, 21–24 June 2004, Singapore, pp. 1489–1496.

Tsai, C. P. & Lee, T. L. 1999 Back-propagation neural network in tidal level forecasting. *Journal of Waterway, Port, Coastal and Ocean Engineering* **125** (4), 195–202.

Ustoorikar, K. & Deo, M. C. 2008 Filling up gaps in wave data with genetic programming. *Marine Structures* **21**, 177–195.

Wu, W., Dandy, G. C. & Maier, H. R. 2014 Protocol for developing ANN models and its application to the assessment of the quality of the ANN model development process in drinking water quality modelling. *Environmental Modelling Software* **54**, 108–127.

Yang, C. T., Marsooli, R. & Aalmi, M. T. 2009 Evaluation of total load sediment transport formulas using ANN. *International Journal of Sediment Research* **24** (3), 274–286.

Yin, J., Zou, Z. & Xu, F. 2013 Sequential learning radial basis function network for real-time tidal level predictions. *Ocean Engineering* **57**, 49–55.

doi: 10.2166/h2oj.2020.041

Yin, J., Li, L., Cao, Y. & Zhao, J. 2016 An adaptive online sequential extreme learning machine for real-time tidal level prediction. In: *Proceedings of ELM-2015 Volume 2. Proceedings in Adaptation, Learning and Optimization* (J. Cao, K. Mao, J. Wu & A. Lendasse, eds), Hangzhou, China, pp. 55–66.
Yitian, L. & Gu, R. R. 2003 Modeling flow and sediment transport in a river system using an artificial neural network. *Environmental Management* **31** (1), 122–134.

First received 30 May 2020; accepted in revised form 19 August 2020. Available online 9 October 2020

Evolutionary Algorithms in Water Resources
doi: 10.2166/h2oj.2020.124

Decision tree-based reduction of bias in monthly IMERG satellite precipitation dataset over India

Shushobhit Chaudhary and C. T. Dhanya*

Department of Civil Engineering, Indian Institute of Technology Delhi, Hauz Khas, New Delhi, India

*Corresponding author. E-mail: dhanya@civil.iitd.ac.in

Abstract

Decision trees are ideally suited for handling huge datasets and modelling non-linear relationships between different variables. Given the relationship between precipitation and bias may be very complex and non-linear, bias-correction of satellite precipitation is a challenge. We examine the applicability of Classification and Regression tree (CART) for bias-correction of the Integrated Multi-satellite Retrievals for Global Precipitation Mission (IMERG) precipitation dataset over India. The gauge-based 0.25° gridded precipitation dataset from India Meteorological Department is considered as the reference. The CART model is trained (2001–2011) and tested (2012–2016) over each 0.25° grids. The training dataset is subjected to 10-fold cross-validation and optimization of the minimum size of leaf node (one of the hyper-parameter). Efficiency of the CART model is evaluated using performance metrics like R^2, RMSE and MAB over the whole of India and different climate and elevation zones in India. CART model is observed to be highly effective in capturing the bias during the training (average $R^2 = 0.77$) and testing (average $R^2 = 0.66$) period. Significant improvement in average monthly MAB (−6.3 to 29.2%) and RMSE (8.7–37.3%) was obtained post bias-correction by CART. Better performance of CART model was observed when compared to two widely adopted bias-correction techniques.

Key words: CART, GPM, IMERG, precipitation bias reduction, regression tree, satellite rainfall

Highlights

- CART is effective in modelling the relationship between satellite precipitation bias and observed precipitation.
- Significant improvement in average monthly MAB and RMSE was obtained post bias-correction by CART.
- CART bias correction algorithm improved the correlation and reduced the MAB and RMSE when compared to linear scaling and CDF matching techniques.
- Though the CART algorithm is found to be robust, over hilly and mountainous regions, the CART model should be cautiously adopted, however.

INTRODUCTION

Recent decades have witnessed a surge in the usage of satellite-based precipitation products, for a variety of hydro-climatic applications and disaster management (Hong *et al.* 2006; Sawunyama & Hughes 2008; Behrangi *et al.* 2014; Koriche & Rientjes 2016). However, the applicability of satellite-based precipitation for deriving reliable hydrological and climatic predictions is still limited, owing to the presence of bias in these measurements. The bias in satellite precipitation estimates may be owing to sampling uncertainty or error in the retrieval algorithm of the satellite estimation process (Hossain & Huffman 2008; Gebregiorgis *et al.* 2017; Prakash 2019). With the advent of multiple satellites for measuring precipitation, the uncertainty among these satellite-based precipitation estimates has also increased (Hossain & Huffman 2008; Abera *et al.* 2016). To minimize the uncertainty among the existing satellite precipitation product, different merged or blended satellite-based precipitation products are proposed, wherein information from multiple sources are combined to provide a high-quality satellite rainfall product (Beck *et al.* 2017; Awange *et al.* 2019; Bhuiyan *et al.* 2019).

The multi-satellite precipitation estimate from the recent Global Precipitation Measurement Mission (GPM), i.e., Integrated Multi-satellite Retrievals for GPM (IMERG) integrates information of various multi-satellite retrievals from NASA Tropical Rainfall Measurement Mission Multi-Satellite Precipitation Analysis (TMPA; inter-satellite calibration and gauge adjustment); NOAA Climate Prediction Center (CPC) morphing technique (CMORPH; Lagrangian time interpolation); Precipitation Estimation from Remotely Sensed Information using Artificial Neural Networks (PERSIANN) – Climate Classification System (Microwave calibrated Infra-Red); and the NASA Precipitation Processing System (input data assembly and processing) (Huffman *et al.* 2014). IMERG satellite precipitation has been available since June 2000 onwards at a very high spatial resolution of $0.1° \times 0.1°$ and multiple temporal resolutions (half-hourly, 3 hourly, one day, 7-day and monthly). IMERG has been extensively compared and validated using ground-based gauges and radar datasets over different regions of the world and its errors and biases are well studied (Tan *et al.* 2016, 2017; Tang *et al.* 2016; Asong *et al.* 2017; Prakash *et al.* 2018). The error or bias in IMERG dataset can be reduced by adopting a variety of statistical techniques like linear monthly scaling, quantile mapping based on an empirical distribution, etc., which are frequently used for bias correction of satellite estimates (Abera *et al.* 2016; Worqlul *et al.* 2018; Chaudhary & Dhanya 2019). Most of the bias correction algorithms applied over satellite precipitation are either parametric or based on the assumptions of linear relationships between predictor (satellite rainfall) and response (bias) (Habib *et al.* 2014; Abera *et al.* 2016; Bhatti *et al.* 2016; Hashemi *et al.* 2017; Worqlul *et al.* 2018). However, the relationship of bias and precipitation may be very complex and non-linear (Gebregiorgis & Hossain 2015). Moreover, the non-linear relationship between the precipitation and bias/error may also be affected by various factors such as landuse/landcover, soil-cover, elevation, climate zones, etc. (Gebregiorgis & Hossain 2012, 2015; Tang & Hossain 2012), which makes it difficult to capture using the usual statistical techniques.

Various machine learning algorithms such as artificial neural networks, k-nearest neighbour, support vector machines, classification and regression trees, etc., have been successfully used in various remote sensing and hydrological applications to effectively model the non-linear associations between multiple predictors and the response variables (Waheed *et al.* 2006; Rasouli *et al.* 2012; Erdal & Karakurt 2013; Tao *et al.* 2016; Mannan *et al.* 2018; Petty & Dhingra 2018; Bhuiyan *et al.* 2019). Moreover, machine learning algorithms are also effective in handling multi-dimensional and complex data (Tan & Beklioglu 2006; Kühnlein *et al.* 2014; Yu *et al.* 2014). In recent years, Classification and Regression tree (CART) based machine learning techniques have been widely adopted in different aspects of hydrology. Abraham *et al.* (2020) adopted CART for the classification of soil into hydrologic groups. Veettil & Mishra (2020) employed the CART for estimating the thresholds associated with climate, catchment, and morphological variables that may potentially influence the hydrological drought characteristic. Pekel (2020) applied the regression tree to estimate soil moisture considering different parameters like air temperature, time, relative humidity, and soil temperature. The applicability of CART for bias correction of satellite-based precipitation estimates remains unexplored till now on both global and regional scales. Given, the potential of CART in handling multi-dimensional and complex data, it would be essential to explore the applicability of CART for the reduction of bias in satellite precipitation.

doi: 10.2166/h2oj.2020.124

In the present study, we explore the possibility of applying Classification and Regression Tree (CART) for possible bias reduction of IMERG precipitation dataset over India. The specific objectives of this study are – (i) to train and test the CART model (wherein, satellite rainfall is the predictor variable and observed rainfall is the response variable) grid-wise over the whole of India for bias reduction of satellite rainfall; (ii) to analyze the performance of CART model over different climate and elevation zones in India; and (iii) to compare the performance of CART-based bias correction techniques with widely adopted and established bias correction techniques. Such an analysis will serve to improve the reliability of IMERG precipitation for applications in water resource and hydrological applications over India.

STUDY AREA

The land region of the Indian sub-continent covering from 66°E to 100°E and from 6°N to 39°N is chosen as the study area. India is a country with varied geophysical and climatic features. The northern part of India is surrounded by the mountainous range of Himalayas, while the southern part is surrounded by sea/oceans (Figure 1, panel a). Western Ghats, located near to the south-western coast receives heavy rainfall. The rainfall during the monsoon months of June, July, August and September, account for more than 75% of the annual rainfall of the country and is highly significant for the agriculture-dominated country. India has seven main Koppen–Geiger climate zones showing different climatic characteristics (Peel *et al.* 2007) (Figure 1, panel b). A brief description about different Koppen–Geiger climate zones over India is shown in Table 1.

DATA USED

Integrated multi-satellite retrievals for GPM (IMERG)

The Integrated Multi-satellite Retrievals for GPM (IMERG) is a U.S. multi-satellite rainfall product from the Global Precipitation Mission (GPM) team. IMERG algorithm estimates precipitation from

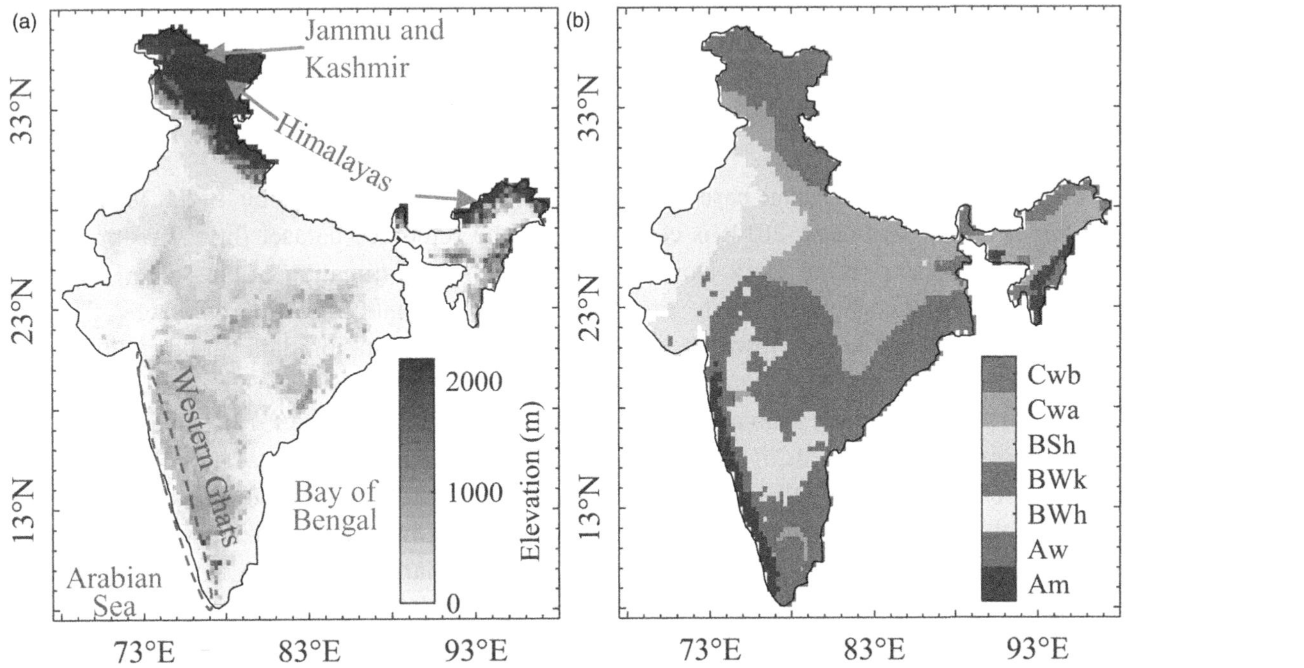

Figure 1 | (a) Digital Elevation Model (DEM) of study region (India) obtained from Shuttle Radar Topography Mission (SRTM) data (*Source*: https://earthexplorer.usgs.gov/). (b) Updated Koppen–Geiger climate zones over India (Peel *et al.* 2007).

doi: 10.2166/h2oj.2020.124

Table 1 | Details of updated Koppen–Geiger climate zones over India

Climate Group	Climate Zone Code	Climate Zone Description	Location
A (Tropical)	*Am*	Tropical Monsoonal	Windward side of Western Ghats and a few parts of North-East India
	Aw	Tropical Savannah	peninsular India
B (Arid)	*BWh*	Arid Hot Desert	North-western desert regions
	BSh	Arid Steepe Hot	Leeward side of Western Ghats and semi-arid region in eastern India
	BWk	Arid Desert Cold	Himalayan regions
C (Temperate)	*Cwa*	Temperate Dry Winter Hot Summer	Gangetic plain region and North-East India
	Cwb	Temperate Dry Winter Warm Summer	Lower Himalayan ranges and few parts of Northern North-East India

Source: Peel *et al.* (2007).

various satellites in the GPM constellation, viz., Tropical Rainfall Measurement Mission Multi-Satellite Precipitation Analysis (TMPA), Precipitation Estimation from Remotely Sensed Information using Artificial Neural Networks- Climate Data Record (PERSIANN), Climate Prediction Center Morphing method (CMORPH), etc. (Hou *et al.* 2014; Huffman *et al.* 2014). GPM is the successor of TRMM and has three noteworthy improvements. Firstly, it has higher global coverage due to the increased orbital inclination from 35° to 65°. Secondly, it can measure very light precipitation (<0.5 mm/hour; Hou *et al.* 2014) owing to increased sensitivity of precipitation radar sensor due to the inclusion of two additional frequencies. Thirdly, solid form of precipitation (snow) can also be measured due to the inclusion of high-frequency channels in the passive microwave sensor imager. The IMERG data is available at a very high spatial resolution of $0.1° \times 0.1°$ and multiple temporal resolutions (half-hourly, 3-hourly, one-day, 7-day and monthly). Version 6 and Level 3 monthly IMERG rainfall data (Research/Final Run) available since June 2000 is used in the present study (https://gpm.nasa.gov/data-access/downloads/gpm). The default units of precipitation in the IMERG datasets is mm/hr, which is transformed to mm/day by accumulating over the entire day (multiplying by 24).

India Meteorological Department (IMD) rainfall

The $0.25° \times 0.25°$ gridded gauge-based daily rainfall dataset developed by India Meteorological Department (IMD) (Pai *et al.* 2014) is considered as the reference dataset (http://imdpune.gov.in/Clim_Pred_LRF_New/Grided_Data_Download.html#), for the estimation of bias. The IMD dataset is constructed by incorporating daily rainfall records information of approximately 6,955 rain gauge stations covering India. However, the availability of gauge station data varied vividly throughout the years. On average, gauge data from 2,600 stations is available per year for construction of data (Pai *et al.* 2014). Information from all the rain gauge data was subjected to various quality checks and then interpolated to $0.25° \times 0.25°$ grids using the Inverse Distance Weighted (IDW) scheme. Orographic dependence of rainfall is accurately captured by the IMD rainfall dataset. For instance, heavy rainfall is observed on the windward side of Western Ghats while low rainfall is observed on the leeward side of Western Ghats (Pai *et al.* 2014; Chaudhary *et al.* 2017). It is noteworthy that the gauge density is low over the hilly regions of northern India including the Himalayas and Jammu and Kashmir, therefore, IMD rainfall is less reliable over those regions.

The $0.1° \times 0.1°$ gridded IMERG dataset was resampled to the $0.25° \times 0.25°$ gridded IMD reference dataset using nearest-neighbour interpolation algorithm for further analysis.

doi: 10.2166/h2oj.2020.124

METHODOLOGY

Bias is defined as the deviation in the magnitude of satellite rainfall from gauge observed rainfall as shown in Equation (1).

$$Bias = P_s - P_g \tag{1}$$

where P_s is satellite (here, IMERG) precipitation; P_g is gauge (here, IMD) precipitation

Additionally, we have adopted the Willmott scheme to decompose the total mean square error (MSE) into its systematic and random component (Willmott 1981; AghaKouchak *et al.* 2012; Prakash *et al.* 2015). The total, systematic and random MSE is derived using Equation (2).

$$\frac{1}{n}\times\left(\sum_{i=1}^{n}(P_s - P_g)^2\right) = \frac{1}{n}\times\left(\sum_{i=1}^{n}(P_s^* - P_g)^2\right) + \frac{1}{n}\times\left(\sum_{i=1}^{n}(P_s - P_s^*)^2\right) \tag{2}$$

where P_s^* is defined using a linear regression error model $P_s^* = a \times P_s + b$, where a being the slope and b being the intercept.

The left-hand side of Equation (2) represents the total MSE which is decomposed into systematic MSE (middle term in Equation (2)) and random MSE (rightmost term in Equation (2)).

Classification and regression tree (CART)

Classification and regression trees (CART), also called decision trees are data structures that are developed by recursively partitioning the input space, and defining a local model in each of the partitioned regions of input space (Murphy 2012). The decision trees are graphically represented with a primary root node at the top, subsequently leading to branches (internal nodes) and finally ending in outer nodes called leaves (terminal or decision nodes). A sample regression decision tree is shown in Figure 2, panel a.

The CART tree is progressed at each internal node by splitting the instance space into two or more subspaces according to a certain discrete function of the input attributes values. To understand the CART algorithm, let $x = x_1, \ldots\ldots x_n$ be the explanatory (predictor, or features; satellite rainfall in the present study) variables and $y = y_1, y_2 \ldots\ldots y_n$ be the target response or dependent (predictand or class; corrected satellite rainfall or observed rainfall in the present study) variables. Here n is the number of predictor and predictand variables (the total number of months in this study). The process of building a CART model involves partitioning the input space into distinct and non-overlapping regions and then predict the responses in each region using the mean of all the response variables in that region. In CART, we divide the predictor space (x) i.e., $x_1, x_2 \ldots\ldots x_n$ into J distinct and non-overlapping regions – ... These regions can take any shape and are formed such that they maximize the reduction in the mean squared error (MSE) over all the splitting candidates. To partition the predictor space into J regions, the top-down recursive binary splitting approach is adopted. The splitting initially begins at the top of the tree, where two regions of partitioned, and then successively the above-partitioned regions are further split and the process continues till the optimization criteria are met (MATLAB 2019; Pekel 2020).

To understand the splitting procedure at any node, let us select a node j that is subjected to binary recursive partitioning. The node j has p predictor variables, i.e., $x_1, x_2 \ldots\ldots x_p$. The MSE of the responses in node j, i.e., $y_1, y_2 \ldots\ldots y_p$ is computed using Equation (3).

$$MSE_j = \sum_{i=1\ to\ p}^{i\in R_j}(y_i - \bar{y}_i)^2 \tag{3}$$

where $\bar{y}_i$ is the mean response for the predictor variables within the jth box.

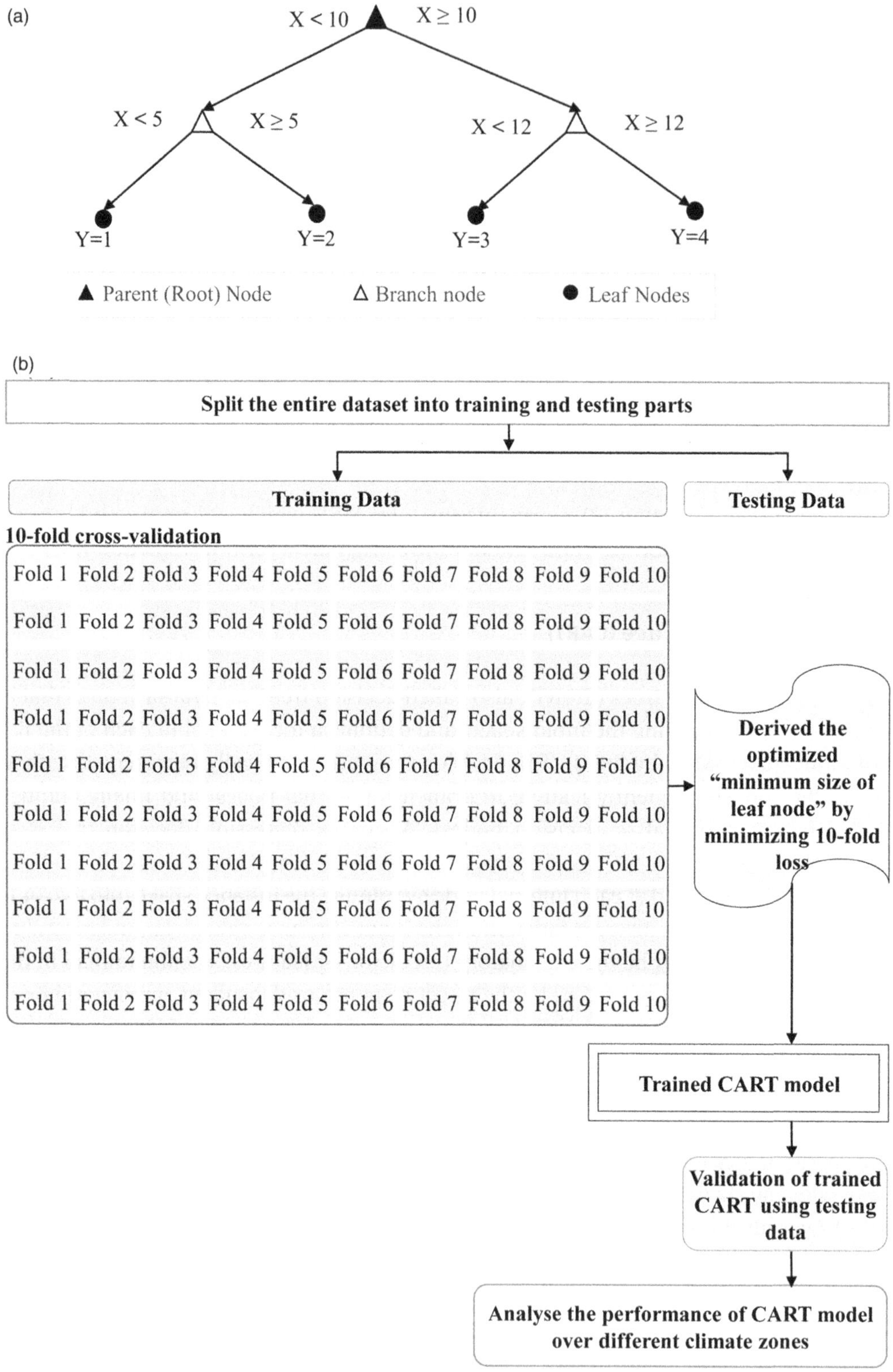

Figure 2 | (a) Illustration of a simple regression decision tree; (b) Methodology adopted for training and testing CART model in the present study.

The objective is to identify a threshold (th_j) of a split of x_j at node j such that two partitioned regions exist, i.e., $\{x|x_j < th_j\}$ (left region) and $\{x|x_j \geq th_j\}$ (right region), where th_j is selected such that it leads to maximizing the reduction in MSE (ΔI) over all the splitting threshold candidates, as shown in

doi: 10.2166/h2oj.2020.124

Equation (4).

$$\Delta I = MSE_j - \{MSE_j\}_{\{x|x_j<th_j\}} - \{MSE_j\}_{\{x|x_j\geq th_j\}} \tag{4}$$

Now this process is recursively repeated until the reduction in MSE is maximized in all the regions (R_1, $R_2 \ldots\ldots R_J$). The splitting of nodes in subsequent nodes is also governed by controlling different hyper-parameters of a decision tree. The maximum number of splits of branch nodes (decision splits), minimum number observations in the branch node (size of parent node) and minimum number of observations in the leaf (size of leaf node) are the generally controlled hyper-parameters of the tree.

To implement CART, we use the statistical classification and regression tree feature available in the Statistics and Machine Learning Toolbox in MATLAB (MATLAB 2019).

Training and testing of CART

The dataset is split into two distinct parts for training and testing of developing the CART model. In this study, 2001–2011 is considered as the training period (132 monthly datasets for training) and 2012–2016 is considered as the testing period (60 monthly datasets for testing). The training dataset is subjected to 10-fold cross-validation (Waheed *et al.* 2006; Bhuiyan *et al.* 2019; Pekel 2020) as shown in Figure 2, panel b. In cross-validation, the training dataset is split into 10 random parts. Out of these 10 parts, one of the parts is retained as the validation data and the remaining 9 data parts are used to train the tree. The cross-validation process is repeated 10 times, till each of the 10 parts serve as the validation data at least once, which yields 10 different decision trees. Cross-validation while training the model leads to improved accuracy of the resulting tree, when it is tested on new data.

In this study, we train the decision trees, by optimizing only one of the hyper-parameter – minimum number observations in the leaf (size of leaf node) at each grid over India. Minimum leaf size puts a limit to split a node when the number of observations in one of the subsequent child nodes is lower than the minimum leaf size. The size of leaf node controls the depth of the decision tree and thus aids in preventing a tree from overfitting the data. We vary the size of leaf node from 0 to 100 (approximately 50% of data size) and train the regression tree for each of the above leaf node size using 10-fold cross-validation. The 10-fold cross-validation loss, which is the total MSE of all the folds is estimated for each of the tree leaf size. Finally, the leaf node size which gives the minimum 10-fold loss is selected as the optimum leaf node size of the training decision tree.

The training of the CART model is performed at each grid of India and the trained CART is subjected over the independent validation period. The performance of the CART model during training and testing is also verified using the four performance metrics – Root Mean Square Error (RMSE), Coefficient of Determination (R^2), adjusted R^2 and Mean Absolute Bias (MAB) as shown in Table 2.

Table 2 | Details of performance metrics used in the present study

Model performance measures	Equation	Range	Ideal Value
Root Mean Square Error (RMSE)	$RMSE = \sqrt{\frac{1}{n}\sum_{i=1}^{n}(y_i - x_i)}$	[0, ∞)	0
Coefficient of Determination (R^2)	$R^2 = 1 - \frac{\sum_{i=1}^{n}(y_i - x_i)^2}{\sum_{i=1}^{n}(y_i - \bar{x}_i)^2}$	(−∞, 1]	1
Mean Absolute Bias (MAB)	$MAB = \frac{1}{n}\sum_{i=1}^{n}\lvert y_i - x_i\rvert$	[0, ∞)	0

n is the total number of values; *y* is the predicted variable; *x* is the observed variable

doi: 10.2166/h2oj.2020.124

RESULTS AND DISCUSSION

Spatial variation of rainfall and bias over India during different seasons

The spatial distribution of 16 years (2001–2016) mean monthly rainfall over India from the observed IMD and satellite-based IMERG dataset during the annual period, winter season (January, February; JF), summer season (March, April, May; MAM), monsoon season (June, July, August, September; JJAS) and post-monsoon season (October, November, December; OND) are shown in Figure 3. The IMERG dataset is effective in capturing the mean annual and seasonal spatial rainfall patterns of the reference IMD dataset. The heavy rainfall over the Western Ghats and North East India during the monsoon season and southern peninsular region during the post-monsoon season is well captured by the IMERG dataset (Figure 3, panels d, e, i, j). Although spatial rainfall pattern is well represented, slight deviations in intensities of IMERG rainfall with respect to IMD rainfall are observed in the majority of grids over India. IMERG over-estimates the all-India mean rainfall as obtained from IMD, especially during the monsoon period. On an annual scale, 3.1 mm/day of all-India mean rainfall is observed in IMERG slightly higher than all-India mean IMD rainfall 3.0 of mm/day (Figure 3, panels a, f). The spatial variation of bias intensity, i.e., deviation in the magnitude of IMERG from IMD (IMERG rainfall - IMD rainfall), is estimated for the annual period and different seasons in Figure 3, panels k–o. High magnitude of bias is evident over the Western Ghats, North-East India and northern state of Jammu and Kashmir. Over the windward side of Western Ghats, a considerably high dry bias (under-estimation) is observed in the IMERG dataset on contrary to the

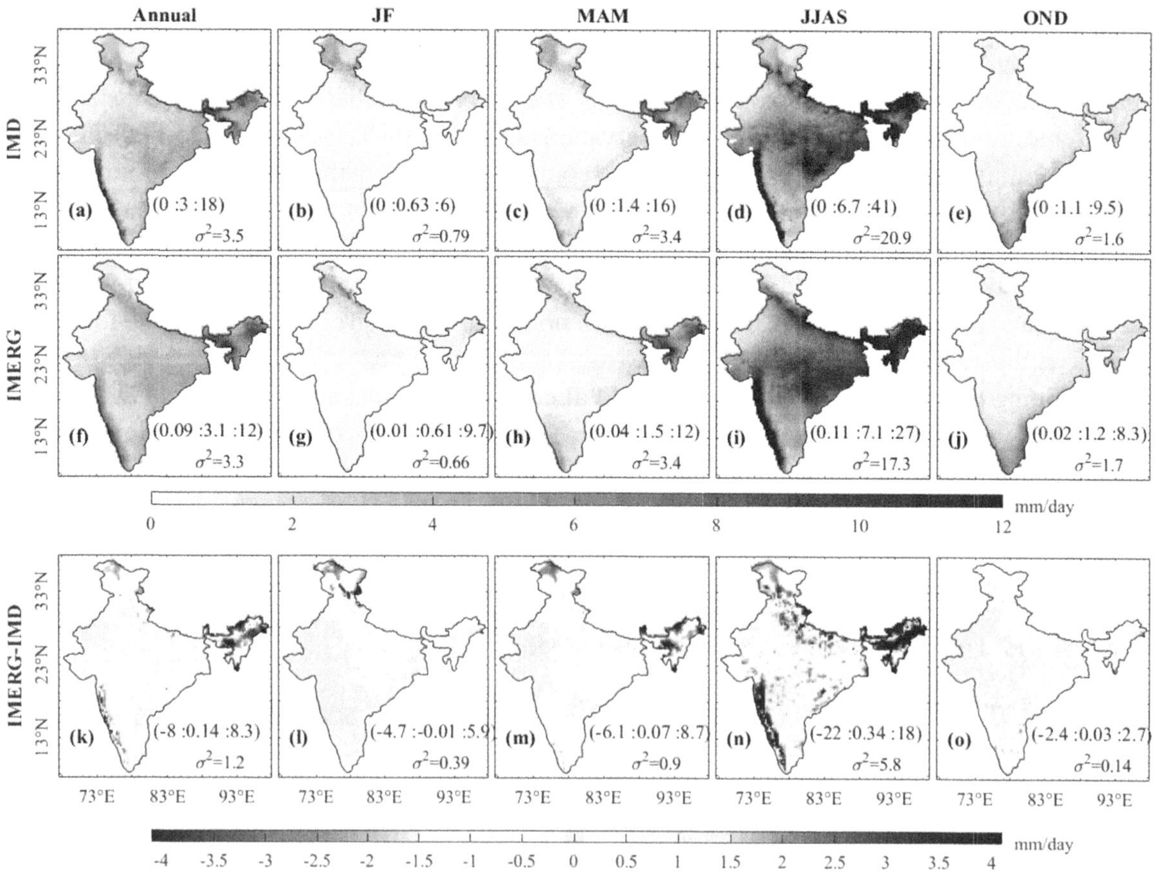

Figure 3 | Spatial variation of (a–e) IMD rainfall, (f–j) IMERG rainfall and (k–o) bias (IMERG – IMD) in rainfall during the annual, winter (JF), summer (MAM), monsoon (JJAS) and post-monsoon (OND) season over India during 16 years duration (2001–2016). All-India minimum, mean and maximum values are given in parentheses (minimum: mean: maximum). Spatial variance (σ^2) is also provided. All units are in mm/day.

doi: 10.2166/h2oj.2020.124

leeward side where significantly high wet bias (over-estimation) is observed (Figure 3, panels k, n). Consistent dry bias is observed in the northern state of Jammu and Kashmir in IMERG dataset over all the seasons. The IMD rainfall is also highly uncertain over these regions as it is highly mountainous and has significantly less gauge network station density (Prakash *et al.* 2019). The spread and intensity of bias are significantly higher during the monsoon season, as evident from Figure 3, panel n.

Further, the MSE in satellite-based IMERG rainfall is decomposed into systematic and random components over India as shown in Figure 4. Larger MSE is generally observed over high rainfall regimes (like Western Ghats and North-East India) and foothills of Himalayas. The spatial variations in systematic and random error components over India expressed in percentages of total MSE are shown in Figure 4(b) and 4(c). Larger systematic error observed over these mountainous zones of North-East India and Jammu and Kashmir can be either attributed to higher uncertainty of satellite rainfall over mountainous zones or uncertainty in observed rainfall due to lack in the required number of gauge station to capture rainfall variability in these zones (Prakash *et al.* 2015). Overall, a higher percentage of random MSE compared to systematic error is observed over India. Systematic error can potentially be corrected using bias correction schemes, whereas random error cannot be possibly reduced but can be statistically quantified (AghaKouchak *et al.* 2012; Prakash *et al.* 2015).

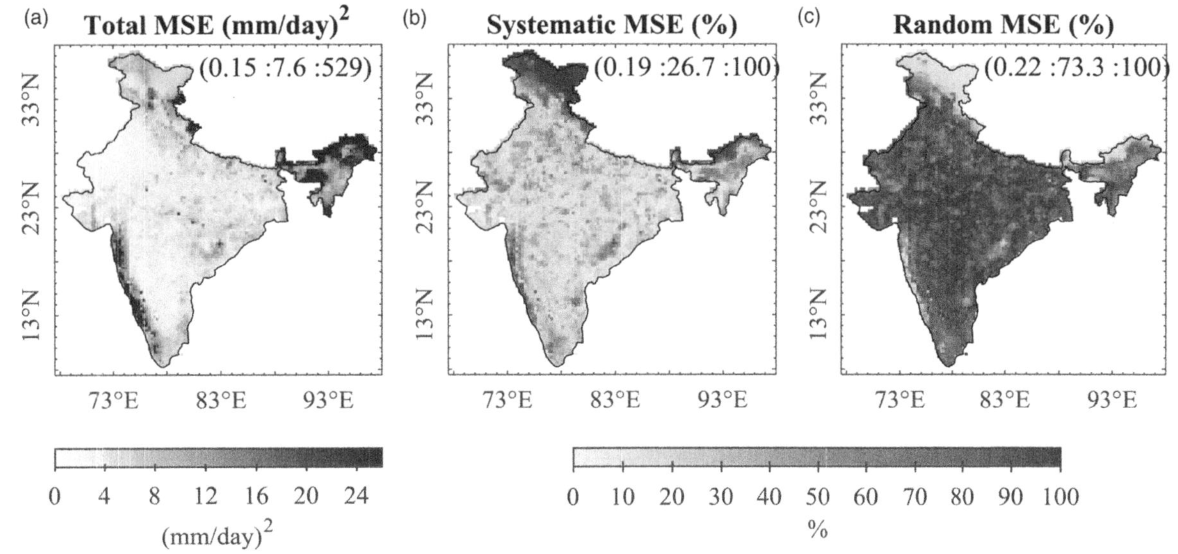

Figure 4 | Spatial variation of (a) Total MSE (b) Systematic MSE and (c) Random MSE in IMERG rainfall with reference to IMD rainfall over India during 16 years duration (2001–2016). All-India minimum, mean and maximum values are given in parentheses (minimum: mean: maximum).

Training and testing of CART over India

The optimum leaf size of the decision tree is estimated at each grid over India by minimizing the total 10-fold cross-validation MSE during the training period. Figure 5(a) shows the variation of total cross-validation loss of 10-fold with different minimum leaf size obtained while training the CART tree on a grid selected (75°E, 15.75°N) over Western Ghats in India. The minimum leaf size of 12, which observes the minimum total 10-folds cross-validation loss of 1.25 (mm/day)^2, is selected as the calibrated model parameter for the tree. The classification tree developed at the selected grid by using the optimized leaf size can be graphically visualized in Figure 5(b). Here, 'r' denotes the satellite rainfall and the values at node indicate the CART corrected satellite rainfall. At the parent node, the value of the rainfall threshold split is observed to be 2.8 mm/day. If the rainfall is greater than 2.8 mm/day,

doi: 10.2166/h2oj.2020.124

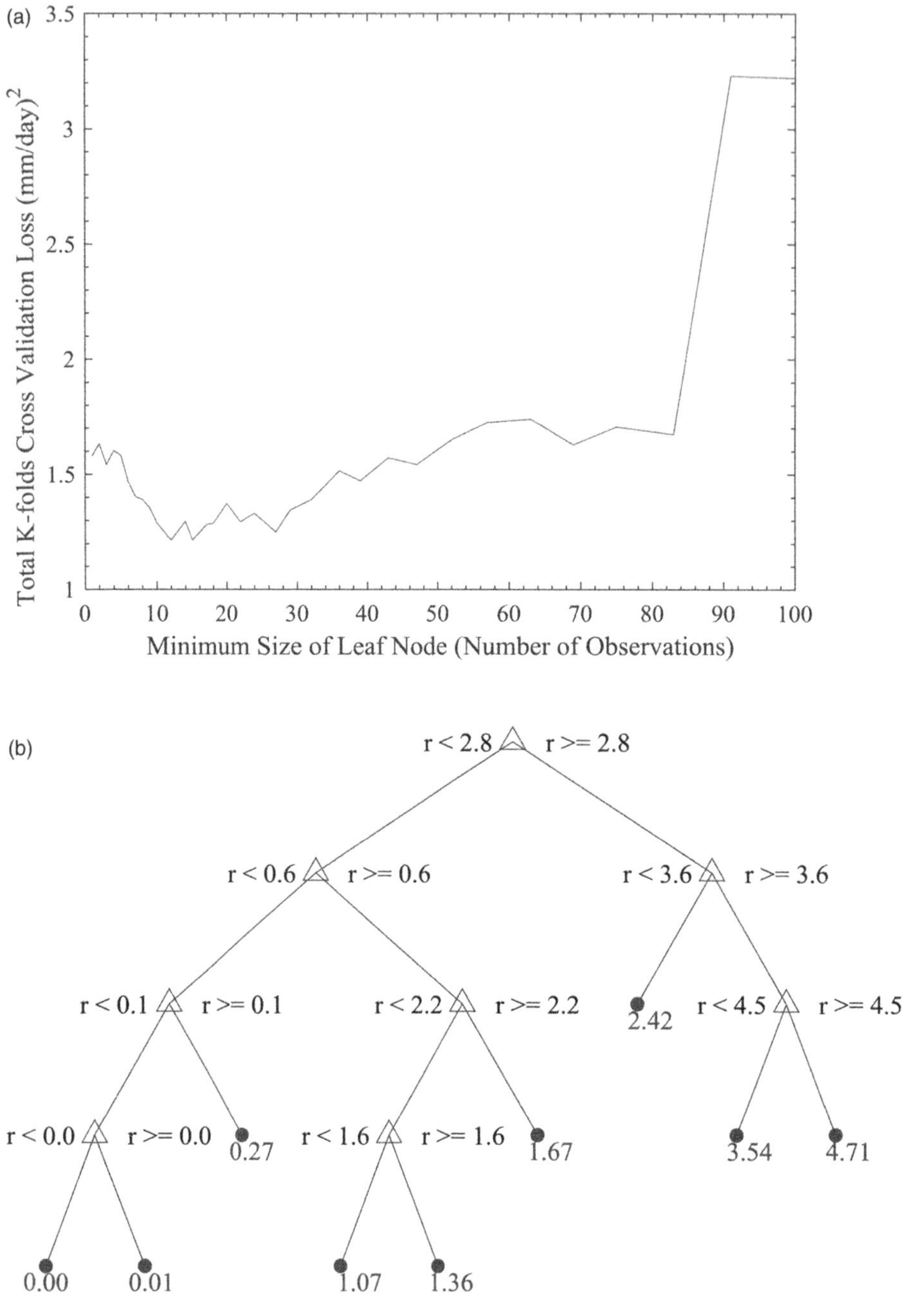

Figure 5 | (a) Variation of total cross-validation loss of 10-folds with different minimum leaf size obtained while training the CART tree; (b) CART decision tree obtained after training the data over a grid selected (75 °E, 15.75 °N) in India. 'r' indicates IMERG rainfall and values at the leaf (in dark blue) indicate the corrected IMERG rainfall.

further if rainfall is greater than 3.6 mm/day, further it is greater than 4.5 mm/day, the corrected satellite rainfall is 4.71 mm/day. Similarly, we can generate rules using the other branches of tree as shown in Figure 5(b).

The optimized minimum leaf size obtained during the training of the CART model at each grid over India is shown in Figure 6, panel a. Higher number (>30) of observations at the leaf nodes (minimum leaf size) were required to train the model in hilly and mountainous regions, where elevation is greater than 3,000 m. On an average minimum, 10 observations were required at the leaf nodes of the CART tree over India. Figure 6, panel b shows the rainfall threshold split at the root node of the CART tree trained over India. Higher (lower) rainfall thresholds were observed at heavy (low) rainfall regions, indicating the possible dependence of root node threshold split on the rainfall pattern over India.

doi: 10.2166/h2oj.2020.124

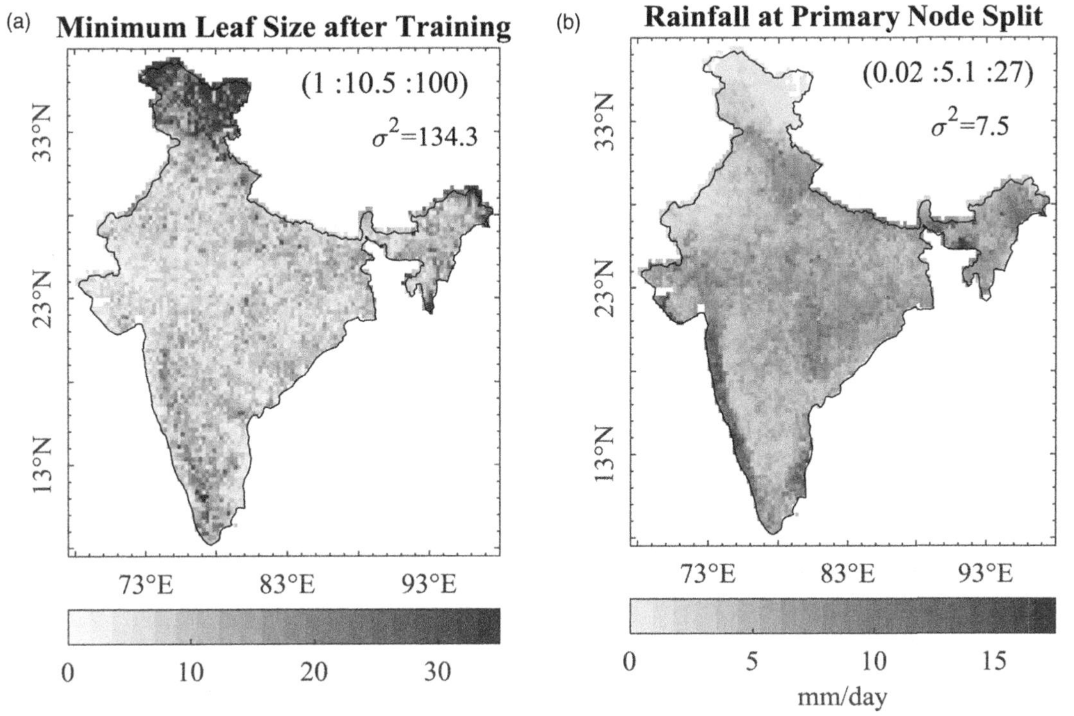

Figure 6 | Spatial variation of (a) minimum leaf size of decision trees obtained during training of CART model and (b) rainfall magnitude obtained at parent node split of CART model developed over the annual period of 2001–2011. All India mean value is indicated in parentheses. All-India minimum, mean and maximum values are given in parentheses (minimum: mean: maximum). Spatial variance (σ^2) is also provided.

The performance of the CART model during the training and testing period is analyzed over India using four performance metrics – Coefficient of Determination (R^2), adjusted R^2, Root Mean Square Error (RMSE) and Mean Absolute Bias (MAB). Figure 7, panel a and b shows the spatial distribution of R^2 during the training and the testing period. Higher R^2 is observed in most of the grids in India except for the states of Jammu and Kashmir, leeward side of Western Ghats and few grids in the southernmost corner of India. The all-India average R^2 of 0.77 and 0.66 is observed during the training and testing period respectively. Higher RMSE error was observed over the leeward side of Western Ghats and hilly regions of North East India (Figure 7, panel c and d). Similarly, high magnitude adjusted-R^2 were observed in both the training and testing period as shown in Figure S1, panel a and b. Lower magnitude of MAB, i.e., all-India average of 0.95 mm/day during the training period and 1.2 mm/day during the testing period were also observed (Figure S1, panel c and d). The above results show that the performance of the CART model is satisfactory and can be used over India for bias correction of monthly satellite estimates. However, over hilly and mountainous regions, the CART model should be cautiously adopted.

Performance of CART over different climate zones of India

The performance of the CART algorithm is evaluated over different climate zones of India. Figure 8 shows box plots that describe the variation of different R^2 and RMSE values in 7 different climate zones of India during the training and testing periods. Least R^2 (<0.4) is observed in the *BWk* climate zone during the training and testing periods. Similarly, moderately lower R^2 values, however, with high variability in R^2 (0.3–0.8) are observed in *Cwb* climate zone. Lower magnitude and higher variability in R^2 observed over *BWk* and *Cwb* climate zone can be attributed to either low reliability of reference dataset (owing to low rain-gauge density in these climatic zones) or possible complexity in capturing the satellite and reference rainfall relationships at very high

doi: 10.2166/h2oj.2020.124

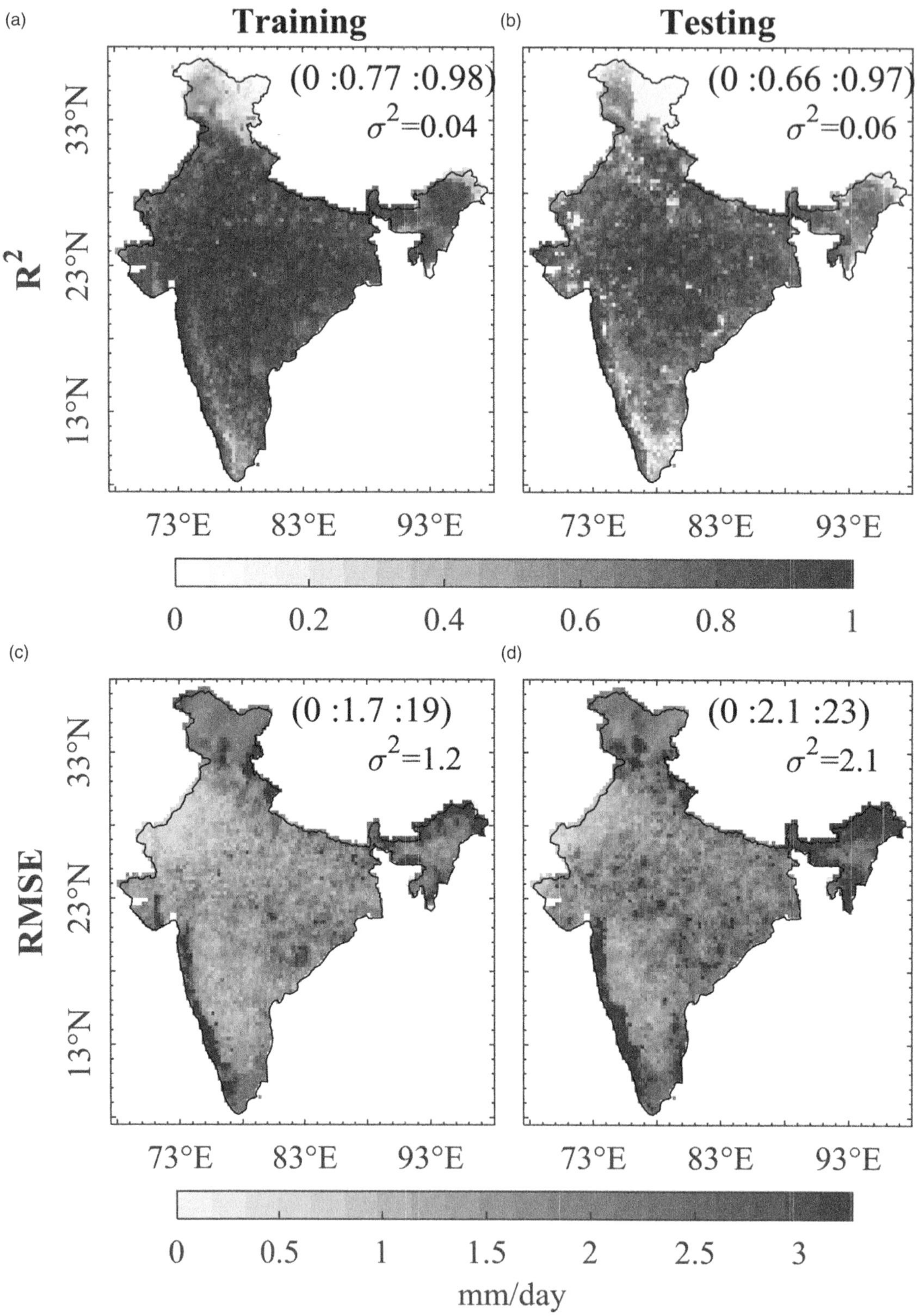

Figure 7 | Spatial variation of (a–b) Coefficient of Determination (R^2) and (c–d) Root Mean Square Error (RMSE) of the trained CART model during the training period (2001–2011) and testing period (2012–2016) over India. All-India minimum, mean and maximum values are given in parentheses (minimum: mean: maximum). Spatial variance (σ^2) is also provided. RMSE units are in mm/day.

altitudes. Over the remaining climate zone, the performance of CART as per R^2 is significantly good and less uncertain. High magnitude and variability of RMSE values are found in *Am* and *Cwb* climate zones.

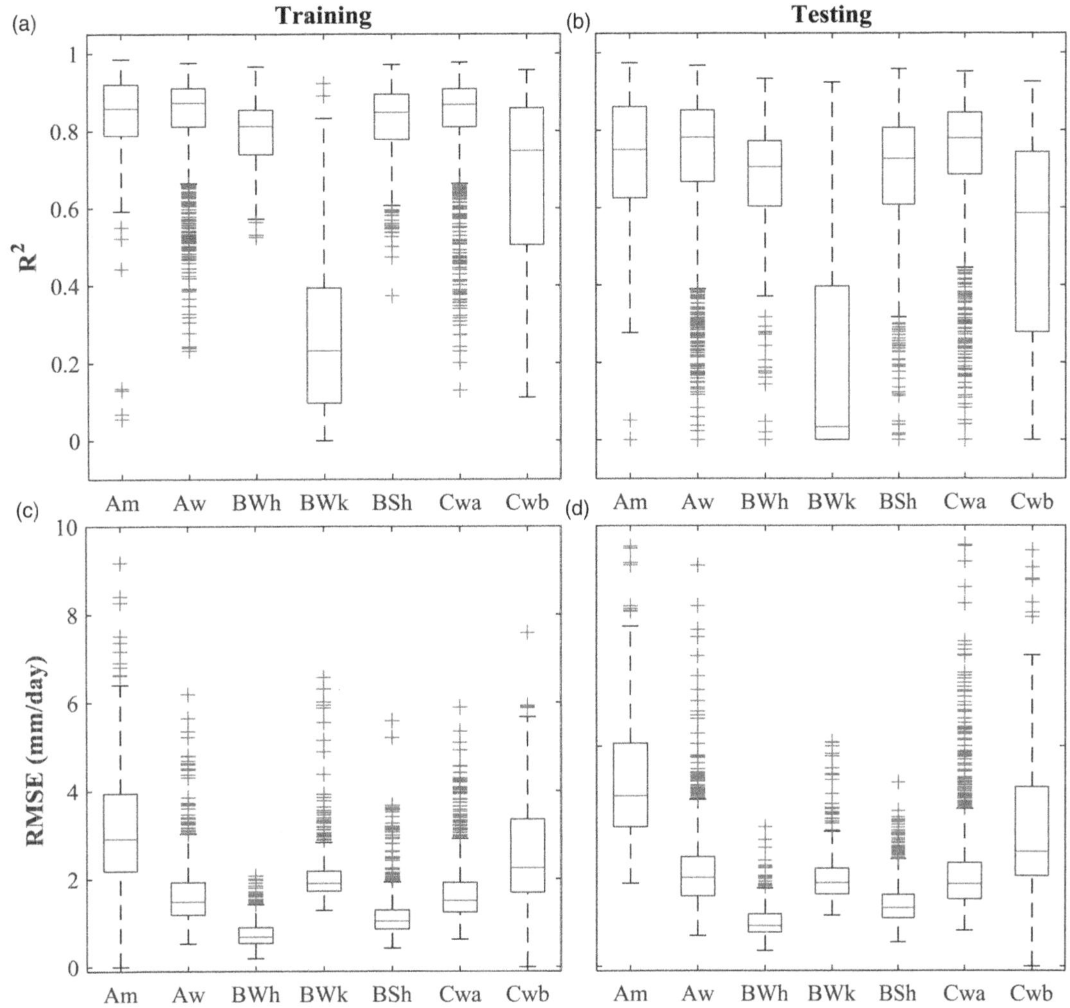

Figure 8 | Box plots showing the variation of (a–b) Coefficient of Determination (R^2) and (c–d) Root Mean Square Error (RMSE) of the trained CART model within different Koppen climate zones over India during the training period (2001–2011) and testing period (2012–2016).

Improvement in IMERG precipitation post application of CART

To quantify the improvement in the IMERG monthly satellite rainfall post-application of CART, we calculated the average monthly MAB and RMSE during the validation period. Table 3 shows the average monthly MAB and RMSE obtained before bias correction (between Original IMERG and IMD) and post bias-correction (CART corrected IMERG and IMD). The percentage of improvement in MAB and RMSE obtained was also quantified as shown in Table 3. High magnitudes of MAB (1.44–2.9) and RMSE (2.26–4.76) are observed during the monsoon season (June, July, August and September). All the months, except December, show a reduction in MAB post bias correction by CART. In December, CART was not able to reduce the absolute bias in the original IMERG dataset as an increase in MAB by 6.3% was observed. However, a significant reduction (17.1%) in RMSE was observed during December. Overall, CART was efficient in reducing the all-India average monthly MAB and RMSE in the IMERG dataset.

The performance of bias correction algorithm was further analysed over different climate and elevation zones over India. Figure 9 shows the variation of monthly MAB in original and CART corrected IMERG dataset spatially averaged over entire India (panel a) and climate zones (panel b–h) during the testing period (2012–2016). CART effectively reduced the all-India MAB over the majority of months. Although the total bias was not removed completely, a significant improvement in MAB ranging from 5.3 to 29.2% was observed. The ability of CART algorithm to reduce the MAB in IMERG dataset varied with varying

Table 3 | Average monthly MAB and RMSE obtained between (i) Original IMERG and IMD and (ii) CART corrected IMERG and IMD for all months during the validation period

	MAB			RMSE		
Months	**Original (mm/day)**	**After CART Correction (mm/day)**	**Improvement (%)**	**Original (mm/day)**	**After CART Correction (mm/day)**	**Improvement (%)**
January	0.46	0.43	5.3	1.04	0.95	8.7
February	0.69	0.53	22.2	1.45	1.12	22.8
March	0.78	0.62	19.6	1.58	1.22	22.5
April	0.98	0.69	29.2	2.02	1.27	37.3
May	1.05	0.83	20.3	1.91	1.45	24.0
June	2.06	1.53	25.6	3.82	2.90	24.2
July	2.90	2.20	24.1	4.76	3.63	23.8
August	2.55	1.93	24.3	4.02	3.11	22.7
September	1.98	1.44	26.9	3.17	2.26	28.8
October	1.00	0.85	14.4	1.70	1.33	22.0
November	0.63	0.54	13.8	1.48	1.13	23.8
December	0.35	0.37	**−6.3**	0.83	0.68	17.1

climate zones. Highest magnitude of MAB was observed in *Am* climate zone which receives heavy rainfall during the monsoon season. CART based bias correction was able to dampen the higher magnitudes of MAB in original data over all the climate zones except *BWk*. Over the *BWk* zone, bias correction with CART algorithm escalated the lower magnitudes of MAB, when compared to the original dataset. Moreover, *Am* zone observed minimum improvement in MAB. The RMSE in the original IMERG dataset was also significantly reduced post-application of CART as shown in Figure S2.

Figure 10 shows the variation of all-India and elevation zones averaged monthly MAB in original IMERG and CART bias-corrected IMERG during the testing period (2012–2016) over India. Seven elevation zones as shown in Table S1 were formulated for evaluating the performance of CART based bias-correction algorithm. Low MAB values were observed in very low elevation zone (<500 m) as well as very high elevation zones (>5,000 m). The performance of CART algorithm was relatively better when elevation was low and deteriorated with increasing elevation. An increase in the magnitude of MAB was observed post bias correction by CART, especially during the months with low MAB in the original IMERG data.

Comparison of CART-based bias correction scheme with other widely adopted schemes

Two widely adopted bias correction schemes – Linear scaling (LS) and Equidistant Cumulative Distribution Matching (EDCDF) are selected for comparison with the proposed CART-based bias correction technique. Detailed formulation of LS and EDCDF technique is given in Table S2. Figure 11 shows the spatial variation of average MAB and RMSE obtained from the original IMERG dataset, CART bias-corrected IMERG, LS corrected and EDCDF matching scheme during the testing period (2012–2016) over India. Over most of the grids over India, least MAB and RMSE were observed in CART-based bias correction followed by EDCDF scheme and LS scheme. Overall, least all-India average MAB of 0.94 mm/day was observed in CART, while the other two techniques, i.e., EDCDF matching and LS corrected method observed the all-India average MAB of 1.1 mm/day and 1.3 mm/day. Figure 12(a) shows the monthly variation of all-India average MAB obtained after CART based bias correction, LS bias correction and EDCDF matching method. CART based method and EDCDF matching method performed best in reducing the all-India average MAB, however, for initial months the performance of CART was much better among the two. Similar inferences were drawn when the bias correction schemes were compared based

doi: 10.2166/h2oj.2020.124

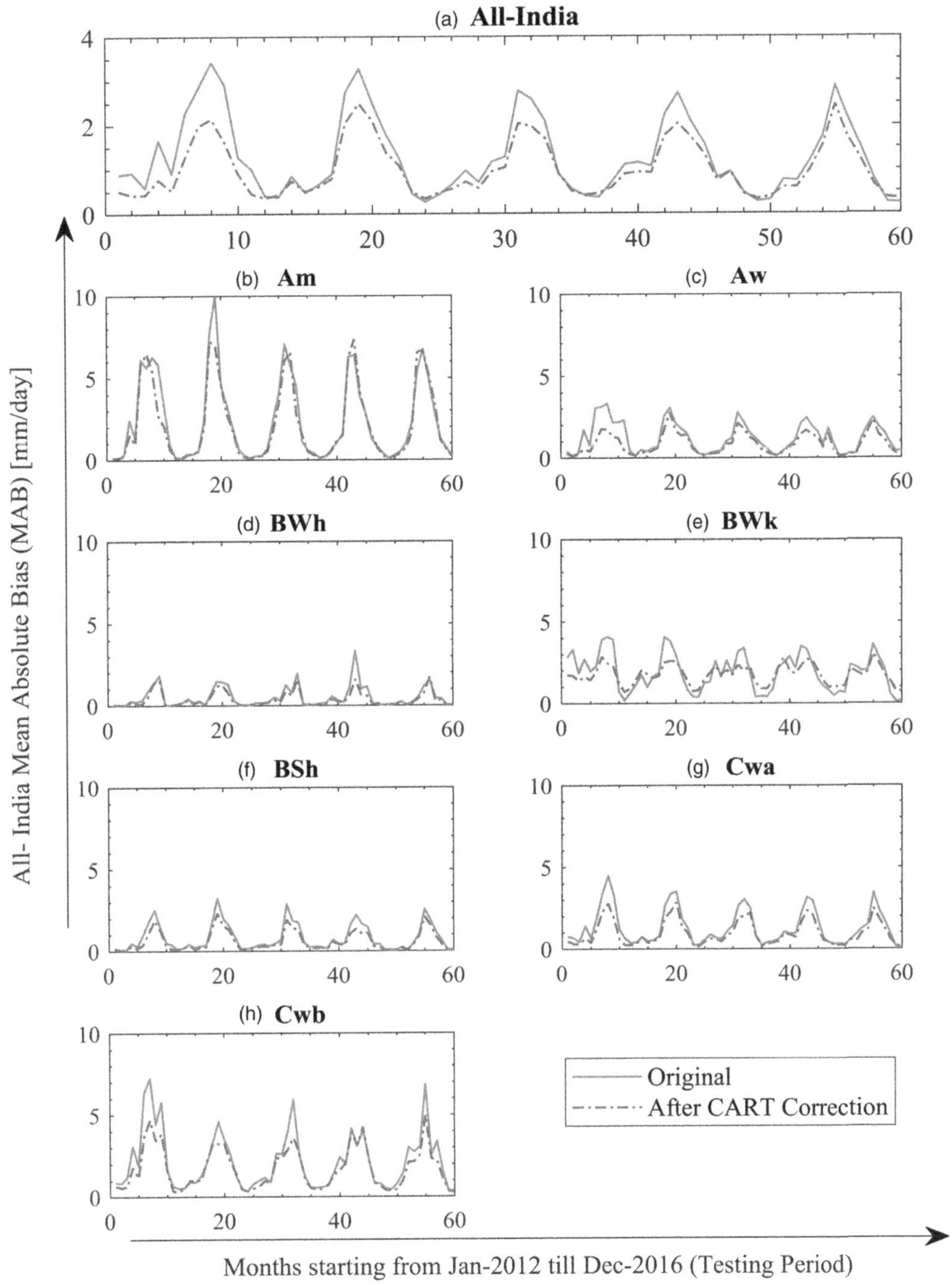

Figure 9 | Monthly variation of Mean Absolute Bias (MAB) in original IMERG and CART bias-corrected IMERG averaged over (a) entire India and (b–h) different climate zone during the testing period (2012–2016) over India.

on their average all-India RMSE values (Figure S3). Better performance of CART model over the other two methods can also be observed in Figure 12(b), wherein CART corrected IMERG dataset shows the highest correlation with IMD reference dataset during the testing period (2012–2016).

CONCLUSIONS

In the present study, we examine the applicability of Classification and Regression tree-based machine learning algorithm for possible bias reduction of the IMERG precipitation dataset over India. The major conclusions observed in the present study are:

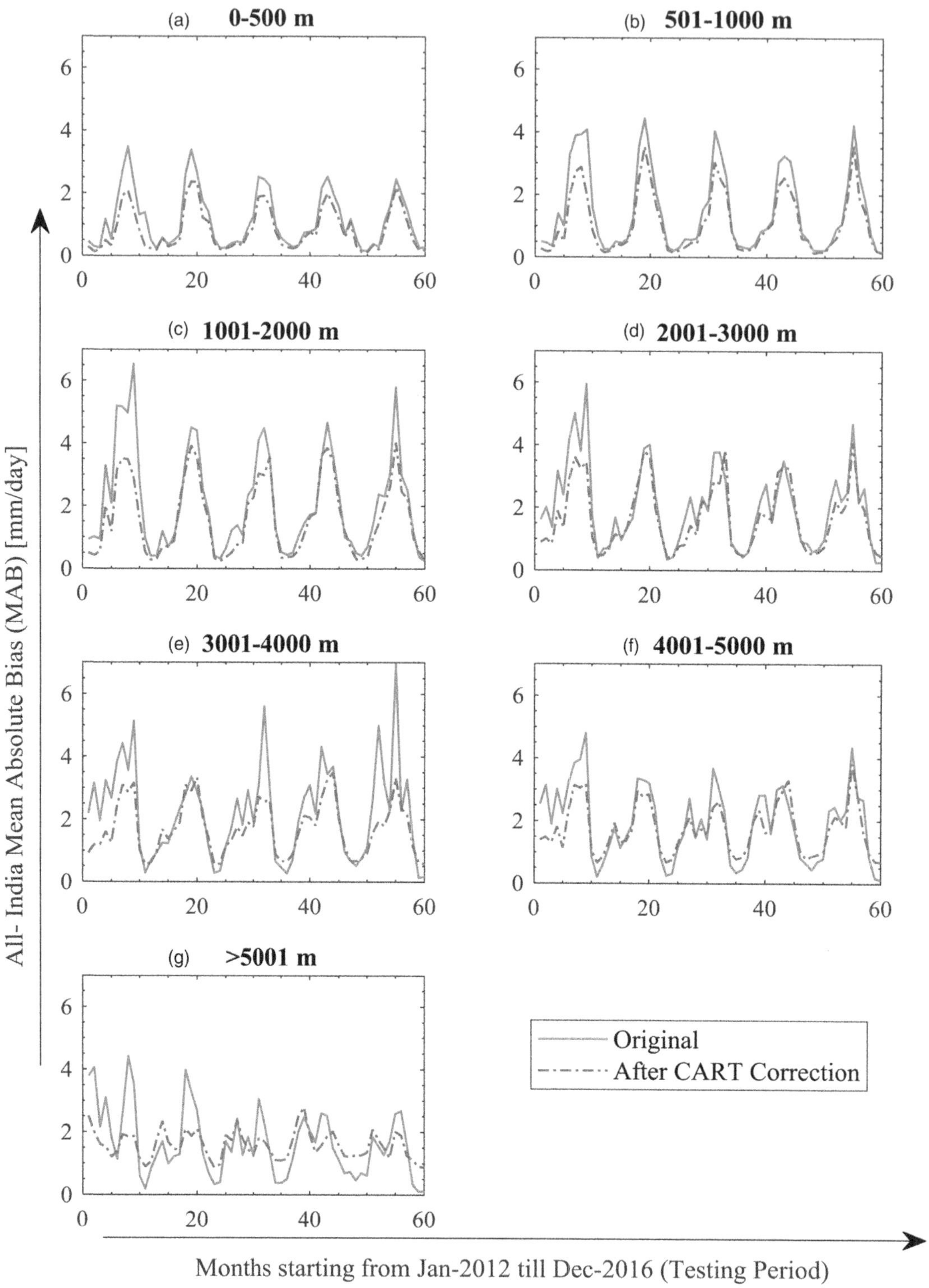

Figure 10 | Monthly variation of Mean Absolute Bias (MAB) in original IMERG and CART bias-corrected IMERG averaged over different (a–g) elevation zone during the testing period (2012–2016) over India.

- The CART model is highly effective in capturing the bias during the training (average $R^2 = 0.77$) and testing (average $R^2 = 0.66$) period. High R^2 (>0.75) and very low variability is observed in all climate zones except for *BWk* and *Cwb* zone.
- Significant improvement in average monthly MAB (−6.3 to 29.2%) and RMSE (8.7–37.3%) was obtained post-application of CART based bias-correction.

doi: 10.2166/h2oj.2020.124

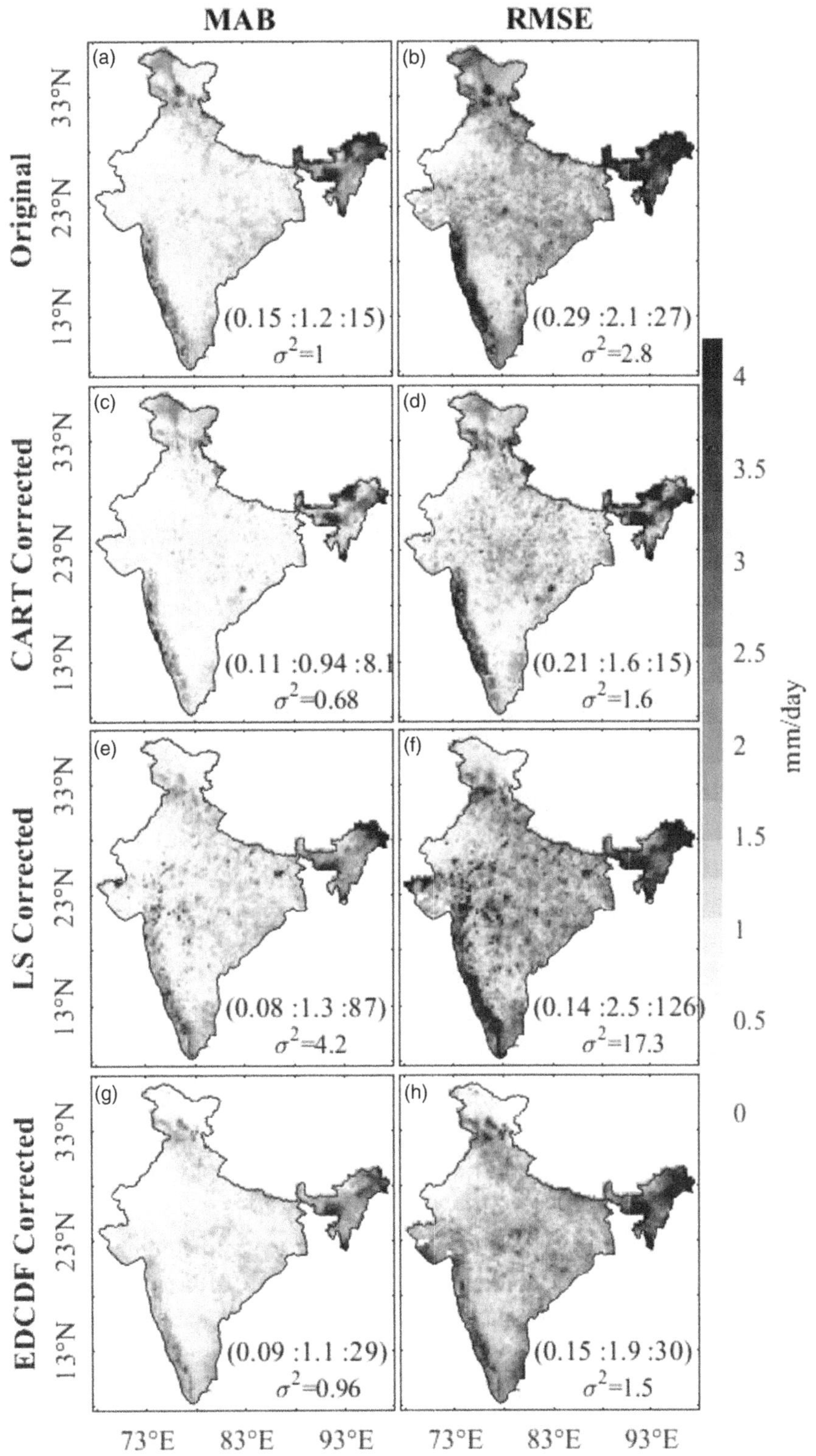

Figure 11 | Spatial variation of MAB (left column) and RMSE (right column) in the (a, b) original IMERG (c, d) CART bias-corrected IMERG (e, f) Linear Scaling (LS) corrected and (g, h) equidistant CDF (EDCDF) matching scheme averaged during the testing period (2012–2016) over India.

- CART based bias correction was able to dampen the higher magnitudes of MAB and RMSE in original data over all the climate zones except *BWk*, where the application of CART algorithm escalated the lower magnitudes of MAB and RMSE, when compared to the original dataset.

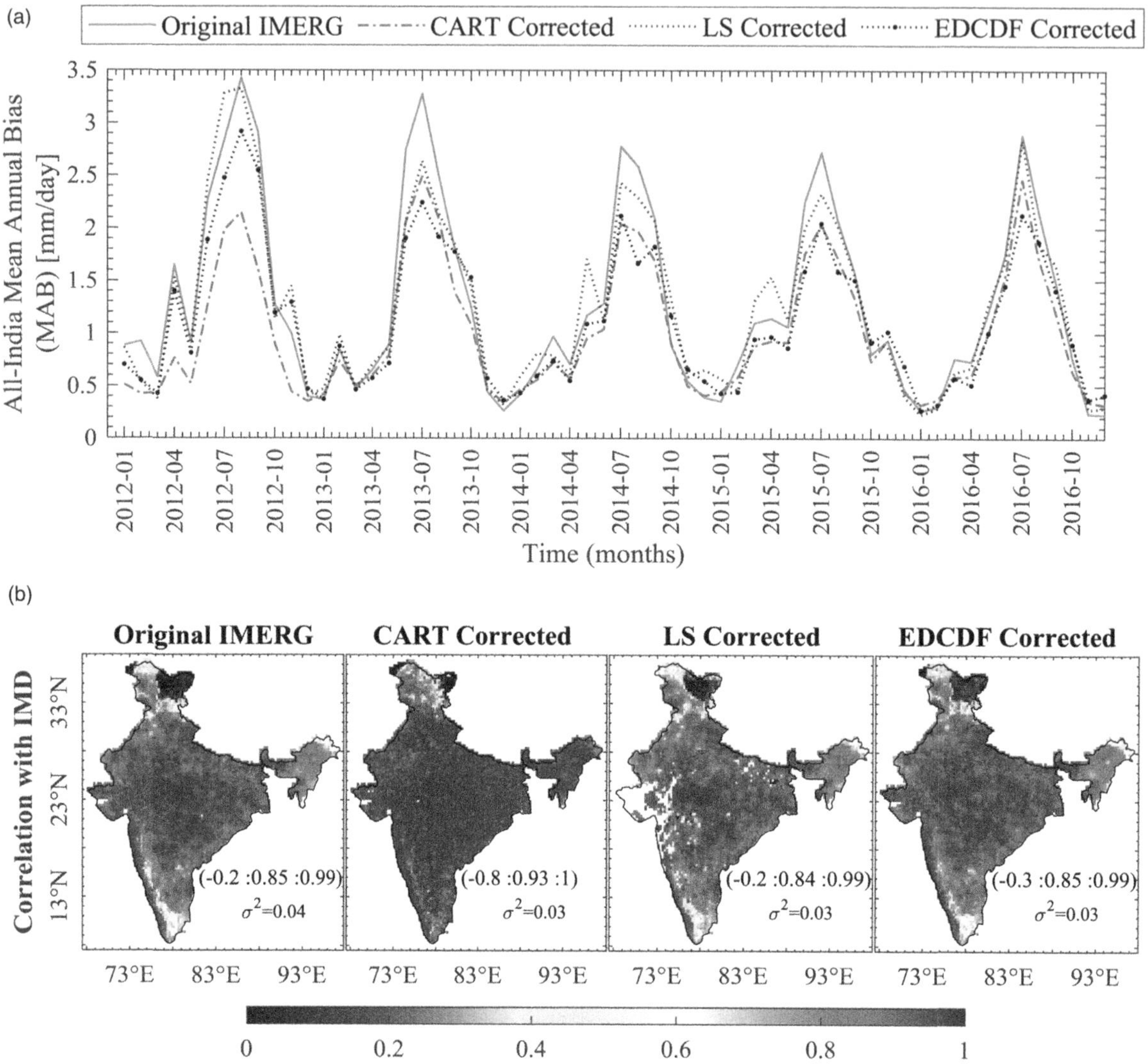

Figure 12 | (a) Monthly variation of Mean Absolute Bias (MAB) in the original IMERG, CART corrected, LS corrected and EDCDF corrected (b) Temporal correlation of original IMERG, CART corrected, LS corrected and EDCDF corrected datasets with IMD (reference) dataset during the testing period (2012–2016) over India.

- The performance of CART algorithm was relatively better when elevation was low and deteriorated with increasing elevation.
- Better performance of CART model over Linear scaling (LS) and Equidistant Cumulative Distribution Matching (EDCDF) was observed. CART corrected IMERG dataset depicted the highest correlation with IMD reference and least average MAB and RMSE dataset during the testing period.

It is important to mention here some of the limitations of the present study. Firstly, we assumed that the reference gauge-based gridded dataset from IMD to be error-free, which may not be true always. Although errors may exist in the gauge-based datasets, their magnitude would be much smaller than those in the satellite-based datasets. For further improvement in bias reduction, station data available from Doppler weather radar or automatic weather stations can be used as a reference in the future. Since over mountainous regions of India, IMD rainfall data is uncertain due to less gauge network density, in the future, a study can be conducted to compare the IMERG datasets directly with gauge station values for these regions. Secondly, we have trained the CART model on 11 years of monthly data, i.e., 132 values which are somewhat less for a data greedy algorithm like CART. Nonetheless, satisfactory fitting of the CART model is observed in the present study which aids in bias

doi: 10.2166/h2oj.2020.124

correction of satellite data. The present study, nevertheless, opens the possibility of applying such a machine learning algorithm for bias correction of the daily or hourly precipitation dataset.

ACKNOWLEDGEMENTS

The authors would like to express sincere gratitude to the Indian Institute of Technology Delhi, India for supporting this work. The authors would also like to thank the (1) Indian Meteorological Department for providing the IMD daily gridded precipitation data and (2) NASA Goddard Earth Sciences Data and Information Services Center (https://disc.gsfc.nasa.gov/) for providing IMERG datasets, and making it available for research. The editor and the three anonymous reviewers are gratefully acknowledged for their valuable comments on our manuscript.

DATA AVAILABILITY STATEMENT

Datasets used in this study is available from India Meteorological Department, from the URL http://www.imdpune.gov.in/Clim_Pred_LRF_New/Grided_Data_Download.html and NASA Goddard Earth Sciences Data and Information Services Center, from the URL https://disc.gsfc.nasa.gov/.

REFERENCES

Abera, W., Brocca, L. & Rigon, R. 2016 Comparative evaluation of different satellite rainfall estimation products and bias correction in the Upper Blue Nile (UBN) basin. *Atmospheric Research* **178**, 471–483.

Abraham, S., Huynh, C. & Vu, H. 2020 Classification of soils into hydrologic groups using machine learning. *Data* **5** (1), 2.

AghaKouchak, A., Mehran, A., Norouzi, H. & Behrangi, A. 2012 Systematic and random error components in satellite precipitation data sets. *Geophysical Research Letters* **39** (9), L09406.

Asong, Z. E., Razavi, S., Wheater, H. S. & Wong, J. S. 2017 Evaluation of integrated multisatellite retrievals for GPM (IMERG) over southern Canada against ground precipitation observations: a preliminary assessment. *Journal of Hydrometeorology* **18** (4), 1033–1050.

Awange, J. L., Hu, K. X. & Khaki, M. 2019 The newly merged satellite remotely sensed, gauge and reanalysis-based multi-source weighted-ensemble precipitation: evaluation over Australia and Africa (1981–2016). *Science of the Total Environment* **670**, 448–465.

Beck, H. E., Van Dijk, A. I., Levizzani, V., Schellekens, J., Gonzalez Miralles, D., Martens, B. & De Roo, A. 2017 MSWEP: 3-hourly 0.25 global gridded precipitation (1979–2015) by merging gauge, satellite, and reanalysis data. *Hydrology and Earth System Sciences* **21** (1), 589–615.

Behrangi, A., Andreadis, K., Fisher, J. B., Turk, F. J., Granger, S., Painter, T. & Das, N. 2014 Satellite-based precipitation estimation and its application for streamflow prediction over mountainous western US basins. *Journal of Applied Meteorology and Climatology* **53** (12), 2823–2842.

Bhatti, H. A., Rientjes, T., Haile, A. T., Habib, E. & Verhoef, W. 2016 Evaluation of bias correction method for satellite-based rainfall data. *Sensors* **16** (6), 884.

Bhuiyan, E. M. A., Nikolopoulos, E. I. & Anagnostou, E. N. 2019 Machine learning–based blending of satellite and reanalysis precipitation datasets: a multiregional tropical complex terrain evaluation. *Journal of Hydrometeorology* **20** (11), 2147–2161.

Chaudhary, S. & Dhanya, C. T. 2019 Investigating the performance of bias correction algorithms on satellite-based precipitation estimates. In remote sensing for agriculture, ecosystems, and hydrology XXI. *International Society for Optics and Photonics* **11149**, 111490Z.

Chaudhary, S., Dhanya, C. T. & Vinnarasi, R. 2017 Dry and wet spell variability during monsoon in gauge-based gridded daily precipitation datasets over India. *Journal of Hydrology* **546**, 204–218.

Erdal, H. I. & Karakurt, O. 2013 Advancing monthly streamflow prediction accuracy of CART models using ensemble learning paradigms. *Journal of Hydrology* **477**, 119–128.

Gebregiorgis, A. S. & Hossain, F. 2012 Understanding the dependence of satellite rainfall uncertainty on topography and climate for hydrologic model simulation. *IEEE Transactions on Geoscience and Remote Sensing* **51** (1), 704–718.

Gebregiorgis, A. S. & Hossain, F. 2015 How well can we estimate error variance of satellite precipitation data around the world? *Atmospheric Research* **154**, 39–59.

Gebregiorgis, A. S., Kirstetter, P. E., Hong, Y. E., Carr, N. J., Gourley, J. J., Petersen, W. & Zheng, Y. 2017 Understanding overland multisensor satellite precipitation error in TMPA-RT products. *Journal of Hydrometeorology* **18** (2), 285–306.

Habib, E., Haile, A. T., Sazib, N., Zhang, Y. & Rientjes, T. 2014 Effect of bias correction of satellite-rainfall estimates on runoff simulations at the source of the Upper Blue Nile. *Remote Sensing* **6** (7), 6688–6708.

Hashemi, H., Nordin, M., Lakshmi, V., Huffman, G. J. & Knight, R. 2017 Bias correction of long-term satellite monthly precipitation product (TRMM 3B43) over the conterminous United States. *Journal of Hydrometeorology* **18** (9), 2491–2509.

Hong, Y., Adler, R. & Huffman, G. 2006 Evaluation of the potential of NASA multi-satellite precipitation analysis in global landslide hazard assessment. *Geophysical Research Letters* **33** (22), L22402.

Hossain, F. & Huffman, G. J. 2008 Investigating error metrics for satellite rainfall data at hydrologically relevant scales. *Journal of Hydrometeorology* **9** (3), 563–575.

Hou, A. Y., Kakar, R. K., Neeck, S., Azarbarzin, A. A., Kummerow, C. D., Kojima, M., Oki, R., Nakamura, K. & Iguchi, T. 2014 The global precipitation measurement mission. *Bulletin of the American Meteorological Society* **95** (5), 701–722.

Huffman, G. J., Bolvin, D. T., Braithwaite, D., Hsu, K., Joyce, R. & Xie, P. 2014 *NASA Global Precipitation Measurement (GPM) Integrated Multi-SatellitE Retrievals for GPM (IMERG)*. Algorithm Theoretical Basis Document (ATBD), Version 4.4, NASA, 30.

Koriche, S. A. & Rientjes, T. H. 2016 Application of satellite products and hydrological modelling for flood early warning. *Physics and Chemistry of the Earth, Parts A/B/C* **93**, 12–23.

Kühnlein, M., Appelhans, T., Thies, B. & Nauss, T. 2014 Improving the accuracy of rainfall rates from optical satellite sensors with machine learning – a random forests-based approach applied to MSG SEVIRI. *Remote Sensing of Environment* **141**, 129–143.

Mannan, A., Chaudhary, S., Dhanya, C. T. & Swamy, A. K. 2018 Regionalization of rainfall characteristics in India incorporating climatic variables and using self-organizing maps. *ISH Journal of Hydraulic Engineering* **24** (2), 147–156.

MATLAB 2019 *Statistics and Machine Learning Toolbox™ User's Guide*. MathWorks. Available from: https://in.mathworks.com/help/stats/.

Murphy, K. P. 2012 *Machine Learning: A Probabilistic Perspective*. MIT press, Cambridge, MA.

Pai, D. S., Sridhar, L., Rajeevan, M., Sreejith, O. P., Satbhai, N. S. & Mukhopadhyay, B. 2014 Development of a new high spatial resolution (0.25 × 0.25) long period (1901–2010) daily gridded rainfall data set over India and its comparison with existing data sets over the region. *Mausam* **65** (1), 1–18.

Peel, M. C., Finlayson, B. L. & McMahon, T. A. 2007 Updated world map of the Köppen-Geiger climate classification. *Hydrology and Earth System Sciences Discussions* **4** (2), 439–473.

Pekel, E. 2020 Estimation of soil moisture using decision tree regression. *Theoretical and Applied Climatology* **139** (3), 1111–1119.

Petty, T. R. & Dhingra, P. 2018 Streamflow hydrology estimate using machine learning (SHEM). *JAWRA Journal of the American Water Resources Association* **54** (1), 55–68.

Prakash, S. 2019 Performance assessment of CHIRPS, MSWEP, SM2RAIN-CCI, and TMPA precipitation products across India. *Journal of Hydrology* **571**, 50–59.

Prakash, S., Mitra, A. K., AghaKouchak, A. & Pai, D. S. 2015 Error characterization of TRMM Multisatellite Precipitation Analysis (TMPA-3b42) products over India for different seasons. *Journal of Hydrology* **529**, 1302–1312.

Prakash, S., Mitra, A. K., AghaKouchak, A., Liu, Z., Norouzi, H. & Pai, D. S. 2018 A preliminary assessment of GPM-based multi-satellite precipitation estimates over a monsoon dominated region. *Journal of Hydrology* **556**, 865–876.

Prakash, S., Seshadri, A., Srinivasan, J. & Pai, D. S. 2019 A new parameter to assess impact of rain gauge density on uncertainty in the estimate of monthly rainfall over India. *Journal of Hydrometeorology* **20** (5), 821–832.

Rasouli, K., Hsieh, W. W. & Cannon, A. J. 2012 Daily streamflow forecasting by machine learning methods with weather and climate inputs. *Journal of Hydrology* **414**, 284–293.

Sawunyama, T. & Hughes, D. A. 2008 Application of satellite-derived rainfall estimates to extend water resource simulation modelling in South Africa. *Water Sa* **34** (1), 1–10.

Tan, C. O. & Beklioglu, M. 2006 Modeling complex nonlinear responses of shallow lakes to fish and hydrology using artificial neural networks. *Ecological Modelling* **196** (1–2), 183–194.

Tan, J., Petersen, W. A. & Tokay, A. 2016 A novel approach to identify sources of errors in IMERG for GPM ground validation. *Journal of Hydrometeorology* **17** (9), 2477–2491.

Tan, J., Petersen, W. A., Kirstetter, P. E. & Tian, Y. 2017 Performance of IMERG as a function of spatiotemporal scale. *Journal of Hydrometeorology* **18** (2), 307–319.

Tang, L. & Hossain, F. 2012 Investigating the similarity of satellite rainfall error metrics as a function of Köppen climate classification. *Atmospheric Research* **104**, 182–192.

Tang, G., Ma, Y., Long, D., Zhong, L. & Hong, Y. 2016 Evaluation of GPM Day-1 IMERG and TMPA version-7 legacy products over Mainland China at multiple spatiotemporal scales. *Journal of Hydrology* **533**, 152–167.

Tao, Y., Gao, X., Hsu, K., Sorooshian, S. & Ihler, A. 2016 A deep neural network modeling framework to reduce bias in satellite precipitation products. *Journal of Hydrometeorology* **17** (3), 931–945.

Veettil, A. V. & Mishra, A. K. 2020 Multiscale hydrological drought analysis: role of climate, catchment and morphological variables and associated thresholds. *Journal of Hydrology* **582**, 124533.

Waheed, T., Bonnell, R. B., Prasher, S. O. & Paulet, E. 2006 Measuring performance in precision agriculture: CART – a decision tree approach. *Agricultural Water Management* **84** (1–2), 173–185.

Willmott, C. J. 1981 On the validation of models. *Physical Geography* **2** (2), 184–194.

Worqlul, A. W., Ayana, E. K., Maathuis, B. H., MacAlister, C., Philpot, W. D., Leyton, J. M. O. & Steenhuis, T. S. 2018 Performance of bias corrected MPEG rainfall estimate for rainfall-runoff simulation in the upper Blue Nile Basin, Ethiopia. *Journal of Hydrology* **556**, 1182–1191.

Yu, J., Qin, X., Larsen, O. & Chua, L. H. C. 2014 Comparison between response surface models and artificial neural networks in hydrologic forecasting. *Journal of Hydrologic Engineering* **19** (3), 473–481.

First received 1 April 2020; accepted in revised form 11 July 2020. Available online 4 August 2020

Evolutionary Algorithms in Water Resources
doi: 10.2166/h2oj.2020.030

Comparative performance evaluation of self-adaptive differential evolution with GA, SCE and DE algorithms for the automatic calibration of a computationally intensive distributed hydrological model

Saswata Nandi[a] and M. Janga Reddy[a,b,*]

[a] Department of Civil Engineering, Indian Institute of Technology Bombay, Mumbai 400076, India

[b] Interdisciplinary Programme in Climate Studies, Indian Institute of Technology Bombay, Mumbai, India

*Corresponding author. E-mail: mjreddy@civil.iitb.ac.in

Abstract

Recently, physically-based hydrological models have been gaining much popularity in various activities of water resources planning and management, such as assessment of basin water availability, floods, droughts, and reservoir operation. Every hydrological model contains some parameters that must be tuned to the catchment being studied to obtain reliable estimates from the model. This study evaluated the performance of different evolutionary algorithms, namely genetic algorithm (GA), shuffled complex evolution (SCE), differential evolution (DE), and self-adaptive differential evolution (SaDE) algorithm for the parameter calibration of a computationally intensive distributed hydrological model, variable infiltration capacity (VIC) model. The methodology applied and tested for a case study of the upper Tungabhadra River basin in India and the performance of the algorithms is evaluated in terms of reliability, variability, efficacy measures in a limited number of function evaluations, their ability for achieving global convergence, and also by their capability to produce a skilful simulation of streamflows. The results of the study indicated that SaDE facilitates an effective calibration of the VIC model with higher reliability and faster convergence to optimal solutions as compared to the other methods. Moreover, due to the simplicity of the SaDE, it provides easy implementation and flexibility for the automatic calibration of complex hydrological models.

Key words: differential evolution, genetic algorithm, parameter estimation, self-adaptive differential evolution, shuffled complex evolution, VIC model

Highlights

- Calibration of a computationally intensive distributed hydrologic model is performed.
- Evaluated the performance of different evolutionary algorithms for automatic calibration of VIC model.
- Self-adaptive scheme avoids tuning of the algorithm parameters.
- SaDE method is capable of finding the optimal parameters quickly and gives superior performance over GA and SCE algorithms.

doi: 10.2166/h2oj.2020.030

Graphical Abstract

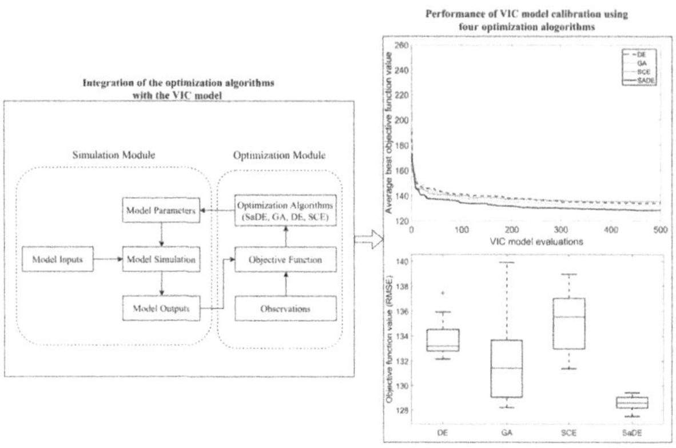

INTRODUCTION

Hydrological models are prevalent tools for representing the hydrological processes of a catchment in mathematical form. Consequently, they are widely used in various water resources management studies such as flood forecasting (Refsgaard *et al.* 1988), drought risk assessment (Sheffield & Wood 2008), integrated river basin management (Cai *et al.* 2003), analyzing impacts of climate change and anthropogenic activities on water resources (Arnell 1999), etc. These hydrological models are broadly classified into conceptual and physically-based models with reference to the interpretations of the dominant physical processes at a catchment. According to the spatial representation of catchment, they are further categorized into lumped, semi-distributed, and distributed models. Physically-based distributed hydrological models (PBDHM) are considered to enhance the hydrological simulation of a watershed by incorporating spatial heterogeneity of catchment variables such as land uses, soil types, topographic features, and meteorological forcing (Chen *et al.* 2016). However, the operational use of PBDHMs is complex and also requires greater computational resources even for a smaller catchment, as it uses different model parameters for its various spatially distinct computational units (Chen *et al.* 2016).

However, regardless of a physical or conceptual model, practically all hydrological models include many parameters, and it cannot be possible to measure all of them in the field directly. Therefore, an effective application of these models requires proper model calibration, which is related to the adjustment of the model parameters to fit the simulated results with the observation data (Tolson & Shoemaker 2007). The calibration of hydrological models can be performed either manually or automatically (Kim *et al.* 2007). Manual calibration involves a trial and error procedure for adjustments of the model parameters, but such calibration is time-consuming, tedious and its efficacy relies on the subjective judgement of a modeller, which includes knowledge of the model and its interaction with the participating watershed (Kim *et al.* 2007). On the contrary, automatic calibration is quick, less subjective, and more efficient for finding the optimal parameter set than that of manual calibration (Kamali *et al.* 2013). With the advancement of computational resources, many studies have chosen automatic calibration over the manual calibration approach for hydrological model applications (Wang 1991; Duan *et al.* 1992). The automatic calibration approach uses an optimization algorithm, which may be either a local or global optimization method, to explore the parameter space according to some specified objective function, which refers to the disparity between the observation and model simulation (Madsen 2000). Local optimization methods were used in the initial studies of automatic calibration for hydrological modelling (Ibbitt & O'Donnell 1971; Sorooshian & Gupta 1983). However, the drawback of such local search methods is that they may get stuck in

doi: 10.2166/h2oj.2020.030

the local optima and never reach the global optimal solution as there may be several local optima for the calibration problem of a hydrological model (Gan & Biftu 1996). Consequently, the overall performance of any local search method mostly depends on its initial solution.

To address such limitations, recent studies have inclined toward the utilization of global search methods for hydrological model calibration. In hydrological modelling, the most commonly used population-based evolutionary algorithms (EAs) are genetic algorithm (GA) (Wang 1991), shuffled complex evolution (SCE) algorithm (Duan *et al.* 1992), particle swarm optimization (PSO) (Scheerlinck *et al.* 2009), and differential evolution (DE) (Guo *et al.* 2014). Even though it is well-known that due to the issue of equifinality, i.e. multiple parameter values leading to the same model efficiency (Beven & Freer 2001), the obtained solution from these algorithms may strongly depend on the model representation, and chosen objective function (Scheerlinck *et al.* 2009; Osuch *et al.* 2015). The GA is one of the earliest evolutionary search based techniques that has been successfully applied for calibration of hydrological models (Wang 1991; Liong *et al.* 1995). Also, few studies on hydrological model calibration have highlighted the dominant performance of the SCE algorithm over other concurrent methods. For example, Arsenault *et al.* (2014) compared several optimization algorithms for three hydrological models in multiple basins over the United States and noted the SCE algorithm would be the most preferred choice, as it outperformed the GA in both convergence speed and computing power. However, the majority of the studies were conducted on simple conceptual hydrological models, which require hardly a few seconds for a model run. Consequently, those studies took advantage of utilizing a large number of function evaluations for model calibration with EAs. For example, Sorooshian *et al.* (1993) noted the convergence of 13 parameter calibration problem for the Sacramento Soil Moisture Accounting (SAC-SMA) model after 23,024 function evaluations, and Gan & Biftu (1996) reported that the EAs required an average of 10,818 function evaluations for estimating 13 parameters of SAC-SMA in eight catchments across the globe; and Qin *et al.* (2016) accomplished the calibration of seven parameters of the Simplified HYDROLOG (SIMHYD) model with an average of 11,153 function evaluations. In the case of modern complex hydrological models, such a large number of function evaluations demands much higher computational resources and time, which makes their calibration unreasonable. For instance, Tolson & Shoemaker (2007) noticed that the calibration of the SWAT model using SCE took a computational time of 14 and 140 days for 10,000 and 100,000 SWAT model evaluations, respectively.

To deal with such issues, there is a need to focus on improving the performance of the algorithms to achieve faster convergence to the optimal solution. In this direction, few optimization algorithms were applied to achieve faster convergence than the others, but the optimization algorithms require sensitivity analysis and fine-tuning of the several algorithm parameters to achieve their best performance. Furthermore, the performance of most of the EAs also depends on the complexity of the problem, which may require huge computational time for a large number of decision variables. Therefore, for the calibration of complex hydrological models, it is desirable to develop an algorithm that would demand nominal user configuration and yield optimal (or near-optimal) solutions with a lesser number of function evaluations (Vrugt *et al.* 2009; Zheng *et al.* 2012).

The primary focus of this study is to present an efficient methodology for automatic calibration of a physically-based complex distributed hydrological model, which will require minimal user configuration and a lesser number of function evaluations. This would facilitate efficient utilization of hydro-meteorological data, which in turn could be extracted by well-calibrated hydrological models. For this purpose, a self-adaptive differential evolution (SaDE) algorithm is presented. The DE algorithm is a global optimization technique proposed by Storn & Price (1997). DE is easy to implement and is reported to have a better (or comparable) performance with respect to other EAs (Reddy & Kumar 2007). However, like other EAs, the performance of the DE is also dependent on the setting of its control parameters, and this may result in a huge computational burden (Zheng *et al.* 2012). The SaDE algorithm is thus an improvement over the basic DE in this regard, where two control parameters of the algorithm are continuously updated during the search process. A few recent studies have also found that self-adaptive variants of DE outperformed other EAs, including GAs and basic DEs, in terms of faster convergence and a higher success rate (Schardong & Simonovic 2015; Pulluri *et al.* 2017; Sirsant & Reddy 2018). However, only a few studies used the self-adaptive DE method for water resources problems, like Zheng *et al.* (2012) used for the optimal design of four water distribution networks; Sirsant & Reddy (2018) applied DE and SaDE methods for the reliability-based design of water distribution networks and noted that SaDE provided quick optimal solutions consistently and performed better than DE, particularly for problems with a large number of decision variables. With this view, the present study examines the potential of the SaDE algorithm for calibration of a complex distributed hydrological model, namely the variable infiltration capacity (VIC) model.

In this study, a three-layer VIC hydrological model (Liang *et al.* 1994) is chosen to simulate the rainfall-runoff process at a medium-size catchment located in the state of Karnataka, India. The VIC model is a physically-based hydrological model, which can simulate most of the important hydrological processes in a catchment. Previously, the VIC model was successfully applied for water resources assessment studies (Schumann *et al.* 2013). The VIC model has several parameters that require rigorous calibration before its application to a catchment for hydrological studies. The calibration of these parameters is a challenging task and requires much computational time and a large number of VIC model simulations. This study presents an effective methodology for automatic calibration of the VIC hydrological model using the SaDE algorithm, and its performance is evaluated by comparing it with the GA, SCE and DE algorithms to verify the competence of the methodology in terms of reliability, variability, and efficiency. The specific objectives of the present study include: (i) to present SaDE based methodology for automatic calibration of the VIC hydrological model; (ii) to evaluate the performance of the SaDE by comparing with the results of the GA, SCE and DE algorithms.

The rest of the paper is organized as follows. The following section presents details of methods used in the study viz., details of the optimization algorithms, and VIC hydrological model. Next, the application of the SaDE method for automatic calibration of the VIC model for the upper Tungabhadra River basin is discussed. The results of the VIC model for hydrological simulations and performance evaluation of the SaDE method by comparing them with the results of GA, SCE and DE methods are presented in the following section. Finally, the conclusions of the study are presented.

METHODOLOGY

VIC hydrological model

In this study, the variable infiltration capacity (VIC) model, a physically-based distributed hydrological model, is selected for simulation of hydrological variables from the basin. The VIC model considers the interaction between the biosphere, vegetation, atmosphere, and soil dynamics for the estimation of water and energy fluxes. In contrast with other comparable hydrological models, the exclusive features of the VIC model are the representation of subgrid variation in vegetation within a single grid, consideration of variable infiltration among multiple soil layers, and estimation of nonlinear baseflow from bottom soil layer (Liang *et al.* 1994).

The Penman–Monteith method and the Horton infiltration curve are used by the model for estimation of evapotranspiration and subgrid variation of soil infiltration capacity, respectively (Liang *et al.* 1994). Total runoff estimation in the model comprises direct runoff (surface flow) and base flow (subsurface flow). Direct runoff is generated from the top two layers by assessing the saturation remainder flow and infiltration surplus flow from those two layers. Baseflow contributed from the bottom-most layer only (Franchini & Pacciani 1991) and is computed according to the ARNO model (ARNO refers to a semi-distributed conceptual rainfall-runoff model, which derives its name

doi: 10.2166/h2oj.2020.030

from its first application to the Arno River, Todini (1996)). In the VIC model, direct runoff and baseflow are mathematically expressed as:

$$Q_d = \begin{cases} P + W - W^{\max}, & P + i_0 \geq i_m \\ P + W - W^{\max} \cdot \left[1 - \left(1 - \dfrac{i_0 + P}{i_m}\right)^{1+b_i}\right], & P + i_0 \leq i_m \end{cases} \quad (1)$$

$$Q_b = \begin{cases} \dfrac{D_S D_m}{W_S W_3^{\max}} \cdot W_3, & 0 \leq W_3 \leq W_S W_3^{\max} \\ \dfrac{D_S D_m}{W_S W_3^{\max}} \cdot W_3 + \left(D_m - \dfrac{D_S D_m}{W_S}\right)\left(\dfrac{W_3 - W_S W_3^{\max}}{W_3^{\max} - W_S W_3^{\max}}\right)^2, & W_3 > W_S W_3^{\max} \end{cases} \quad (2)$$

where Q_d is the direct runoff (mm), P is the precipitation (mm), W and $W^{\max}$ are the initial and maximum soil moisture for the upper soil layers (mm), i_0 and i_m (mm) are the initial and maximum infiltration rate (mm), b_i is a parameter controlling the shape of the variable infiltration curve (unitless), Q_b denotes the base flow (mm), D_m represents maximum daily base flow (mm), D_s is a fraction (unitless) of D_m whereby nonlinear baseflow begins, W_3 and $W_3^{\max}$ are initial and maximum soil moisture at the bottom soil layer (mm), and W_s is the fraction (unitless) of $W_3^{\max}$ where nonlinear baseflow begins.

The VIC model produces the runoff individually in model grid cells. Therefore, a separate routing model is selected to transfer simulated runoff from the individual model grids to the outlet of the study area. This study employed the routing method of Nijssen *et al.* (1997), which uses the unit hydrograph theory for transporting the flow inside a grid to the outlet of that grid and linearized Saint-Venant equations for the channel routing of the flow from the grid outlet to the basin outlet through the river network.

Evolutionary algorithms

This study uses the GA, SCE, DE, and SaDE algorithms for calibration of the parameters of the VIC hydrological model. Brief details of these EAs are given below.

Genetic algorithm

Genetic algorithms (GA) are population-based heuristic search algorithms inspired by natural selection and survival of the fittest concept (Goldberg 1989). The GA algorithm starts by initializing a population of candidate solutions that have to be drawn randomly from a feasible parameter space. The individual members are chosen in each generation by fitness values, in which better fit chromosomes in the population are chosen to replicate new promising offspring. Next, the new population is produced by applying crossover and mutation operations to those selected members. The crossover operator selects parent solutions to generate new offspring solutions by sharing important characteristics of both parent solutions. In order to increase the diversity of the new population, the offspring solutions are randomly mutated. In this way, the generation of new candidate solutions is repeated until the stopping criterion is satisfied. More details of the algorithm can be found in Goldberg (1989) and Reddy & Kumar (2012). The pseudo-code of the GA algorithm is given in Appendix A.1.

Shuffled complex evolution algorithm

The shuffled complex evolution (SCE) algorithm was proposed as a global optimization technique for parameter estimation of hydrological models (Duan *et al.* 1992). The SCE algorithm begins by randomly initializing the population within its feasible range and then partitioning the population into different communities (or complexes). All such communities are allowed to evolve (generate offspring) freely for guiding the search in a reformed direction. All the members of a particular community have the chance to take part in the offspring generation. However, one better member of the community has a higher possibility of getting engaged in the generation of the offspring, which ensures the competitiveness of evolution. After a few generations, the entire population is shuffled, and new communities are formed. This process helps in sharing information between various communities to explore the search space thoroughly. The pseudo-code of the SCE-UA algorithm is given in Appendix A.2. For more details of the method, one can refer to Duan *et al.* (1992) and Vrugt *et al.* (2003).

Differential evolution

Differential evolution (DE) is one of the widely used population-based global optimization algorithms which is based on the process of natural evolution and has outperformed many other EAs (Storn & Price 1997). The unique trial vector generation strategy (such as mutation and crossover operation) distinguished DE from other EAs, although it has similar basic operations (mutation, crossover, and selection) to the other EAs. In DE, the mutation operator is applied first to produce the mutant vectors by modifying the initial vectors, and then the crossover operator is employed to those mutant vectors to generate the trial vectors. Finally, the selection operator is used to select the trial vectors which will survive to the next iteration. DE is an iterative procedure; after the initialization, it enters into an iterative loop of mutation, crossover, and selection operations and continues until the terminal condition is satisfied. Among the various operators, the mutation is the most important operator in DE and is highly dependent on the mutation factor (F). The crossover rate (CR) in the crossover operation greatly influences the generation of the trial vector and the convergence characteristics of the algorithm. Therefore, the performance of the DE algorithm is more reliant on the choice of F and CR values. The pseudo-code of the DE algorithm is given in Appendix A.3.

Moreover, like the GA and SCE algorithms, the performance of DE is also highly dependent on the chosen parameter settings, and their incorrect selection may lead to the poor performance of the algorithm (Das *et al.* 2016). So, proper sensitivity analysis is required before its application to a given problem.

Self-adaptive differential evolution algorithm

The self-adaptive differential evolution (SaDE) algorithm abstains from the substantial computational costs of traditional DE for finding the appropriate control parameters of the algorithm. Here, the control parameters are progressively self-adapted by means of their past involvement in producing favourable solutions. This study adopts the SaDE algorithm with changes to the parameter selection of the crossover rate (CR) and the mutation factor (F) in contrast to the one presented in Qin *et al.* (2009). Here, the mean value of the F and CR parameters are updated after every ten generations based on the successful values of the parameters over the last ten iterations, and then a new set of mutation and crossover rates are generated for each individual using these updated mean values. More details about the steps involved in the SaDE methodology are explained below.

1. Generate initial population $X_i^G = \{x_{i,1}^G, x_{i,2}^G, \ldots, x_{i,D}^G\}$ by randomly choosing individuals within the search space. Here $x_{i,j}^G$ represents the initial value of the jth parameter in the ith individual at the initial population, G represents the current generation number, and D denotes the dimension of the problem.
2. Initial values of mutation factor F and crossover rate CR are randomly generated using their specified mean and standard deviation values as follows:

$$\left.\begin{aligned} F_i &= \mu_F + Randn_1 \times \sigma_F \\ CR_i &= \mu_{CR} + Randn_2 \times \sigma_{CR} \end{aligned}\right\} \tag{3}$$

doi: 10.2166/h2oj.2020.030

where $F_i \in [0, 2]$ and $CR_i \in [0, 1]$ are the initial values of mutation factor and crossover rate for the ith population; $randn_1$ and $randn_2$ are random numbers generated from the standard normal distribution; μ_F and σ_F are the mean and standard deviation of the mutation factor; μ_{CR} and σ_{CR} are mean and standard deviation of the crossover rate. The mean and standard deviation values are updated based on the following recommendation from Qin *et al.* (2009) and also after performing initial sensitivity analysis, the μ_F and σ_F are initialized with values of 0.5 and 0.3, while μ_{CR} and σ_{CR} are set at values of 0.5 and 0.1, respectively.

3. Calculate the objective function values for each member of the population in the current generation.
4. Mutation is performed to every individual (X_i^G) in the population. These X_i^G are considered as the target vector, and their corresponding mutant vectors $V_i^G = \{V_{i,1}^G, V_{i,2}^G, \ldots, V_{i,D}^G\}$ are generated by adding the weighted difference between two randomly chosen individuals from the current population to a third randomly selected individual from the current population. Mathematically it can be written as:

$$V_i^G = X_{r_1}^G + F_i(X_{r_2}^G - X_{r_3}^G) \quad \forall i = 1, 2, \ldots, NP \tag{4}$$

where V_i^G = mutant vector associated with the ith population at the Gth generation. $X_{r_1}^G$, $X_{r_2}^G$ and $X_{r_3}^G$ are randomly selected vectors from the current generation ($r_1 \neq r_2 \neq r_3$), F_i is the mutation factor for target vector i, and NP represents the population size.

5. Next, crossover is implemented for the generation of trial vectors $U_i^G = \{U_{i,1}^G, U_{i,2}^G, \ldots, U_{i,D}^G\}$ by choosing between either a mutant vector (V_i^G) or the corresponding target vector (X_i^G) using a crossover factor. Thus, the trial vector (U_i^G) is generated as:

$$u_{i,j}^G = \begin{cases} v_{i,j}^G, & \text{if } \mathrm{rand}^j\,[0, 1] \leq CR_i \text{or } (j = j^{rand}) \\ x_{i,j}^G, & \text{otherwise} \end{cases} \quad \forall i = 1, 2, \ldots, NP \tag{5}$$

where $u_{i,j}^G$, $v_{i,j}^G$, $x_{i,j}^G$ are the trial vector, mutant vector, and target vectors associated with the jth decision parameter of the ith member at generation G. The $rand^j$ is a uniform random number in between 0 to 1, and j^{rand} can take any random value in the range 1 to D, D is the dimension of the target vector.

6. Subsequently, the objective function value $f(U_i^G)$ for the individual trial vector is computed and then compared with the objective function value of the corresponding target vector X_i^G. The vector which holds the lower value of the objective function continues for the next generation (X_i^{G+1}) (for minimization type of objective function). In other words:

$$X_i^{G+1} = \begin{cases} U_i^G, & \text{if } f(U_{i,G}) \leq f(X_{i,G}) \\ X_i^G, & \text{otherwise} \end{cases} \quad \forall i = 1, 2, \ldots, NP \tag{6}$$

The above steps of mutation, crossover, and selection operations in Equations (4)–(6) are performed for all members in the population (NP).

7. To update the values of two control parameters (i.e. F and CR) in the SaDE algorithm, the values of F and CR contributing to trial vectors effectively arriving for the next generation are noted after a certain number of generations (say, ten generations). Then, averaging those successful values of the F and CR provides the updated mean values of mutation factor (μ_F) and crossover rate (μ_{CR}), respectively. With these updated values of μ_F and μ_{CR}, a new set of F and CR values are generated as described previously in Equation (3) (i.e. assuming normal distributions for F and CR with the specified standard deviation values).

8. The above steps are repeated in a loop until the convergence criterion is satisfied. The convergence criteria employed in the present study are either there is no improvement in the objective function value for ten successive generations, or the algorithm has reached the maximum number of allowable generations ($I_{\max}$).

More details of the SaDE method can be found in Qin *et al.* (2009) and Sirsant & Reddy (2018).

APPLICATION

Study area

The study area is situated in the upper Tungabhadra River basin in Karnataka, India. The location map of the study area is shown in Figure 1. The spatial extent of the study area ranges from 75°7′E to 75°57′E and 13°7′N to 14°15′N, with the basin outlet at Honnali gauging station. The basin has a catchment area of 7,075 km^2 with elevations ranging from 542 to 1,890 m above mean sea level. The study area receives an average annual rainfall of 1,016 mm. The average annual temperature is 26.7 °C in the basin. A semi-arid climatic condition prevails in the study area, and the majority of rains occur from June to November during the south-west and north-east monsoons. It is predominantly covered by agricultural land with loamy to red sandy soil as the dominant soil type. Chikamagalur, Shimoga, and Devanagari are the three major districts falling under the study area and they are greatly dependent on water from the Tunga River. Accessibility of different observed hydro-meteorological datasets and nonexistence of large control structures are the main reasons for the selection of this study area.

Data used

Daily gridded precipitation data ($0.25 \times 0.25°$) collected from the Indian Meteorological Department (IMD) are used in this study as the primary forcing variable for all hydrological simulations. Daily temperature data was also taken from the IMD at a spatial resolution of $1 \times 1°$. Wind speed data were collected from the National Center for Environmental Prediction (NCEP) reanalysis data at $2.5 \times 2.5°$ scale. This study used soil information from the Food and Agricultural Organization (FAO) harmonized world soil database at 30 arc-second (~1 km) spatial resolution. Land-use and land-cover data for the study were derived from MODIS MOD12Q1 Land Cover Products (1 km), which was obtained from NASA's data distribution portal. SRTM (90 m) DEM was collected from the National Remote Sensing Centre of India. The data of daily observed streamflow at the outlet of the catchment, Honnali Hydro-observation site, were obtained from WRIS-India.

VIC model calibration

VIC model settings

Based on previous studies, 13 flow-related parameters from the VIC model were chosen for the automatic calibration (Demaria *et al.* 2007; Xie *et al.* 2007). These 13 parameters, along with their descriptions, model response, ranges, and units, are presented in Table 1. Among these 13 parameters, one governs direct runoff response, eight govern drainage responses, and four control the baseflow responses. During the calibration process, these selected 13 parameters shared the same value for the whole model domain. The other VIC model parameters are obtained from the field measurement, which is varied across the model domain. The calibration could have considered independent spatial values for all the calibration parameters, but that would have increased the complexity of the

doi: 10.2166/h2oj.2020.030

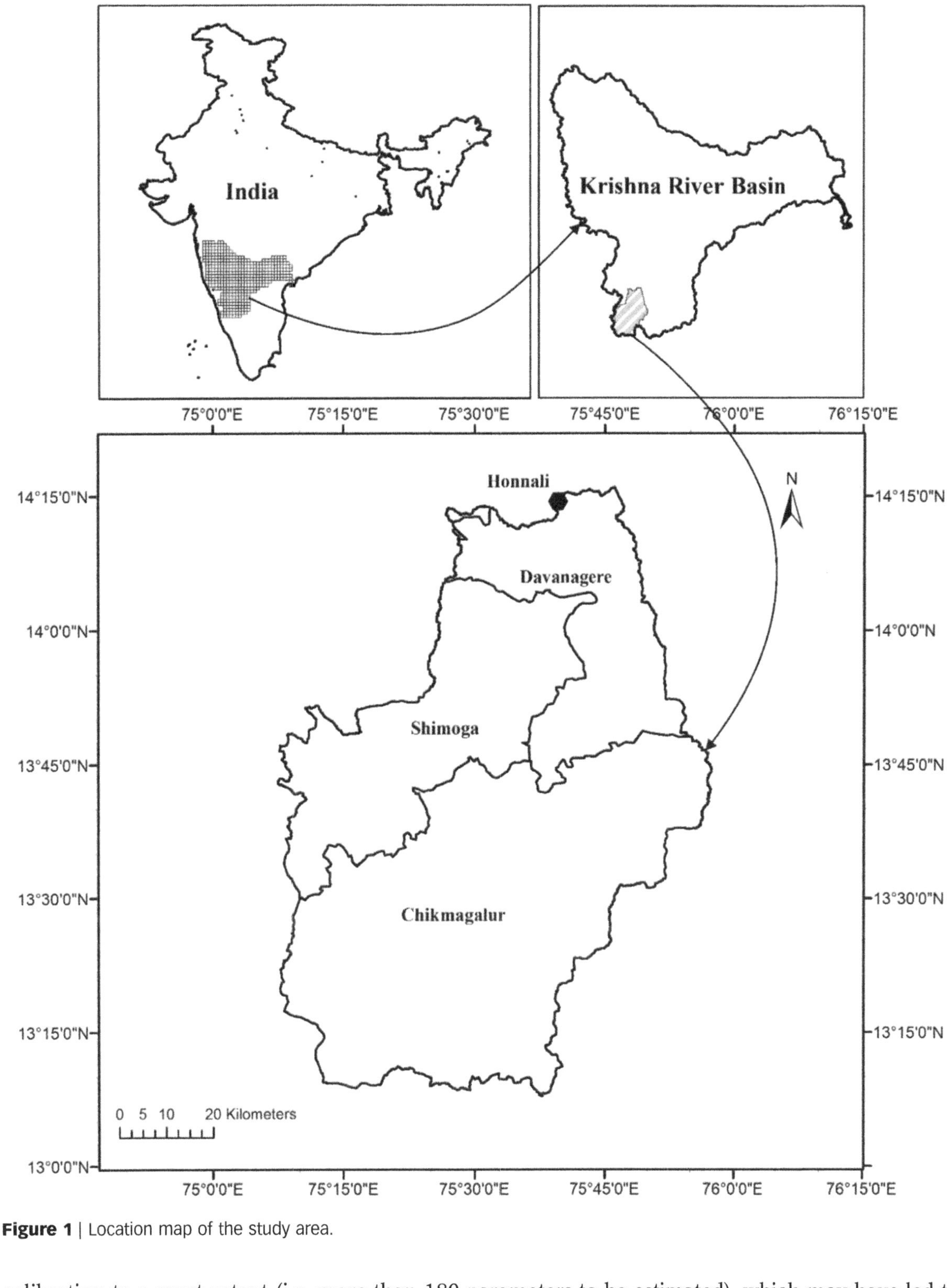

Figure 1 | Location map of the study area.

calibration to a great extent (i.e. more than 180 parameters to be estimated), which may have led to impractical longer optimization time with the available computational resources.

The VIC model and aforementioned routing model source code are first compiled in the Linux environment using the GNU Compiler Collection (GCC). The integration of the SaDE optimization algorithm with the VIC model is shown in Figure 2. The simulation-optimization framework is implemented in the MATLAB environment. For this integration it first requires creation of all the input files (i.e. meteorological forcing files, soil parameter file, vegetation parameter file, and vegetation

Table 1 | Details of the VIC model parameters selected for calibration

Parameter	Description	Model response	Range	Units
b_i	Infiltration parameter	Direct flow	1×10^{-5}–0.4	–
D_m	Maximum baseflow velocity	Base flow	0–30	mm/day
D_s	Fraction of max baseflow velocity for nonlinear baseflow	Base flow	0–1	–
W_s	Fraction of max soil moisture for nonlinear baseflow	Base flow	0–1	–
d_1	Thickness of soil layer 1	Drainage	0.01–0.5	m
d_2	Thickness of soil layer 2	Drainage	0.1–2	m
d_3	Thickness of soil layer 3	Base flow	0.1–2	m
Ks_1	Saturated hydraulic conductivity for first layer	Drainage	1–10,000	mm/day
Ks_2	Saturated hydraulic conductivity for second layer	Drainage	1–10,000	mm/day
Ks_3	Saturated hydraulic conductivity for third layer	Drainage	1–10,000	mm/day
exp_1	Exponent (Brooks-Corey drainage eq.) for first layer	Drainage	8–30	–
exp_2	Exponent (Brooks-Corey drainage eq.) for second layer	Drainage	8–30	–
exp_3	Exponent (Brooks-Corey drainage eq.) for third layer	Drainage	8–30	–

library file) and a global parameter file (i.e. model configuration file) for the simulation of the VIC model. We also need to define the required input files (i.e. flow direction file, area mask file, area fraction file, outlet station file, and a model configuration file) for the routing model. Next, the SaDE algorithm is implemented in the MATLAB, as discussed above under 'Self-adaptive differential evolution algorithm' (called the SaDE module). The SaDE module calls the simulation module whenever it needs to compute the fitness values of the member solutions of the initial population or the recent trial vectors. The simulation module first updates the appropriate VIC input files using the feedback received from the SaDE module and then runs the VIC model to produce the hydrological fluxes at each model grid. Next, the routing model is executed via the simulation module to obtain the simulated discharge at the basin outlet. Subsequently, the root-mean-square error (RMSE) is computed as the fitness measure by comparing the simulated discharge with the observed discharge data and sends this information back to the SaDE module. The optimization using the SaDE module progresses with its continuous interconnection with the simulation module until certain stopping conditions are met. Similarly, the GA, SCE and DE algorithms are also implemented and linked to the VIC model in the MATLAB environment. The study used four years of data (1 January 1995–31 December 1998) for calibration, and three years of data (1 January 1999–31 December 2001) for validation of the VIC model.

Calibration criteria and performance measures

Evaluation of any hydrological model is usually conducted by pairwise comparisons of model-simulated flow values with the observed ones. The RMSE between observed and simulated flows is used as the calibration criteria (objective function) to determine optimal parameters of the VIC hydrological model (Madsen 2000; Kavetski & Clark 2010). RMSE calculates the average squared errors for all simulated flow values and is defined as follows:

$$RMSE = \sqrt{\frac{1}{n}\sum_{i=1}^{n}(S_i - O_i)^2} \tag{7}$$

where O_i and S_i are observed and simulated flow values for the ith time step, and n is the number of time steps.

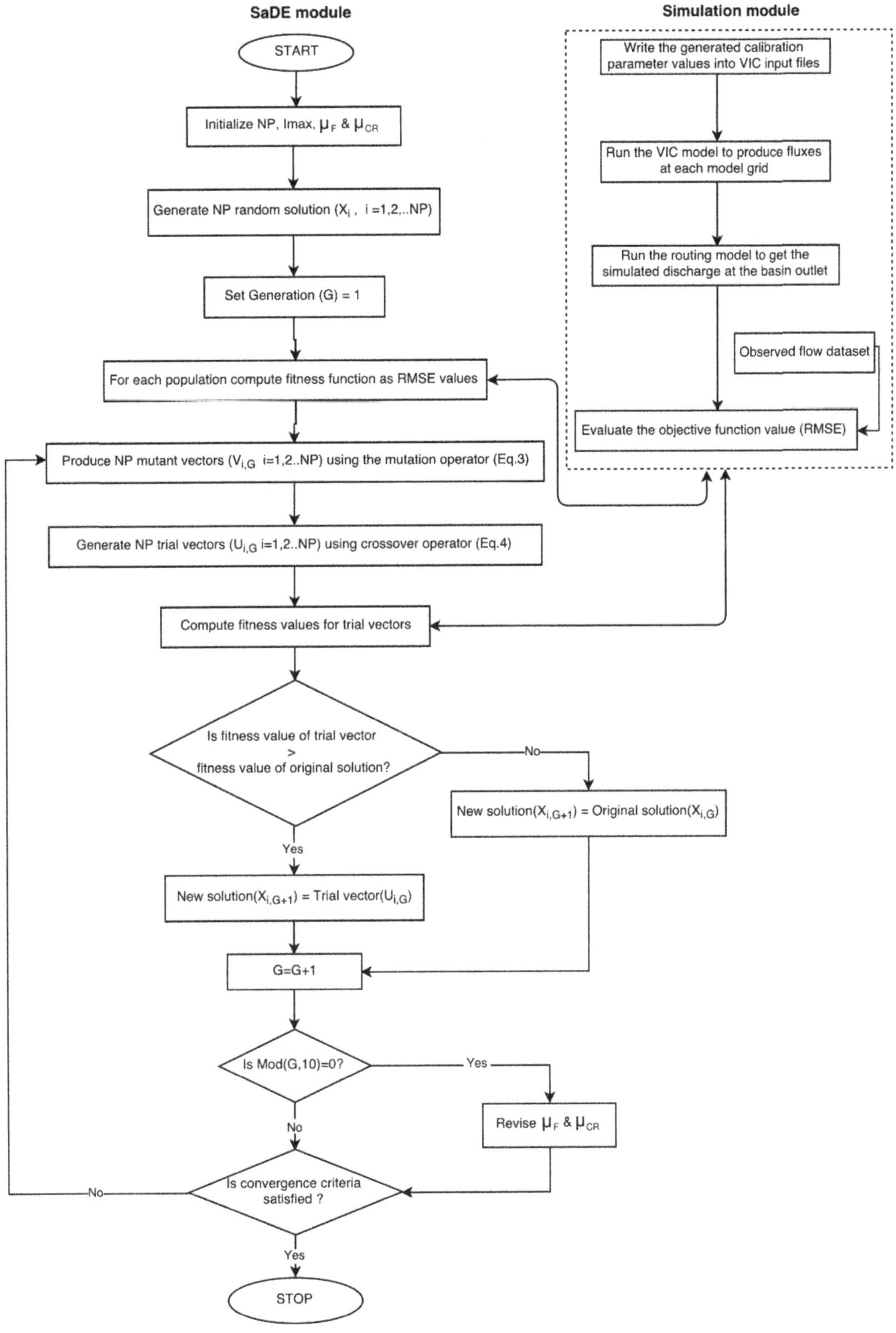

Figure 2 | Schematic layout of SaDE based automatic calibration of VIC hydrological model.

Then, the performance of the calibration algorithms is assessed by four commonly used evaluation criteria, i.e. Nash–Sutcliffe efficiency (E_{ns}), determination coefficient (R^2), percent bias ($PBIAS$), and mean of simulated flow ($\overline{Q_s}$). These E_{ns}, R^2, and $PBIAS$ are defined in Appendix B. The E_{ns} assesses the degree of fitness between simulated and observations hydrograph by a normalized statistic based on the ratio of the residual variance to observed data variance. Values of E_{ns} can range between

$-\infty$ to 1, while a value of $E_{ns} = 1$ suggests a perfect match between simulated and observed data, and a value of $E_{ns} < 0$ indicates that the observed mean would produce better accuracy than the model simulated streamflows. The R^2 is the ratio of explained variation to total variation, which reveals the strength of association between the simulated and observed data. Values of R^2 can vary between 0 and 1, while a higher value designates a better agreement. The *PBIAS* provides a view of the average tendency of under- or overprediction of the simulated discharge with respect to the observation data.

Parameter settings of the algorithms

All the algorithms used in this study need some parameter settings to be defined by the user. For the SaDE algorithm, all the parameter settings except for the population size were discussed previously. With a limited number of function evaluations, a relatively lower population size is found to be more effective than the larger population size for calibration of hydrological models by EAs (Zhang *et al.* 2009). Therefore, a population size of 30 was selected for GA, DE and SaDE algorithms. The mutation factor and crossover rate for DE were chosen as 0.5 and 0.9, respectively, following suggestions from past studies (Storn & Price 1997; Brest *et al.* 2006) and after performing some preliminary sensitivity analysis for the parameters. For the implementation of GA, the MATLAB genetic algorithm toolbox is used with crossover fraction and probability rate of mutation values set at 0.8 and 0.01, respectively, which are selected after some preliminary sensitivity analysis. In the SCE algorithm, the number of complexes (*p*) was the most important parameter. In this study, for the SCE algorithm the value of *p* is set equal to the number of calibration parameters (i.e. 13), and the other parameter values are selected following suggestions from past studies (Duan *et al.* 1992; Kuczera 1997) as well as after carrying out some preliminary sensitivity analysis.

RESULTS AND DISCUSSION

As one execution of the VIC hydrological model took an average of 65 seconds on a personal computer (PC) with a configuration of Intel Core i7-4790 3.60 GHz processor with 8 GB RAM, performing a large number of model evaluations during the calibration process is cumbersome. This is particularly true for EAs, which need to be executed for multiple trial runs. Therefore, the calibration of the VIC model is performed with a perspective of a limited number of model evaluations. The maximum number of model evaluations was fixed at 500 for each optimization run (except for checking the global convergence, where 5,000 function evaluations were permitted). Results from the proposed SaDE based calibration were compared with the results from GA, SCE and DE algorithms. This comparison consists of ten optimization runs for each optimization algorithm as all four algorithms are stochastic in nature, and therefore their comparative performance must be evaluated for several runs originating from the independent random population. The variation of the average best solution attained in ten optimization runs with the number of function evaluations was presented, which provided a comparison of average performance for the four algorithms. For further evaluation of the algorithm performance, additional tests such as stability (or variability in multiple trials), reliability, and statistical tests were also conducted in the study. It also examined the ability of the SaDE algorithm to achieve stable convergence to the global optimal solution with additional computational expenses.

Comparison of algorithms' performance

To check the ability of the algorithms to reach the global optimal solution, ten separate optimization trial runs were performed (with a computational cost of 5,000 model evaluations in each trial run) for

doi: 10.2166/h2oj.2020.030

the GA, SCE, DE, and SaDE algorithms. Each trial run of an EA required almost 4 days to execute for the current problem. Therefore, such comparisons were accomplished only after a massive computational cost, which can help in understanding the evolution skill of the SaDE and other algorithms in finding a global optimal solution, if possible, later in a restricted computational cost. The results showed that most of the time, SaDE and other EAs (i.e. GA, SCE, DE) attained optimal solutions at the end of 5,000 model evaluations. However, their performance at different stages of evolution was highly variable and dissimilar. For instance, the SaDE found its best solution after the 2,005 model evaluation, while GA and SCE took 2,982 and 4,474 model evaluations, respectively, to find the optimal solution. Even though the SaDE took 2,005 model evaluations for its complete convergence, it was noteworthy that more than 95% of its maximum fitness was achieved within 346 model evaluations. For the SCE algorithm, it took 2,632 model evaluations to attain the same 95% maximum fitness. Overall, it seems clear from these results that the SaDE algorithm can converge to a global optimal solution if an adequate computational budget allowed, and converges much faster than other EAs.

For computationally intensive complex hydrological models, the number of model evaluations required to achieve better objective values is a vital feature for choosing an optimization algorithm. With a detailed representation of a large catchment area, other physically-based hydrological models similar to the VIC model may also take a longer time for their executions. In such a case, running the model for a large number of times becomes very challenging. Therefore, preference should be given to the algorithm which is able to obtain superior objective function value with a limited computational budget (Duan *et al.* 1992; Tolson & Shoemaker 2007). In this direction, the performance of the SaDE was further assessed with limited model evaluations. To achieve this, the maximum number of model evaluations was fixed at 500 for each optimization run, which took about 8 hours of computational time. Ten such trial runs were performed for each of the GA, SCE, DE and SaDE algorithms and the plot of average best RMSE as the objective function value against the number of model evaluations is shown in Figure 3. From Figure 3, it is to be noted that the convergence speed of SaDE is faster than the other three algorithms. Initially, the SCE algorithm exhibited better convergence than the GA and DE algorithms, but finally, the GA algorithm reached lower objective function values than the SCE and DE algorithms. Overall, the SaDE algorithm consistently performed better than the other three algorithms. Even at half of the total model evaluations (250), the SaDE average objective function value (130.85) was still better than the final average objective function values (i.e. after 500 model evaluations) of GA (132.1), DE (138.25) and SCE (136.2). This shows the dominant performance of SaDE over the other algorithms, even with a smaller number of model evaluations.

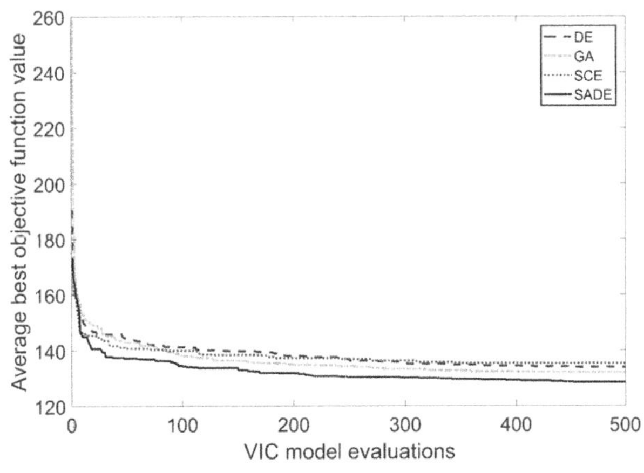

Figure 3 | Comparison of convergence characteristics of average best objective function values (RMSE) for the GA, SCE, DE and SaDE algorithms with a limited function evaluation over ten optimization trial runs.

doi: 10.2166/h2oj.2020.030

Nevertheless, the results reveal that four optimization algorithms showed diverse performance at a different number of model runs (Figure 3). Although the results confirmed that SaDE was the most efficient optimization algorithm for the calibration of the VIC model, as it had resulted in the lowest objective function value compared to the GA, SCE and DE algorithms, there was still a need to assess the significance of the results as well. Therefore, the statistically significant difference between the mean and median of the best objective function values from the four algorithms were examined using the well-known two-sample *t*-test and Wilcoxon rank-sum test, respectively, and the results are presented in Table 2. These two tests with the null hypothesis of identical mean were carried out at the 5% significance level. A *p*-value lower than the reference significance level suggested a rejection of the null hypothesis. The *p*-values obtained for the comparison of SaDE with those other algorithms were 1.2×10^{-5} (*t*-test for DE), 0.012 (*t*-test for GA), 9.7×10^{-5} (*t*-test for SCE), 1.8×10^{-4} (Wilcoxon Rank-Sum-test for DE), 0.017 (Wilcoxon Rank-Sum-test for GA), and 1.8×10^{-4} (Wilcoxon Rank-Sum-test for SCE). Therefore, results from both the two-sample t-test and the Wilcoxon rank-sum test were consistent, which suggested that the SaDE performed significantly better than the other three algorithms.

In the case of expensive model evaluation, it is of vital importance to select an optimization algorithm that can generate satisfactory solutions repeatedly, i.e. produce smaller variability in the objective function values for multiple trial runs. Smaller variability indicates greater stability of an algorithm. Therefore, a little variability with a smaller mean objective function value (RMSE) recommends a more robust algorithm for calibration. For practical calibration of the hydrological model with a lengthy simulation run, one will most likely prefer an optimization algorithm that needs only a single run for its calibration. Consequently, the algorithm that gives consistently good results is evidently a better choice than an algorithm which delivers a good solution with an equal probability to produce a poor solution. Conventionally, variability is measured by the standard deviation and the corresponding interquartile ranges. Therefore, a boxplot is used to represent the variability, which displays the lower quartile, median, upper quartile, whiskers (dashed lines spanning from top and bottom of the box for showing the stretch of the leftover data), and also the outliers with a '+' symbol. Figure 4 shows the boxplot of the best objective function values obtained from ten trial runs for each of the four algorithms. It can be noted from the figure that the SaDE algorithm is subjected to lowest variability (interquartile range) and best objective function value (lowest median) in the ten different trial runs, suggesting a robust performance of SaDE as compared to the other EAs. The GA was associated with a higher spread in the final objective function values. The DE algorithm is found to have a lower spread but higher objective function values (RMSE). The SCE algorithm was associated with higher variability and worst optimal values. Earlier successful applications of the SCE

Table 2 | The results of *p*-values obtained for statistical significance tests to assess the comparative performance of SaDE with the GA, SCE and DE algorithms via the two-sample *t*-test and Wilcoxon rank-sum tests

Statistical test	DE	GA	SCE	SaDE
Two-sample t-test				
DE	–			
GA	0.221	–		
SCE	0.153	*0.005*	–	
SaDE	*1.2×10^{-5}*	*0.012*	*9.7×10^{-5}*	–
Wilcoxon rank-sum test				
DE	–			
GA	0.140	–		
SCE	0.212	*0.045*	–	
SaDE	*1.8×10^{-4}*	*0.017*	*1.8×10^{-4}*	–

Note: the values in italics indicate statistical difference exists at the 5% significance level.

doi: 10.2166/h2oj.2020.030

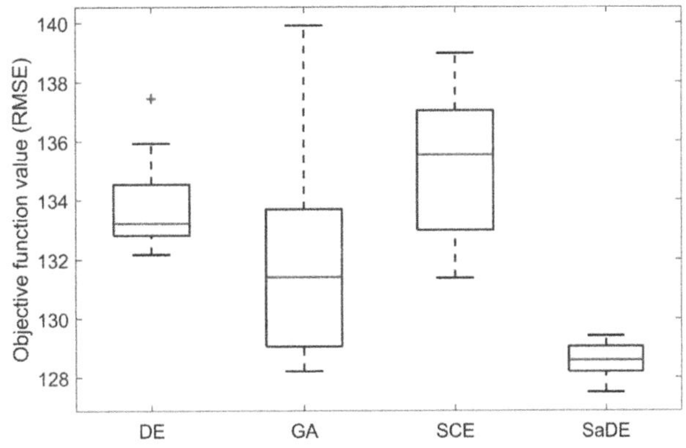

Figure 4 | Boxplot of best objective function values (RMSE) obtained using GA, SCE, DE, and SaDE algorithms for ten trial runs.

algorithm for hydrological model calibration had used very large numbers (often in excess of 10,000) of model evaluations to obtain the final solution (Tolson & Shoemaker 2007). However, with a limited number of model evaluations, the SCE algorithm could not reach the global (or near-global) optimal solution, as reported by previous studies (Tolson & Shoemaker 2007; Arsenault *et al.* 2014; Qin *et al.* 2016). Thus, restrictions on the number of model evaluations may have led to the poor performance of the SCE algorithm in the current application.

Figure 5 shows a comparison of the empirical cumulative distribution function plots between the four algorithms. The empirical cumulative distribution function (as used by Devore (2004)) for the different algorithms provides a useful comparison of the measure of the reliability of the algorithms. It is an approximation of the original cumulative distribution function and states the occurrence frequency of best objective function values below some reference value. In Figure 5, the empirical probabilities (or the occurrence frequencies) are the plotting positions attributed to the best objective function values of the four algorithms from their ten trial runs using $r/(n+1)$, where r and n stand for the rank of ordered best objective value and the number of data points respectively. Therefore, for ten trial runs, there will be ten data points, and the empirical cumulative distribution will be in the range of 0.09–0.91. The algorithm with a higher probability (or approximately vertical empirical cumulative distribution plot) at lower objective function values would be preferred in the case of the minimization

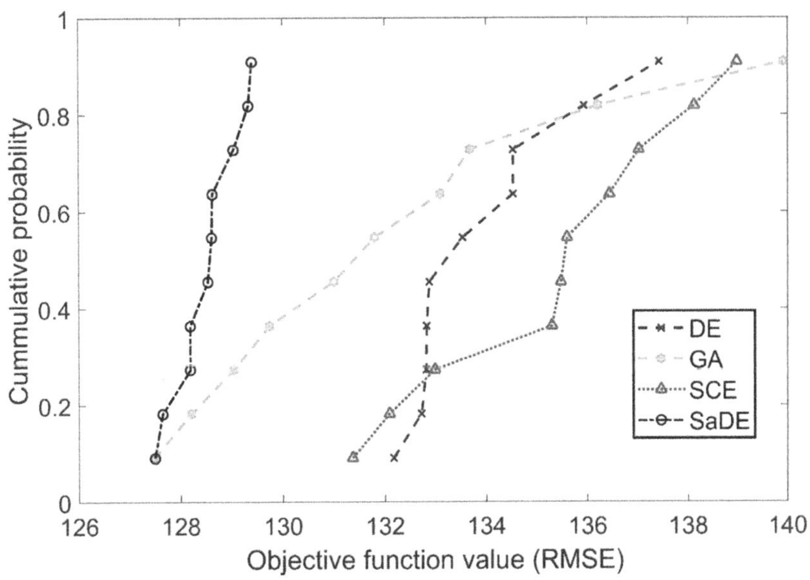

Figure 5 | Comparison of the empirical cumulative distribution for the best objective function values (RMSE) resulted from 10 trial runs using GA, SCE, DE, and SaDE algorithms.

problem. Based on such criteria, the SaDE algorithm is the best among the four algorithms, as it holds a nearly steep slope with a sense of smaller spread, and also as it occupies the position farthest left thus indicating its capability of producing smaller objective function values than other algorithms. Both the SCE and DE algorithms are positioned farther right to SaDE and have flattened out curves, suggesting that they are less reliable. It can be noted that GA starts at the farthest left but flattens out for higher cumulative probabilities, indicating that it is less reliable as it produces poor results in some trial runs.

This study also analysed the parameter equifinality (which concerns the principle that in open systems a given end state can be reached by several possible ways) of different EAs in their application to VIC model calibration. All the parameter sets within 1% of the optimal RMSE values are considered equifinal (Nemri & Kinnard 2019), see Figure 6. From the dispersion of the equifinal parameters in Figure 6, it can be noted that different parameters are associated with different degrees of equifinality.

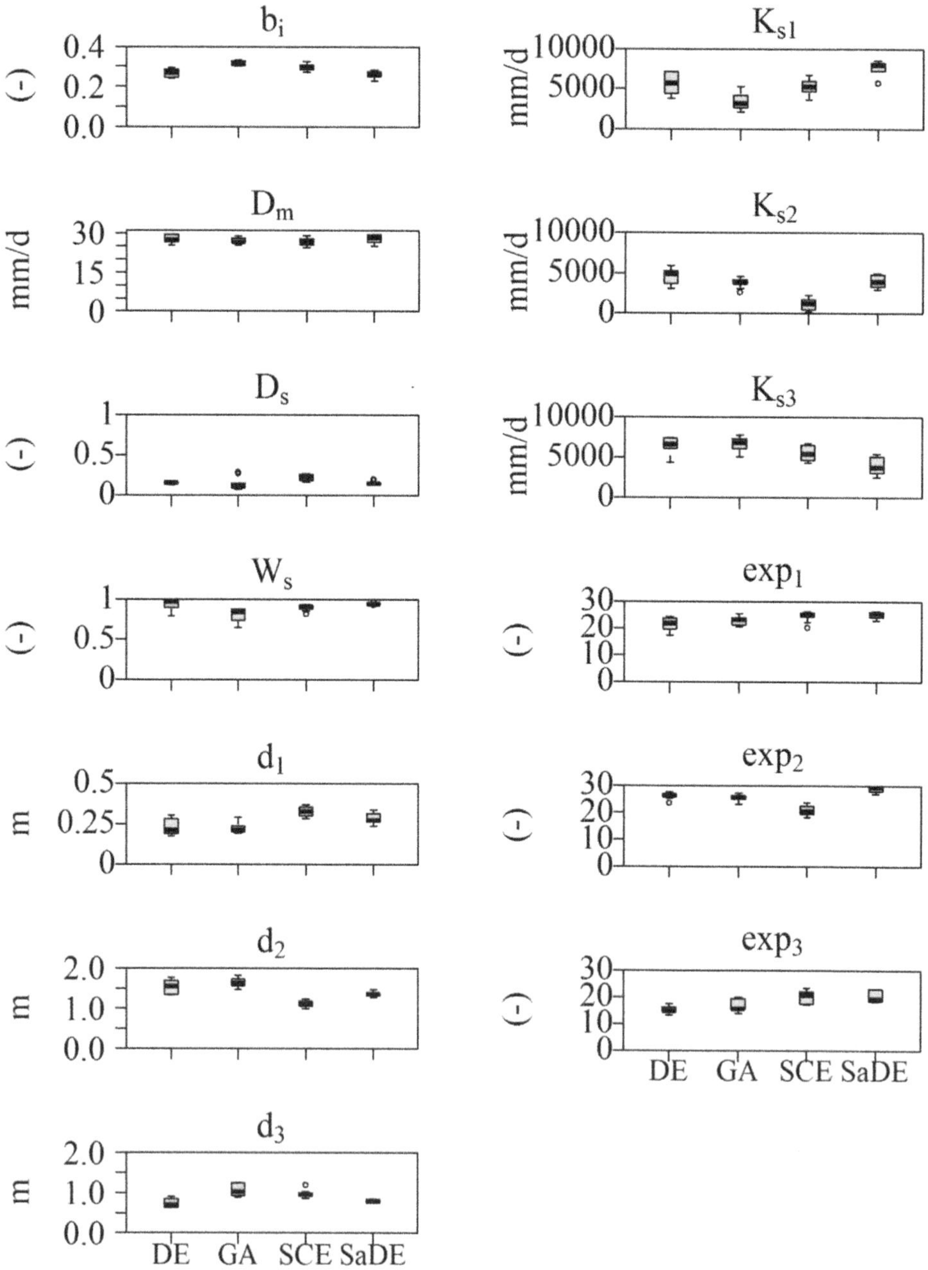

Figure 6 | Boxplots of equifinal parameters set for the VIC model calibration using GA, SCE, DE, and SaDE algorithms.

Evolutionary Algorithms in Water Resources
doi: 10.2166/h2oj.2020.030

For example, parameters such as b_i, D_m and D_s are more identifiable due to their smaller variability as compared to the parameters such as K_{s1}, K_{s2}, K_{s3}, exp_1 and exp_3. Different algorithms also have shown varying degrees of equifinality, as can be seen from Figure 6. Overall, the SaDE and GA methods have formed a more homogenous parameter distribution with reduced dispersion in their equifinal parameter sets.

Runoff estimation

The earlier assessment of the algorithmic performance was carried out in terms of efficiency, variability, and reliability with a limited number of function evaluations after examining their ability to attain an optimal solution. In this section, an effort was made to evaluate the performance of different calibration algorithms for reliable runoff simulation. Four performance measures (E_{ns}, R^2, *PBIAS*, and mean of simulated flow) were computed by adopting the best parameter values obtained from the ten optimization trial runs with 500 function evaluations. It can be seen from Table 3 that the mean daily runoff was better estimated with the SaDE algorithm during both calibration and validation periods. The SaDE overestimated the mean daily flow by 1.55% during calibration, while GA, SCE and DE overestimated it by 2.07, 5.18, and 2.59%, respectively. For validation, the GA, SCE and DE underestimated the mean daily flow by 7.35, 5.39, and 6.37%, respectively, whereas SaDE overestimated it by 1.47%. The E_{ns}, R^2, and *PBIAS* for SaDE during calibration were 0.87, 0.93, and −1.75, respectively, which were more favourable than the other three algorithms. For the validation period, E_{ns} and R^2 were nearly the same for all the algorithms, but *PBIAS* pointed towards the marginally better performance of the SaDE. Earlier studies recommended that for a satisfactory model simulation its E_{ns} must be greater 0.5, R^2 should be higher than 0.6, and *PBIAS* less than or equal to 25% (Santhi *et al.* 2001; Moriasi *et al.* 2007). Based on the recommendations from past studies, it can be inferred that all the algorithms performed satisfactorily in streamflow simulation, although the overall performance of the SaDE was comparatively better than the other three algorithms.

Table 3 | Statistical performance measures for the simulated streamflow with the best DE, GA, SCE, and SaDE estimated parameters of the VIC model

Simulation period	Algorithm	$\overline{Q_s}$ (m³/s)	E_{ns}	R^2	*PBIAS*
Calibration (mean observed flow = 193 m³/s)	GA	197	0.86	0.92	−2.10
	SCE	198	0.82	0.90	−2.63
	DE	203	0.82	0.91	−5.15
	SaDE	*196*	*0.87*	*0.93*	*−1.75*
Validation (mean observed flow = 204 m³/s)	GA	189	0.86	0.92	7.28
	SCE	191	0.86	0.93	2.46
	DE	193	0.85	0.93	5.32
	SaDE	*207*	0.86	0.93	*−1.51*

Note: Values in italics signify best among all the algorithms.

The calibrated values of the VIC model parameters (using the SaDE algorithm) are given in Table 4. These parameter values were obtained from the best solution of the SaDE algorithm at the end of 500 evaluations from ten optimization trial runs. The VIC model simulated flows using those calibrated parameter values is compared with the observed flows in Figure 7. It is clear from the figure that most of the time the simulated flows closely match the observed flows. Different streamflow characteristics are generally represented by the different flow signals (i.e. high flow, low flow, and medium flow) from the hydrograph. Therefore, to obtain a comprehensive idea about the different streamflow characteristics, the effect of SaDE based automatic calibration on different flow signals was analysed.

Table 4 | Details of the VIC model parameters selected from calibration using SaDE

Parameter	Optimal value	Parameter	Optimal value
b_i	0.24	Ks_1	5,936.10
D_m	30.00	Ks_2	3,239.10
D_s	0.16	Ks_3	3,403.70
W_s	0.95	exp_1	21.58
d_1	0.28	exp_2	29.72
d_2	1.47	exp_3	16.46
d_3	0.84		

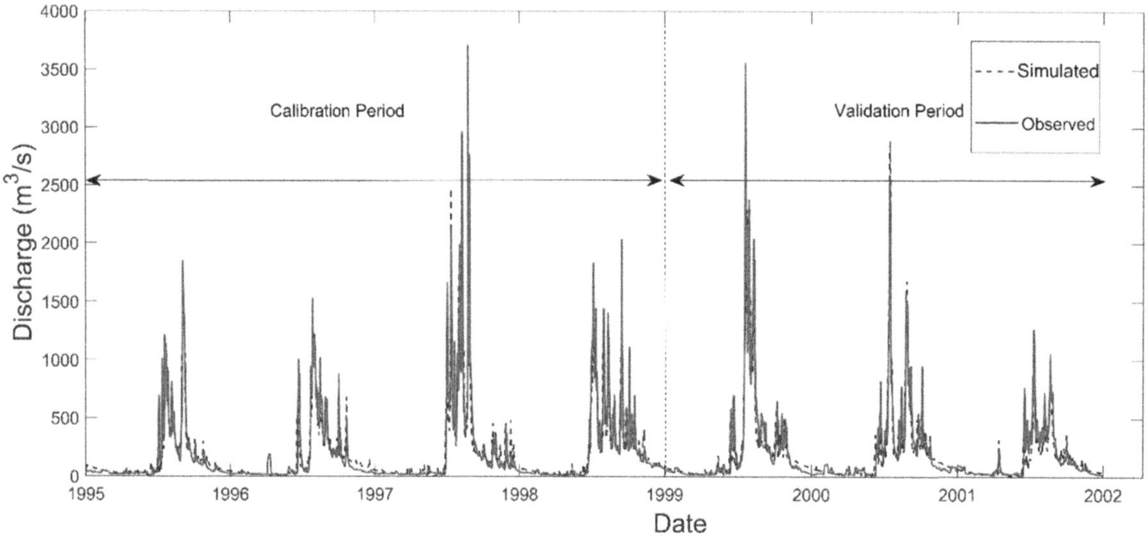

Figure 7 | Comparison of SaDE calibrated VIC model simulated flows with observed flows during calibration and validation periods.

Based on the exceedance probability, the flow signals were categorised into the low flow (exceedance probability $\geq$90%), medium flow (10% < exceedance probability <90%), and high flow (exceedance probability $\leq$10%) signals. Figure 8 shows the comparison of VIC model simulated flows (with calibration parameters obtained using the SaDE) with the observed flows for different flow signals. It is clear from Figure 8(a) that the daily observed and simulated flows during low flow periods were very poorly associated ($R^2 = 0.14$) as the model tends to overestimate the majority of low flow signals.

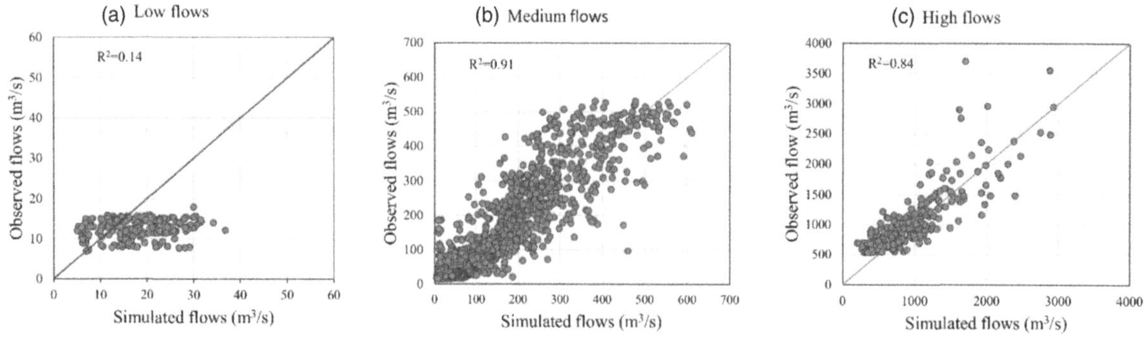

Figure 8 | Scatter plots of observed vs simulated flows from the SaDE calibrated VIC model, for (a) low flows, (b) medium flows, and (c) high flows.

In contrast, for the medium flows, the agreement between the simulated and observed flows was high ($R^2 = 0.91$), but both had some over- and underestimations (Figure 8(b). The simulated high flows were also predicted with good accuracy ($R^2 = 0.84$) along with a few underestimations (Figure 8(c). Overall, the VIC model simulations for high and medium flows showed good agreement with the observed flows, while the low flows were poorly estimated by the VIC model.

To further investigate the behaviour of the modelled flow values at different flow levels, performance measures of the modelled streamflow were computed after partitioning the streamflow series into wet and dry season flows. The months of June–November, where the observed monthly flows had exceedance probability less than 50%, were considered as the wet season, and the rest of the months were considered as the dry season. The wet season contained more than 91% of the annual average streamflow in the study area. Table 5 represents the comparison of three performance measures (E_{ns}, R^2, and *PBIAS*) for wet and dry seasons. The wet season exhibited a similar performance to the whole time period (Table 3) in terms of E_{ns} and R^2 during both calibration and validation periods. In Table 5, small positive biases were found for the wet season. In contrast, a large negative bias was found for calibration during the dry season. Overall, the performance measured during the dry season was very poor compared to those for the wet season and also for the entire period. Similar low performances of the estimated streamflows during dry periods (even after the rigorous model calibration) were reported in several past studies (Gan *et al.* 1997; Croke *et al.* 2004; Kim 2015). However, the limitation of accurate streamflow estimation in dry conditions may not be a consequence of the particular choice of a calibration algorithm, rather it may be linked to the inadequate model structure and parameterization of the hydrological model, selection of calibration data length, and in many instances due to the choice of the objective function (Kim 2015).

Table 5 | Performance of the SaDE based calibrated streamflows for dry and wet periods during VIC model calibration and validation

	Calibration		**Validation**	
	Wet period	**Dry period**	**Wet period**	**Dry period**
E_{ns}	0.87	0.09	0.88	–0.21
R^2	0.93	0.63	0.94	0.43
PBIAS	3.12	–24.99	5.02	–1.84

CONCLUSIONS

Hydrological models mimic the behaviour of hydrological processes in catchments, and their reliable results help in gaining enhanced knowledge of various environmental processes. Calibration of a physically-based hydrological model is a challenging task because of the large number of unknown parameters, equifinality of parameter sets, and the computational cost for their simulation that restricts the number of model runs that are required for the calibration. In this study, the SaDE algorithm is presented for the automatic calibration of a physically-based VIC hydrological model in the upper Tungabhadra river basin of India. The SaDE eliminates the need for fine-tuning of control parameters (i.e. mutation factor F and crossover rate CR) of DE by implementing an automatic self-adapting process, which helps in saving a considerable amount of time and computational budget in finding optimal parameters. Further, apart from SaDE, three more evolutionary algorithms, namely GA, SCE, and DE algorithms, are applied for calibration of VIC model parameters, and the results of SaDE are compared with those from the GA, SCE and DE algorithms. The performance of algorithms was evaluated using standard performance measures subject to a limited computational budget (or a number of function evaluations) and assessed the usefulness of the SaDE approach.

Analysis from the results of the study revealed that the SaDE is a powerful tool for hydrological model calibration and it can produce significantly better performance than the other three algorithms. Here, the main advantage of the SaDE method is that it eliminates the tedious and computationally demanding task of parameter fine-tuning. The empirical CDF plots prepared for reliability assessment of the algorithm clearly indicated the advantage of the SaDE over the other algorithms during the calibration and validation of the VIC model. Moreover, the SaDE algorithm leads to the smallest standard deviation and interquartile range (evaluated based on different trial runs of fitness/objective values), which reveals its consistent precedence over the other three algorithms. Although the problem of 'equifinality' is found in the model calibration, it can be expected that the equifinality in parameter space will be less pronounced when using the SaDE method. The results of this study also demonstrate the superior performance of SaDE in the case of a limited number of model evaluations. If an adequate number of function evaluations are allowed, then SaDE is found to achieve global optimal solutions like other EAs, but it achieves much faster than the GA and SCE algorithms. A lesser number of function evaluations is a major advantage in hydrological modelling applications where time and computational constraints prevent higher model evaluations and recurrent runs of the algorithm. Results from this study also showed that SaDE produces a slightly better performance overall (in terms of E_{NS}, R^2, and $PBIAS$ statistics) during both model calibration and validation periods for streamflow simulation.

Thus, from the results of the study, it can be inferred that obtaining 'good' solutions with the SaDE is straightforward and efficient since it does not require a prior parametric study and it is also computationally inexpensive compared to the other EAs. Moreover, the SaDE algorithm is very simple and so it can be easily implemented in any programming language consistent with the operating environment of the model subjected to calibration. Therefore, the study recommends SaDE as a potential tool for calibration of complex hydrological models.

CONFLICT OF INTEREST

The authors declare that there is no conflict of interest.

ACKNOWLEDGEMENTS

The authors would like to thank the various organizations that provided support for the study: Department of Science and Technology (SPLICE–Climate Change Programme), Government of India, Project #DST/488/CCP/CoE/140/2018 for funding support; the Indian Meteorological Department (IMD), National Remote Sensing Centre (NRSC) of India, India-WRIS, Central Water Commission (CWC) of India, National Center for Environmental Prediction (NCEP), Food and Agricultural Organization (FAO), and the MODIS science and development team of NASA for providing various datasets used in this study.

DATA AVAILABILITY STATEMENT

Data cannot be made publicly available; readers should contact the corresponding author for details.

REFERENCES

Arnell, N. W. 1999 The effect of climate change on hydrological regimes in Europe: a continental perspective. *Glob. Environ. Chang.* **9** (1), 5–23. https://doi.org/10.1016/S0959-3780(98)00015-6.

Arsenault, R., Poulin, A. & Brissette, F. 2014 Comparison of stochastic optimization algorithms in hydrological model calibration. *J. Hydrol. Eng.* **19** (7), 1374–1384. https://doi.org/10.1061/(ASCE)HE.1943-5584.0000938.

Beven, K. & Freer, J. 2001 Equifinality, data assimilation, and uncertainty estimation in mechanistic modelling of complex environmental systems using the GLUE methodology. *J. Hydrol.* **249** (1–4), 11–29. https://doi.org/10.1016/S0022-1694(01)00421-8.

Brest, J., Greiner, S., Boskovic, B., Mernik, M. & Zumer, V. 2006 Self-adapting control parameters in differential evolution: a comparative study on numerical benchmark problems. *IEEE Trans. Evol. Comput.* **10** (6), 646–657. https://doi.org/10.1109/TEVC.2006.872133.

Cai, X., McKinney, D. C. & Lasdon, L. S. 2003 Integrated hydrologic-agronomic-economic model for river basin management. *J. Water Resour. Plan. Manag.* **129** (1), 4–17. https://doi.org/10.1061/(ASCE)0733-9496(2003)129:1(4).

Chen, Y., Li, J. & Xu, H. 2016 Improving flood forecasting capability of physically based distributed hydrological models by parameter optimization. *Hydrol. Earth Syst. Sci.* **20** (1), 375–392. https://doi.org/10.5194/hess-20-375-2016.

Croke, B. F., Merritt, W. S. & Jakeman, A. J. 2004 A dynamic model for predicting hydrologic response to land cover changes in gauged and ungauged catchments. *J. Hydrol.* **291** (1–2), 115–131. https://doi.org/10.1016/j.jhydrol.2003.12.012.

Das, S., Mullick, S. S. & Suganthan, P. N. 2016 Recent advances in differential evolution–an updated survey. *Swarm. Evol. Comput.* **1** (27), 1–30. https://doi.org/10.1016/j.swevo.2016.01.004.

Demaria, E. M., Nijssen, B. & Wagener, T. 2007 Monte Carlo sensitivity analysis of land surface parameters using the Variable Infiltration Capacity model. *J. Geophys. Res. Atmos.* **16** (112), D11113. https://doi.org/10.1029/2006JD007534.

Devore, J. L. 2004 *Probability and Statistics for Engineering and the Sciences*, 6th edn. Cengage Learning, Ontario.

Duan, Q., Sorooshian, S. & Gupta, V. 1992 Effective and efficient global optimization for conceptual rainfall-runoff models. *Water Resour. Res.* **28** (4), 1015–1031. https://doi.org/10.1029/91WR02985.

Franchini, M. & Pacciani, M. 1991 Comparative-analysis of several conceptual rainfall runoff models. *J. Hydrol.* **122** (1–4), 161–219. https://doi.org/10.1016/0022-1694(91)90178-K.

Gan, T. Y. & Biftu, G. F. 1996 Automatic calibration of conceptual rainfall-runoff models: optimization algorithms, catchment conditions, and model structure. *Water Resour. Res.* **32** (12), 3513–3524. https://doi.org/10.1029/95WR02195.

Gan, T. Y., Dlamini, E. M. & Biftu, G. F. 1997 Effects of model complexity and structure, data quality, and objective functions on hydrologic modelling. *J. Hydrol.* **192** (1–4), 81–103. https://doi.org/10.1016/S0022-1694(96)03114-9.

Goldberg, D. E. 1989 *Genetic Algorithms in Search, Optimization, and Machine Learning.* Addison-Wesley, Reading, MA.

Guo, J., Zhou, J., Lu, J., Zou, Q., Zhang, H. & Bi, S. 2014 Multi-objective optimization of empirical hydrological model for streamflow prediction. *J. Hydrol.* **511**, 242–253. https://doi.org/10.1016/j.jhydrol.2014.01.047.

Ibbitt, R. P. & O'Donnell, T. 1971 Fitting methods for conceptual catchment models. *J. Hydraul. Div.* **97** (9), 1331–1342.

Kamali, B., Mousavi, S. J. & Abbaspour, K. C. 2013 Automatic calibration of HEC-HMS using single-objective and multi-objective PSO algorithms. *Hydrol. Process.* **27** (26), 4028–4042. https://doi.org/10.1002/hyp.9510.

Kavetski, D. & Clark, M. P. 2010 Ancient numerical daemons of conceptual hydrological modeling: 2. Impact of time stepping schemes on model analysis and prediction. *Water Resour. Res.* **46** (10), W10511. https://doi.org/10.1029/2009WR008896.

Kim, H. S. 2015 Effectiveness of a regional model calibrated to different parts of a flow regime in regionalisation. *Hydrol. Earth Syst. Sci. Disc.* **12** (7), 7057–7098. https://doi.org/10.5194/hessd-12-7057-2015.

Kim, S. M., Benham, B. L., Brannan, K. M., Zeckoski, R. W. & Doherty, J. 2007 Comparison of hydrologic calibration of HSPF using automatic and manual methods. *Water Resour. Res.* **43** (1), W01402. https://doi.org/10.1029/2006WR004883.

Kuczera, G. 1997 Efficient subspace probabilistic parameter optimization for catchment models. *Water Resour. Res.* **33** (1), 177–185. https://doi.org/10.1029/96WR02671.

Liang, X., Lettenmaier, D. P., Wood, E. F. & Burges, S. J. 1994 A simple hydrologically based model of land surface water and energy fluxes for general circulation models. *J. Geophys. Res.* **99** (D7), 14415–14428. https://doi.org/10.1029/94JD00483.

Liong, S. Y., Chan, W. T. & ShreeRam, J. 1995 Peak-flow forecasting with genetic algorithm and SWMM. *J. Hydraul. Eng.* **121** (8), 613–617. https://doi.org/10.1061/(ASCE)0733-9429(1995)121:8(613).

Madsen, H. 2000 Automatic calibration of a conceptual rainfall–runoff model using multiple objectives. *J. Hydrol.* **235** (3), 276–288. https://doi.org/10.1016/S0022-1694(00)00279-1.

Moriasi, D. N., Arnold, J. G., Van-Liew, M. W., Bingner, R. L., Harmel, R. D. & Veith, T. L. 2007 Model evaluation guidelines for systematic quantification of accuracy in watershed simulations. *Trans. ASABE* **50** (3), 885–900. https://doi.org/10.13031/2013.23153.

Nemri, S. & Kinnard, C. 2019 Comparing calibration strategies of a conceptual snow hydrology model and their impact on model performance and parameter identifiability. *J. Hydrol.* **582**, 124474. https://doi.org/10.1016/j.jhydrol.2019.124474.

Nijssen, B., Lettenmaier, D. P., Liang, X., Wetzel, S. W. & Wood, E. F. 1997 Streamflow simulation for continental-scale river basins. *Water Resour. Res.* **33** (4), 711–724. https://doi.org/10.1029/96WR03517.

Osuch, M., Romanowicz, R. J. & Booij, M. J. 2015 The influence of parametric uncertainty on the relationships between HBV model parameters and climatic characteristics. *Hydrol. Sci.* **60** (7–8), 1165–1173. https://doi.org/10.1080/02626667.2014.967694.

Pulluri, H., Naresh, H. R. & Sharma, V. 2017 An enhanced self-adaptive differential evolution based solution methodology for multiobjective optimal power flow. *Appl. Soft Comput.* **54**, 229–245. https://doi.org/10.1016/j.asoc.2017.01.030.

Qin, A. K., Huang, V. L. & Suganthan, P. N. 2009 Differential evolution algorithm with strategy adaption for global numerical optimization. *IEEE Trans. Evol. Comp.* **13**, 398–417.

Qin, Y., Kuczera, G. & Kavetski, D. 2016 Comparison of Newton-type and SCE optimisation algorithms for the calibration of conceptual hydrological models. *Australas. J. Water Resour.* **20** (2), 169–176. https://doi.org/10.1080/13241583.2017.1298180.

Reddy, M. J. & Kumar, D. N. 2007 Multiobjective differential evolution with application to reservoir system optimization. *J. Comput. Civ. Eng.* **21** (2), 136–146. https://doi.org/10.1061/(ASCE)0887-3801(2007)21:2(136).

Reddy, M. J. & Kumar, D. N. 2012 Computational algorithms inspired by biological processes and evolution. *Curr. Sci.* **103** (4), 1–11.

Refsgaard, J. C., Havnø, K., Ammentorp, H. C. & Verwey, A. 1988 Application of hydrological models for flood forecasting and flood control in India and Bangladesh. *Adv. Water Resour.* **11** (2), 101–105. https://doi.org/10.1016/0309-1708(88)90043-7.

Santhi, C., Arnold, J. G., Williams, J. R., Dugas, W. A., Srinivasan, R. & Hauck, L. M. 2001 Validation of the SWAT model on a large river basin with point and nonpoint sources. *J. Am. Water Resour. Assoc.* **37** (5), 1169–1188. https://doi.org/10.1111/j.1752-1688.2001.tb03630.x.

Schardong, A. & Simonovic, S. P. 2015 Coupled self-adaptive multiobjective differential evolution and network flow algorithm approach for optimal reservoir operation. *J. Water Resour. Plann. Manag.* **141** (10), 4015015. https://doi.org/10.1061/(ASCE)WR.1943-5452.0000525.

Scheerlinck, K., Pauwels, V., Vernieuwe, H. & Baets, B. D. 2009 Calibration of a water and energy balance model: recursive parameter estimation versus particle swarm optimization. *Water Resour. Res.* **45** (10), W10422. https://doi.org/10.1029/2009WR008051.

Schumann, G. J. P., Neal, J. C., Voisin, N., Andreadis, K. M., Pappenberger, F., Phanthuwongpakdee, N., Hall, A. C. & Bates, P. D. 2013 A first large-scale flood inundation forecasting model. *Water Resour. Res.* **49** (10), 6248–6257. https://doi.org/10.1002/wrcr.20521.

Sheffield, J. & Wood, E. F. 2008 Projected changes in drought occurrence under future global warming from multi-model, multi-scenario, IPCC AR4 simulations. *Clim. Dyn.* **31** (1), 79–105. https://doi.org/10.1007/s00382-007-0340-z.

Sirsant, S. & Reddy, M. J. 2018 Reliability-based design of water distribution networks using self-adaptive differential evolution algorithm. *ISH J. Hydraul. Eng.* **24** (2), 198–212. https://doi.org/10.1080/09715010.2017.1408038.

Sorooshian, S. & Gupta, V. K. 1983 Automatic calibration of conceptual rainfall-runoff models: the question of parameter observability and uniqueness. *Water Resour. Res.* **19** (1), 260–268. https://doi.org/10.1029/WR019i001p00260.

Sorooshian, S., Duan, Q. & Gupta, V. K. 1993 Calibration of rainfall-runoff models: application of global optimization to the Sacramento Soil Moisture Accounting Model. *Water Resour. Res.* **29** (4), 1185–1194. https://doi.org/10.1029/92WR02617.

Storn, R. & Price, K. 1997 Differential evolution-a simple and efficient heuristic for global optimization over continuous spaces. *J. Glob. Optim.* **11** (4), 341–359. https://doi.org/10.1023/A:1008202821328.

Todini, E. 1996 The ARNO rainfall – runoff model. *J. Hydrol.* **175** (1–4), 339–382. https://doi.org/10.1016/S0022-1694(96)80016-3.

Tolson, B. A. & Shoemaker, C. A. 2007 Cannonsville reservoir watershed SWAT2000 model development, calibration and validation. *J. Hydrol.* **337** (1), 68–86. https://doi.org/10.1016/j.jhydrol.2007.01.017.

Vrugt, J. A., Gupta, H. V., Bouten, W. & Sorooshian, S. 2003 A shuffled complex evolution metropolis algorithm for optimization and uncertainty assessment of hydrologic model parameters. *Water Resour. Res.* **39** (8), 1201.

Vrugt, J. A., Robinson, B. A. & Hyman, J. M. 2009 Self-adaptive multimethod search for global optimization in real-parameter spaces. *IEEE Trans. Evol. Comput.* **13** (2), 243–259. https://doi.org/10.1109/TEVC.2008.924428.

Wang, Q. J. 1991 The genetic algorithm and its application to calibrating conceptual rainfall-runoff models. *Water Resour. Res.* **27** (9), 2467–2471. https://doi.org/10.1029/91WR01305.

Xie, Z., Yuan, F., Duan, Q., Zheng, J., Liang, M. & Chen, F. 2007 Regional parameter estimation of the VIC land surface model: methodology and application to river basins in China. *J. Hydrometeorol.* **8** (3), 447–468. https://doi.org/10.1175/JHM568.1.

Zhang, X., Srinivasan, R., Zhao, K. & Liew, M. V. 2009 Evaluation of global optimization algorithms for parameter calibration of a computationally intensive hydrologic model. *Hydrol. Process.* **23** (3), 430–441. https://doi.org/10.1002/hyp.7152.

Zheng, F., Zecchin, A. C. & Simpson, A. R. 2012 Self-adaptive differential evolution algorithm applied to water distribution system optimization. *J. Comput. Civ. Eng.* **27** (2), 148–158. https://doi.org/10.1061/(ASCE)CP.1943-5487.0000208.

First received 10 April 2020; accepted in revised form 21 June 2020. Available online 8 August 2020

Evolutionary Algorithms in Water Resources
doi: 10.2166/h2oj.2020.035

Quantifying natural organic matter concentration in water from climatological parameters using different machine learning algorithms

Sina Moradi[a,b], Anthony Agostino[a], Ziba Gandomkar[c], Seokhyeon Kim[d], Lisa Hamilton[e], Ashish Sharma[d], Rita Henderson[a] and Greg Leslie[b,*]

[a] Algae & Organic Matter Laboratory, School of Chemical Engineering, University of New South Wales, Sydney 2052, Australia

[b] UNESCO Centre for Membrane Science & Technology, School of Chemical Engineering, University of New South Wales, Sydney 2052, Australia

[c] Discipline of Medical Imaging Sciences, Faculty of Medicine and Health, University of Sydney, Sydney 2006, Australia

[d] School of Civil and Environmental Engineering, University of New South Wales, Sydney 2052, Australia

[e] Water and Catchment Protection, WaterNSW, Sydney 2150, Australia

*Corresponding author. E-mail: g.leslie@unsw.edu.au

Abstract

The present understanding of how changes in climate conditions will impact the flux of natural organic matter (NOM) from the terrestrial to aquatic environments and thus aquatic dissolved organic carbon (DOC) concentrations is limited. In this study, three machine learning algorithms were used to predict variations in DOC concentrations in an Australian drinking water catchment as a function of climate, catchment and physical water quality data. Four independent variables including precipitation, temperature, leaf area index and turbidity ($n = 5{,}540$) were selected from a large dataset to develop and train each machine learning model. The accuracy of the multivariable linear regression, support vector regression (SVR) and Gaussian process regression algorithms with different kernel functions was determined using adjusted R-squared (adj. R^2), root-mean-squared error (RMSE) and mean absolute error (MAE). Model accuracy was very sensitive to the time interval used to average climate observations prior to pairing with DOC observations. The SVR model with a quadratic kernel function and a 12-day time interval between climate and water quality observations outperformed the other machine learning algorithms (adj. $R^2 = 0.71$, RMSE = 1.9, MAE = 1.35). The area under the receiver operating characteristic curve method (AUC) confirmed that the SVR model could predict 92% of the elevated DOC observations; however, it was not possible to estimate DOC values at specific sampling sites in the catchment, probably due to the complex local geological and hydrological changes in the sites that directly surround and feed each sampling point. Further research is required to establish potential relationships between climatological data and NOM concentration in other water catchments – especially in the face of a changing climate.

Key words: climate, machine learning, natural organic matter, water quality

Highlights

- The application of different machine learning algorithms to quantify NOM in an Australian catchment.
- Investigate potential relationships between climatological factors and water quality parameters.

doi: 10.2166/h2oj.2020.035

- The estimation of DOC concentration of water from simple measurable water quality and climatological variables.
- The detection of high DOC concentration events using the machine learning technique.

Graphical Abstract

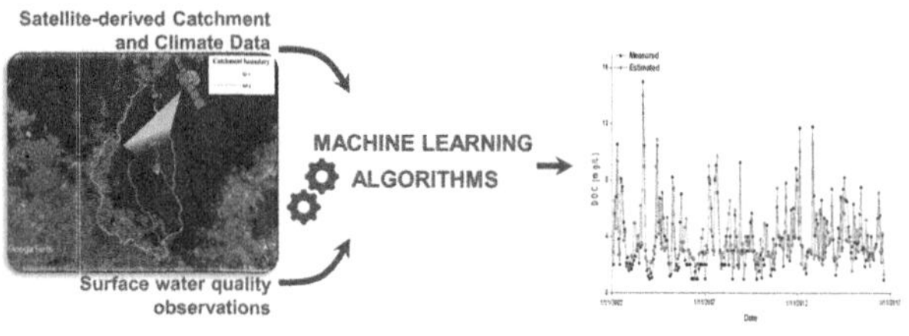

INTRODUCTION

A warming climate is driving a suite of changes in global surface waters, from nutrient, temperature and sediment profiles in Chesapeake Bay, USA (Najjar *et al.* 2010) to the timing and intensity of monsoons in Southeast Asia (Loo *et al.* 2015). Higher temperatures impact water quality in lakes by reducing oxygen concentrations, increasing the release of phosphorus from sediments (Arnell *et al.* 2015) and increasing the concentration of dissolved organic matter (El-Jabi *et al.* 2014). Intense, but less frequent rainfall events will increase sediment loads while decreasing water security in drinking water catchments. These changes will have profound effects on the potable water supplies of current and future generations (Hansen *et al.* 2013).

The character and composition of natural organic matter (NOM) in freshwater ecosystems also appears to respond to variations in climatic conditions (Hejzlar *et al.* 2003; Delpla *et al.* 2009; Zhu *et al.* 2017; Gavin *et al.* 2018; Parr *et al.* 2019). This trend is also observed in Australian catchments with deleterious consequences for the performance of existing treatment infrastructure on more highly variable potable water supplies (Mohiuddin *et al.* 2014). Observed increases in dissolved organic carbon (DOC) concentrations have been linked to changes in climatic conditions (Tranvik & Jansson 2002; Freeman *et al.* 2004; Evans *et al.* 2006). Potential climatic drivers of the upward trends in DOC concentrations have included temperature (Evans *et al.* 2006), soil moisture (SM) (Hudson *et al.* 2003) and precipitation (Erlandsson *et al.* 2008). However, the present understanding of how changes in climatic conditions will continue to impact the flux of NOM from the terrestrial to aquatic environments and thus aquatic DOC concentrations is limited. Hence, establishing potential relationships between climatological data and NOM concentration in water catchment will be critical in planning for the delivery of potable water in a warmer climate.

The link between climate, catchment and water quality is dynamic, complex and with some exceptions not well understood. The expanded use of satellite-derived observations and increases in the type and frequency of water quality measurements has created more data but not necessarily better outcomes for the management of water quality. Consequently, there is an opportunity to apply advanced data-driven modelling techniques to catalogue and interrogate these complex datasets to identify the trends and interdependencies between the variables, and to perform dimensionality reduction of the variables (Lary *et al.* 2016; Ruescas *et al.* 2018). Supervised data-driven machine learning algorithms can potentially map DOC concentration in a catchment to climatological data without attempting to accurately model underlying processes. Machine learning methods have been widely used and achieved promising results in different environmental settings (Kim *et al.* 2014; Park *et al.* 2015; Granata *et al.* 2017; Ruescas *et al.* 2017), with the dataset size varying depending on the type of observation from $n = 63$ (Kim *et al.* 2014) to $n = 2{,}255$ (Park *et al.* 2015) for analytical measurements and $n > 20{,}000$ for remote sensing applications (Ruescas *et al.* 2017). The application of data-driven modelling techniques

doi: 10.2166/h2oj.2020.035

to estimate DOC concentrations has been demonstrated in other studies (Clair *et al.* 1999; Snauffer *et al.* 2018). For example, a neural network model was developed to estimate DOC concentration using hydrological, and climatic variables such as temperature, and total precipitation (Clair *et al.* 1999). In another application, the concentration of coloured dissolved organic matter was determined from remote sensing signals by comparing different machine learning regression approaches such as Gaussian process regression (GPR), support vector regression (SVR), random forest and kernel ridge regression (Ruescas *et al.* 2017, 2018).

The low computational cost and interpretability makes the multivariate linear regression (MLR) algorithm desirable as there are clear trade-offs between model complexity and interpretability. In many cases, particularly where limited data are available, more complex algorithms can achieve high levels of generalisation and prediction accuracy (Khalil *et al.* 2005). Among non-linear approaches, SVR is a nonparametric (i.e. not limited by a functional form) supervised learning algorithm that simultaneously can minimise prediction error and model complexity (Vapnik 2000). In contrast to linear and non-linear machine learning algorithms that are trained from exact values for every parameter in a function, GPR uses a (Gaussian) probability distribution over all possible values that fit the data. Exponential kernel function and squared exponential are widely used choices for covariance kernel functions (Jiang *et al.* 2019). GPR has been applied in regression problems in different environmental fields of studies including prediction of stream water temperature (Grbić *et al.* 2013), groundwater salinity (Lal & Datta 2018) or streamflow forecasting (Sun *et al.* 2014). However, the approach has not been used to explore correlations between DOC and catchment and climate variables.

The prospective application of an appropriate machine learning algorithm to a sufficiently large dataset may enable water utilities to predict what climate conditions are conducive to excursions in NOM in water catchments. The first challenge is to select the appropriate machine learning algorithm and develop a sufficiently large dataset to train and evaluate the performance of each machine learning algorithm. In most cases, the data required can be sourced from routinely collected water quality data, which are often fragmented by water utilities and catchment management authorities and temporally disconnected. An ever-growing library of climatological data is freely available from multiple weather and geographic satellite sources. However, as surface water is influenced by climate conditions through the transfer of momentum and matter, a time gap exists between climatological and water quality data. Moreover, as the catchment area increases, the lag time (here the optimal lagged day) increases (Rostami *et al.* 2018). Therefore, the second challenge is to explore the effects of different statistical pairing of observed DOC concentrations with averaged climatological variables from multiple time steps prior to the water quality observation. Thus, by combining and harmonising temporal differences between fragmented datasets, and training and validating an appropriate machine learning algorithm, it may be possible to identify potential relationships between the climatological and water quality data.

The objective of this paper was to evaluate three commonly used machine learning algorithms, namely MLR, SVR and GPR for the estimation of DOC concentration of water from simple measurable water quality and climatological variables using datasets from an Australian catchment. The secondary aim of this study was to establish new knowledge about the potential relationships between climatological factors and water quality parameters by using the exhaustive feature selection technique. This may enable water utilities to minimise the impact of changes in climate conditions on treatment plant performance and potable water quality.

METHODOLOGY

Study area

This study was conducted in the Nepean catchment in Australia (386 km^2) over a spatial domain of 33.81°S to 35.09°S and 149.97°E to 151.16°E with an elevation of 130–720 m above sea level

doi: 10.2166/h2oj.2020.035

(Figure 1), The dominant climate class is temperate, characterised by no dry season, a warm summer and 800–1,600 mm of annual rainfall. The major land cover are trees (82%) with the remaining areas used for rural residential and agricultural purposes, including pasture and cropping (Peel *et al.* 2007b; Sixsmith *et al.* 2015; Office of Environment and Heritage 2017).

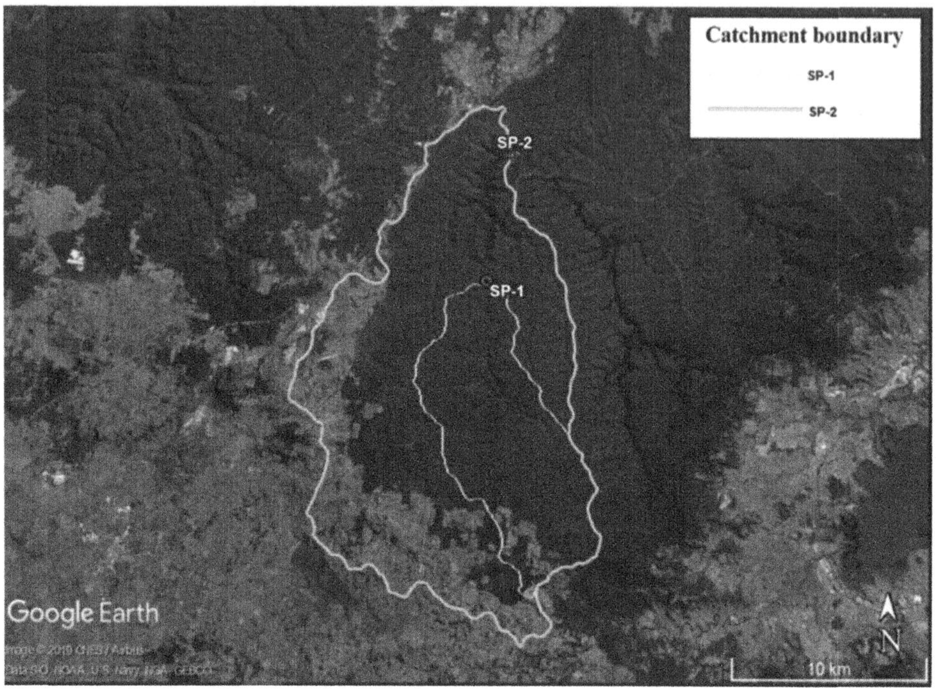

Figure 1 | Nepean dam catchment and two sampling points (SP-1 and SP-2).

Dataset

The dataset ($n = 5{,}540$) was compiled from surface water quality observations and climatological and catchment variables derived from satellites, and a land surface model with a temporal range of 15 years and 2 months (1 November 2002 to 1 January 2018) (Table S1, Supporting Information). A pre-processing step was employed to remove inaccuracies, including reading errors, abnormal data and missing values.

Water quality data

Water quality data were collected at the Burke River (SP-1), one of three main tributaries to Lake Nepean and Lake Nepean (SP-2) in the Nepean catchment (Figure S1, Supporting Information). Parameters obtained from routine water quality monitoring included pH, water temperature (°C), turbidity (NTU), alkalinity (mg $CaCO_3$/L), true colour (UTC) and DOC concentration (mg/L).

Catchment and climate data

The catchment boundary (DEM) was obtained from the Shuttle Radar Topography Mission (SRTM) and the Hydrologically Enforced Digital Elevation Model (DEM-H) Version 1.0 (Read *et al.* 2011); land cover (LC) was obtained from the Dynamic Land Cover Dataset Version 2.1 (Sixsmith *et al.* 2015) in 2012–2013, consisting of 22 LC classes; and climate class (CZ) was obtained from the updated Köppen–Geiger climate classification (Peel *et al.* 2007a) (Table S1). Daily rainfall (mm/day) and air temperature (°C) were obtained from the Australian Water Availability Project

doi: 10.2166/h2oj.2020.035

(AWAP) dataset, gridded at 0.05° (Raupach *et al.* 2009). Mean temperatures were expressed as arithmetic averages of daily maximum and minimum temperatures.

Daily SM (%) and actual evapotranspiration (ET) (mm/day) were extracted from the Australian Water Resource Assessment Landscape (AWRA-L) dataset, gridded at 0.05° (Frost *et al.* 2016). Leaf area index (LAI) (m^2/m^2) defined as the one-sided green leaf area per unit ground surface area was calculated from 8-day composite Moderate Resolution Imaging Spectroradiometer (MODIS) measurements from Terra and Aqua satellites products (Didan *et al.* 2015). Raw LAI data were re-projected and resampled to 0.05° × 0.05° grids (identical to the gridded rainfall data) by using the MODIS re-projection tool (Dwyer & Schmidt 2006). Outliers were removed using the Savitzky–Golay filter (Savitzky & Golay 1964) in the TIMESAT software package (Jönsson & Eklundh 2004). The filtered 8-day LAI maps were linearly interpolated to a daily time scale for consistency with the other parameters in the dataset.

Data processing

Water quality data were collected on different time steps, and the dates of sample collection were not consistent. When multiple samples were taken within a day, the corresponding measurements were averaged to produce a single representative daily value. The climatological data series was complete and void of missing values, so additional data processing was not necessary. The climatological and water quality data series were merged with each other such that only the dates with both climatological and water quality data available could be considered for further data processing.

Establishing the machine learning algorithms

The workflow to establish, train and assess the machine learning algorithms is presented in Figure 2. To evaluate predictive models, the original dataset was randomly partitioned into 10 equal size non-overlapping subsets to enable the 10-fold cross-validation technique. Nine subsets were used for

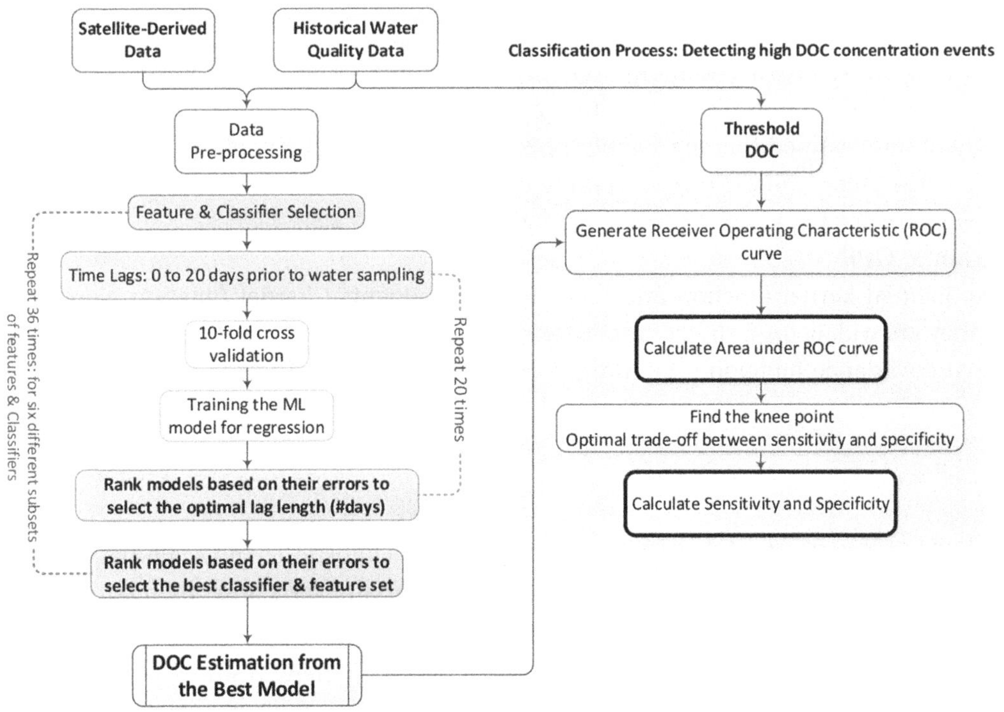

Figure 2 | The overall machine learning framework to establish, train and assess the machine learning algorithms.

training and one set for validation. The cross-validation process was repeated 10 times such that each of the 10 subsets was used exactly once as the validation data. Simple regression analysis was used to identify the potential correlations between the observed DOC concentrations (Target) and other water quality, catchment and climate data (Predictors). After understanding the associations between the predictors and the target, the three algorithms were programmed, trained and validated using MATLAB 2018b (Mathworks, Inc.).

Algorithm 1. MLR

The MLR model learns the linear function to map the climatological parameters to the DOC concentrations in the catchment. Equation (1) represents the relationship between the predicted DOC for the kth sample point, $\widehat{\mathrm{DOC}}_k$, using the MLR model:

$$\widehat{\mathrm{DOC}}_k = a_0 + \sum_{n=1}^{M} (a_n x_i) \tag{1}$$

where a_0 is the intercept and a linear term for each predictor (x_i). The MLR is the simplest algorithm and was used to establish the baseline response between target and predictors. The ordinary least square approximation by QR decomposition was used to estimate the algorithm parameters.

Algorithm 2. SVR

The SVR can be expressed as Equation (2):

$$\widehat{\mathrm{DOC}}_k = a_0 + \sum_{i=1}^{M} (a_i - a_i^*) \cdot K(x_i, x) \tag{2}$$

where K is the kernel function which transforms the data to map into a high-dimensional space. The effectiveness of several kernel functions, including linear, quadratic and cubic, to estimate DOC concentration using climatological parameters were tested. MATLAB allows the setting of certain parameters along with the kernel function as the SVR hyper parameters. Default MATLAB values were used for building the model. Briefly, the epsilon parameter, which represents the half width of epsilon-insensitive band and set to the interquartile range of DOC divided by 13.49 (i.e. an estimator for 10% of the standard deviation). When training an SVR, no penalty is associated with the training points inside the epsilon-insensitive band. The box constraint parameter, which controls the penalty of misclassified training samples, hence limiting model complexity, was set to a multiple of 10 times the epsilon value. Prior to training the model, the predictors were standardised, and the kernel scale parameter was set automatically by MATLAB.

Algorithm 3. GPR

The exponential kernel function and the squared exponential kernel function were used in this study, as they are widely used choices for covariance kernel functions (Jiang *et al.* 2019). The squared exponential covariance function used in the GPR algorithm was expressed as Equation (3):

$$k(x, x') = \sigma^2 \times e^{\left[\frac{-(x-x')^2}{2l^2}\right]} \tag{3}$$

where σ is the noise standard deviation and l is the characteristic length-scale which controls the effect of distance between x and x'. Intuitively, when x and x' are very close, then k (covariance function) leads to a higher value, which means $f(x)$ is highly correlated with $f(x')$. On the other hand, for two distant points, $k(x,x')$ is near to zero. This property of the squared exponential function is also highly desirable as for predicting the target variable at a new x, negligible effects from distant observations are anticipated.

Evolutionary Algorithms in Water Resources
doi: 10.2166/h2oj.2020.035

Time-lag function

As the system time lag was unknown, the performance of the machine learning algorithms as measured by different statistical criteria (highlighted in this section) was evaluated as a function of the lag (T_D). Equation (4), where $\hat{y}$ and θ indicate the predicted value and model parameters, respectively, summarises the dominant delay estimation process. $Perf(\hat{y}, y)$ shows a performance metric describing the accuracy of the model. As implied by Equation (4), the largest spike (arg max) in the performance metric specified the dominant lag time.

$$\arg\max_{T_D} \; Perf(\hat{y}, y) \text{ where } \hat{y} = f(x_i, \theta) \tag{4}$$

Performance and accuracy assessment of the DOC models

Three summary statistics describing the accuracy of the models were used to assess the models' performances: adjusted R-squared (adj. R^2), the root-mean-squared error (RMSE) and mean absolute error (MAE). The adj. R^2 indicates the proportion of the total sum of squares explained by each model adjusted for the number of coefficients. To quantify the level of statistical significance in the performance of selected machine learning algorithms, the paired t-test as a commonly used statistical hypothesis test for comparing machine learning algorithms was applied into the 10-fold cross-validation results. The paired t-test results that are often a test statistic and a p-value can be interpreted to confirm whether there is a real or statistically significant difference between the machine learning algorithms (Nadeau & Bengio 2003; Bouckaert & Frank 2004). Although predicting the DOC as a continuous variable is highly desirable, ultimately predicting the events with high DOC concentrations might be more useful in practice. To do so, thresholding the predicted DOC value to produce a binary outcome was required.

To visualise and quantify the trade-off between the false alarms and missed events, the receiver operating characteristic (ROC) curve was generated for the best performing model (the classification process in Figure 2). The ROC curve plots true positive rate (that is known as sensitivity) against false positive rate (that is also known as one minus specificity), obtained by varying the probability threshold on the predicted output (DOC concentration) (Fawcett 2006). The sensitivity measures the accuracy of the model in predicting high DOC concentration events, while the specificity measures the accuracy of the model in predicting instances with below-threshold DOC concentration. The probability threshold was used to find out how accurately the model could determine if there exists a high DOC event or not. Since increases by as much as 100% in DOC concentrations were observed during wet weather events in other researches (Hinton *et al.* 1997; Schoenheinz & Grischek 2011; Whitworth *et al.* 2012; Dhillon & Inamdar 2013), a high DOC event was defined in this study when the DOC concentration was two times higher than the averaged observed historical DOC concentration in the dataset. To provide an aggregate measure to quantify the model performance across all possible threshold values for $\widehat{DOC}$, the area under the ROC curve (AUC) was calculated. The AUC value of 1.0 indicates a perfect model performance, while a value of 0.5 represents the chance-level (Kordestani *et al.* 2019).

RESULTS AND DISCUSSION

Removal of irrelevant data

Table 1 enumerates the basic statistics of the monitored water quality data at selected sampling points (SP-1 and SP-2 in Figure 1) including the mean and standard deviation of each monitored water

Table 1 | Summary of the basic statistical assessments of selected measured water quality variables in selected sampling sites

	SP-1			SP-2		
Variable	**Mean**	**Standard deviation**	**R^2 with DOC**	**Mean**	**Standard deviation**	**R^2 with DOC**
pH	6.53	0.50	0.00	7.31	0.38	0.00
Temperature (°C)	16.00	5.30	0.26	14.81	3.52	0.17
Turbidity (NTU)	7.72	4.94	0.30	2.15	1.86	0.10
Alkalinity (mg $CaCo_3$/L)	5.17	3.54	0.00	11.07	1.08	0.00
True colour (HU)	23.12	10.40	0.22	7.14	1.86	0.25
DOC concentration (mg/L)	4.30	1.81	1.00	3.7	0.99	1.00

quality data and their R^2 with DOC concentration for each site. pH and alkalinity were dropped from the analysis, as they were variables exhibiting an insignificant relationship with DOC concentration. It should be noted that some studies suggested rising DOC concentrations may have been linked to increases in alkalinity or pH, associated with recovery from acidification (Evans *et al.* 2005; Winterdahl *et al.* 2014). However, there are some studies that challenged the role of pH (Couture *et al.* 2012; Pärn & Mander 2012). Although the temporal correlations were not strong between pH and DOC concentrations in this dataset, the possibility that as pH increases the DOC concentrations increase in this catchment is not dismissed. True colour was infrequently measured in our dataset ($n = 1{,}758$ compared with $n = 10{,}761$ for turbidity at SP-1), thus it was excluded as a parameter in DOC modelling. Temperature and turbidity were retained as water quality variables for the modelling of the DOC concentration.

Selection of correct predictor variables

Climatological data including precipitation (Pre.), temperature (Temp.), SM, ET and LAI measured at SP-1 from November 2002 to January 2018 are shown in Figure S2. To select the correct climatological variables for DOC estimation, Pearson's correlation coefficient of each climatological variable and selected water quality parameters with DOC concentration in selected sampling sites were determined (Table 2 and Table S2). Strong correlations were observed between SM and precipitation data ($r = 0.91$ at SP-1) and between temperature and ET data ($r = 0.80$ at SP-1) (Table 2 and Table S2). To ensure parameters used to estimate DOC concentration were independent from each other, parameters with high correlation were excluded. Therefore, two pathways arise for potential machine learning model input variables, one considering precipitation and temperature and the other including SM and ET. The effects of including more input variables on estimated DOC concentration were also studied by increasing the number of model input variables from only two variables

Table 2 | Pearson's correlation coefficient of climatological variables and selected water quality parameters with DOC concentration of SP-1 dataset

MP1	DOC	Turb.	Pre.	Temp.	SM	ET	LAI
DOC	1.00						
Turb.	0.29	1.00					
Pre.	0.45	0.19	1.00				
Temp.	0.26	0.06	0.07	1.00			
SM	0.41	0.18	0.91	0.07	1.00		
ET	0.29	0.02	0.34	0.80	0.38	1.00	
LAI	0.12	0.00	0.11	0.15	0.05	0.18	1.00

doi: 10.2166/h2oj.2020.035

(Pre. and Temp. versus SM and ET) to three and four input variables by adding turbidity and LAI, respectively.

Impact of considering time lags for the climatological data on estimated DOC concentrations

To investigate the effects of considering time lags for the climatological data on estimated DOC concentrations, time lags of up to 20 days prior to water sampling for the climatological data were investigated in this study. Adj. R^2 and RMSE values of estimated DOC concentrations at SP-1 using MLR as well as SVR and GPR with different kernel functions are presented in Figure 3(a)–3(f) and Figure S3(a)–S3(f),

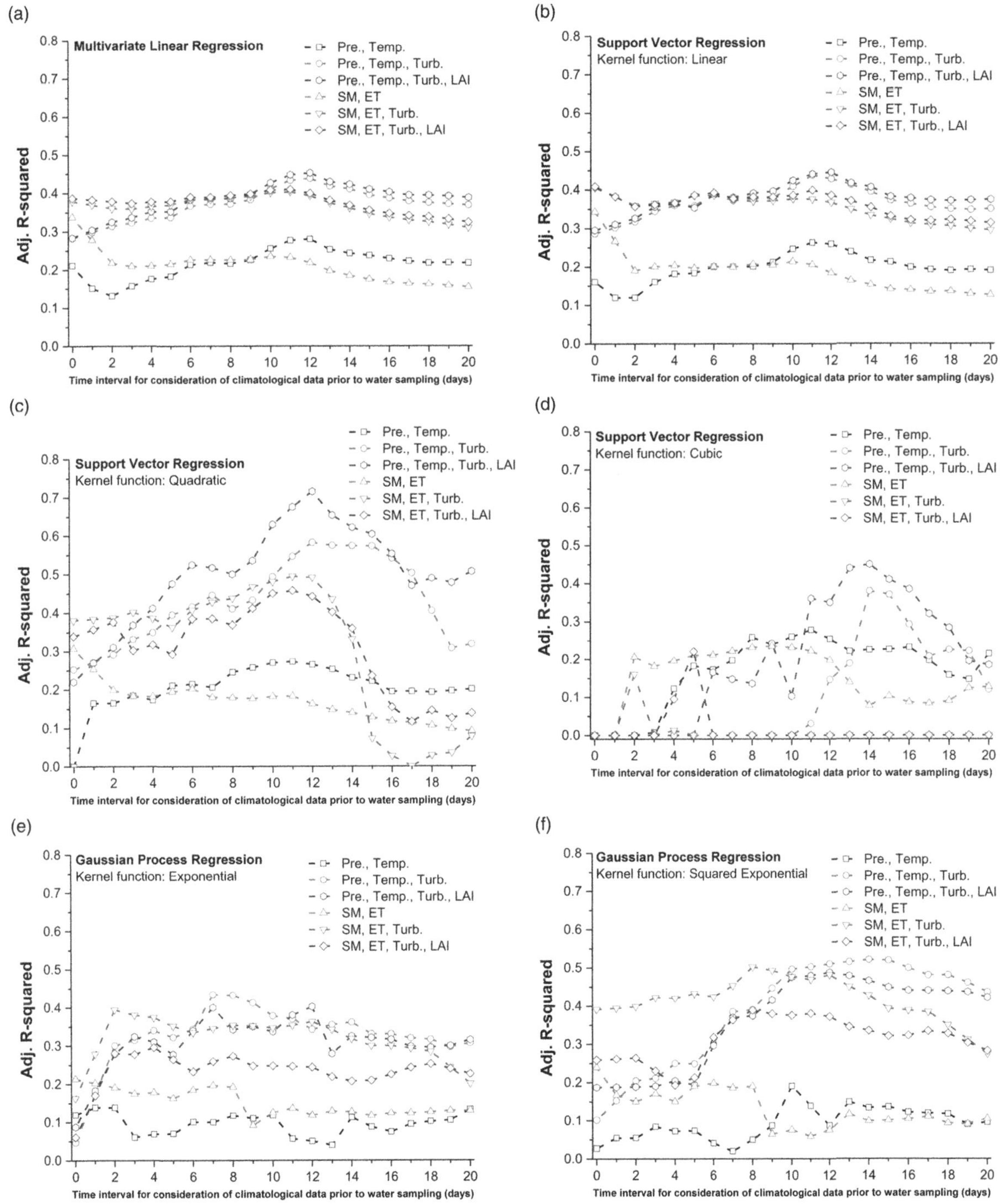

Figure 3 | Adj. R^2 values of estimated DOC concentration at SP-1 using MLR, SVR and GPR with different kernel functions.

respectively. Figure S4(a)–S4(f) represents how MAE values of estimated DOC concentration at SP-1 varied by increasing the time interval for consideration of climatological data prior to water sampling. The results in Figure 3(a)–3(f) showed that increasing the number of inputs gave better results, especially with the more sophisticated machine learning approaches. The model performance was improved by considering turbidity and LAI as input variables, and the first scenario (Pre. and Temp.) resulted better than the second one (SM and ET) (Figure 3(a)–3(f)). SVR with a linear kernel function (Figure 3(b)) resulted in slightly better performance than MLR in adj. R^2 values when the time lag between climatological data and water quality data was increased. Changing the kernel function from linear to quadratic also improved the model estimation results (adj. R^2 of the model was increased) (Figure 3(c)). However, applying SVR with a cubic kernel function was unsuccessful for this dataset for DOC concentration estimation (Figure 3(d)). Similarly, GPR with kernel functions either exponential (Figure 3(e)) or squared exponential (Figure 3(f)) exhibited poor performance with maximum adj. R^2 values on average less than 0.5. The adj. R^2 increased when the time lag between the climatological data and the day of water sampling was increased from 1 to 12 days, beyond which it continued to decrease (Figure 3(a)–3(d)).

Therefore, using SVR with a quadratic kernel function with Pre., Temp., LAI and Turb. as model input variables with a 12-day time lag between climatological and water quality data outperformed the other selected machine learning algorithms in this study in terms of statistical indices (adj. $R^2 = 0.71$, RMSE = 1.9, MAE = 1.35). To estimate whether the differences between the performance of SVR with a quadratic kernel function and other selected machine learning algorithms are true and reliable or are just due to statistical chance, a paired *t*-test was conducted on 10-fold cross-validation results, while a 12-day time lag between climatological and water quality data was considered. The paired *t*-test results between SVR with a quadratic kernel function and other machine learning algorithms rejected the null hypothesis at the 5% significance level, and the averaged *p*-value was 1.37% which is smaller than the considered significance level (i.e. 5%). Hence, the results statistically provided convincing evidence that machine learning algorithms performed differently and any observed difference in the performance of SVR with a quadratic kernel function is likely due to a difference in the models. Therefore, the SVR showed a better capability than MLR and GPR to handle the nonlinearity and complexity of climatological characteristics of the catchment. However, it should be acknowledged that the performance of the GPR model was investigated with only two kernel functions, the exponential and the squared exponential. Exploring other kernels could be a potential future work to this study and might result in a higher performance metrics for the GPR.

The trend of RMSE (Figure S3(a)–S3(f)) was opposite to that of adj. R^2, as the objective function with RMSE was to minimise the estimation error. It should be noted that by using the entire dataset in the hyperparameter selection process, a bias is introduced, which likely results in an overly optimistic performance metrics. The hyperparameters were the best model types (i.e. SVR with quadratic kernel function), the best set of features (Pre., Temp., LAI and Turb.) and the most appropriate value time delay (i.e. 12 days). It should be noted that the 12-day time interval would not be a universal value for all the other catchments, as the time interval could be related to some physical characteristics of the catchment such as catchment area/slope or the LC.

Machine learning model performance assessment

One potential avenue is to explore the performance of the best model for a completely blinded dataset, which was not used for the feature and model selection. Therefore, a 12-day time lag was chosen to consider the impacts of climatological data prior to water sampling, as it led to the highest adj. R^2 value and lowest RMSE value in this study (Figure 3(c)). The measured and estimated DOC concentrations and their scatter plot using the SVR model with the quadratic kernel function at SP-1 are presented in Figure 4. As seen from the figure, the plotted data points revealed a good agreement

doi: 10.2166/h2oj.2020.035

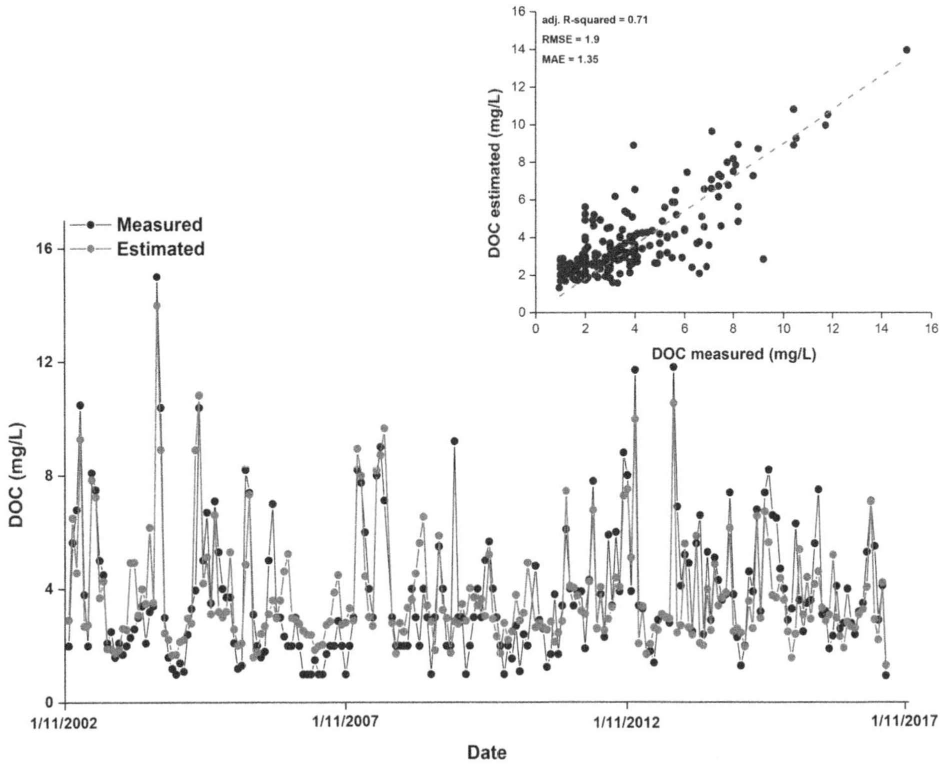

Figure 4 | The measured and estimated DOC concentrations and the scatter plot using the SVR model with the quadratic kernel function at SP-1.

between measured and estimated DOC concentrations, as they generally correlated close towards the 1:1 sloped line.

ROC analysis

To investigate if the developed SVR model with a quadratic kernel function and a 12-day time lag between climatological and water quality data could detect high DOC concentration events, the ROC analysis was carried out. The averaged measured DOC concentration at SP-1 from November 2002 to January 2018 was 3.4 mg/L (Figure 4). In this study, the DOC concentration threshold for such events was set to be two times higher than the averaged observed DOC concentrations over the past 15 years (6.8 mg/L at SP-1). Figure 5 shows the ROC curve for the high DOC concentration events using the SVR model with the quadratic kernel function with Pre., Temp., LAI and Turb. as model input variables and a 12-day time lag between climatological and water quality data at SP-1. Different cut-off values can be selected from the ROC curve. Here, as high sensitivity will guarantee that most of the high DOC concentration events will be detected, a lax operating point is desirable. Therefore, a knee point in the ROC curve, which ensures a sensitivity of 90%, was selected as the operating point. The corresponding sensitivity and specificity were 0.89 and 0.88, respectively (denoted by a red circle in Figure 5). The confidence intervals in each point were computed by generating 100 bootstrap replicas. It was found that the AUC value for the SVR model with the quadratic kernel function and a 12-day time lag was 0.92 which is above the random level (i.e. 0.5), indicating that the developed SVR model, which uses climatological variables as model inputs, is capable of being used as an alarm system, indicating a possible high DOC event.

Adj. R^2, RMSE and MAE values for the estimated DOC concentrations at SP-2 with the same machine learning algorithms are shown in Figure S5(a)–S5(f), Figure S6(a)–S6(f) and

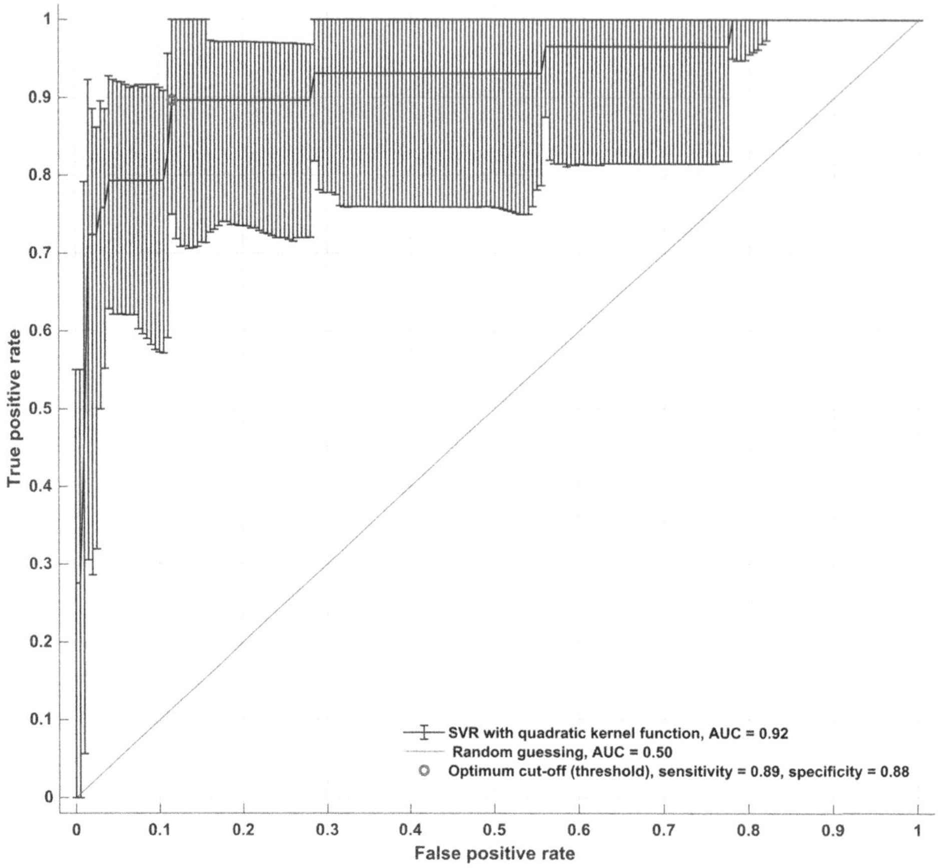

Figure 5 | The ROC curve and the AUC corresponding to the high DOC concentration events using the SVR model with the quadratic kernel function with Pre., Temp., LAI and Turb. as model input variables and a 12-day time lag between climatological and water quality data at SP-1.

Figure S7(a)–S7(f), respectively. Unlike SP-1 that is a river-based sampling point, SP-2 is located at a lake whose water comes from three main tributaries (SP-1 is at one of these three main tributaries into the lake). The results show that none of the applied machine learning algorithms could estimate the DOC concentration at SP-2 (adj. R^2 <0.3). As shown in Figure S5(a)–S5(f), Figure S6(a)–S6(f) and Figure S7(a)–S7(f), estimated DOC concentrations at SP-2 did not improve by including more input parameters to machine learning algorithms, or by increasing the time lag between climatological and water quality datasets. This could largely be due to the complex local geological and hydrological changes in the sites that directly surround and feed each sampling point or because the lake water combines water from its tributaries with potentially different physiochemical and climatological characteristics. One future approach to deal with lake-based sites could be to consider data from all input tributaries. Another possible reason, that would make it hard to understand how water quality parameters in a lake are affected by climatological parameters, could be the inherent variability in lake characteristics such as lake size, shape and depth compared to the rivers. It could also be due to the complex nature of lake–atmosphere interactions (O'Reilly *et al.* 2015; Winslow *et al.* 2015) or water clarity in the lake as it influences the depth range over which heat can be absorbed (Rose *et al.* 2016) or the strength of lake stratification (Winslow *et al.* 2015, 2017). Hence, the response of the lake to changing climatological parameters was complex, and considering only the climatological parameters could not estimate the DOC concentration at the lake sampling point in our study (SP-2).

doi: 10.2166/h2oj.2020.035

CONCLUSION

This research utilised machine learning and satellite data to develop and train a model to estimate DOC concentration in water in an Australian catchment. Given the ease of accessing high-resolution satellite data, it is feasible to use climatological data derived from the satellite datasets and link them with water quality indicators to meet the current and future challenges in large-area water quality monitoring. The analysis of the results showed that precipitation, temperature, LAI and turbidity yielded the optimal results using the SVR model with the quadratic kernel function. Considering the impact of time lag on climatological data prior to water sampling showed an impact on model accuracy, and a 12-day time lag between climatological and water quality data brought far better results for the datasets used in this study in terms of statistical indices (adj. $R^2 = 0.71$, RMSE = 1.9, MAE = 1.35). To show the usefulness of the proposed method compared with other traditional model and kernel-based models, different machine learning models were constructed on the dataset. Experimental results show that the forecasting capability of the SVR model outperforms those of other kernel-based models, thereby generating more accurate results.

From the ROC analysis, the developed SVR model with the quadratic kernel function can be successfully used to indicate possible high DOC events (AUC = 92%). Although the results showed that the SVR model works well for the river-based sampling point, the response of the lake to changing climatological parameters was complex and none of the applied machine learning algorithms could estimate the DOC concentration at lake-based sampling point, probably due to higher water residence times in lakes than rivers or diverse variability in lake characteristics such as lake size, shape and depth. Further studies would be needed to test and clarify the effect of considering other potential lake characteristics as model input variables for the DOC estimation at lake-based sampling points and the potential for utilising the approach to river-based modelling used in this study as inputs to DOC estimation models for lake-based sites.

ACKNOWLEDGEMENTS

This research was supported via the Australian Research Council's Linkage Projects funding scheme (project number LP160100620) which included support from Sydney Water and WaterNSW. In addition, the authors thank Water Research Australia for the PhD top-up scholarship provided to Anthony Agostino. The authors also acknowledge the support of the UNESCO Centre for Membrane Science and Technology.

DATA AVAILABILITY STATEMENT

All relevant data are included in the paper or its Supplementary Information.

REFERENCES

Arnell, N. W., Halliday, S. J., Battarbee, R. W., Skeffington, R. A. & Wade, A. J. 2015 The implications of climate change for the water environment in England. *Progress in Physical Geography: Earth and Environment* **39**, 93–120.

Bouckaert, R. R. & Frank, E. 2004 Evaluating the replicability of significance tests for comparing learning algorithms. In: *Advances in Knowledge Discovery and Data Mining* (Dai, H., Srikant, R. & Zhang, C., eds). Springer Berlin Heidelberg, Berlin, Heidelberg, pp. 3–12.

Clair, T. A., Ehrman, J. M. & Higuchi, K. 1999 Changes in freshwater carbon exports from Canadian terrestrial basins to lakes and estuaries under a 2 × CO_2 atmospheric scenario. *Global Biogeochemical Cycles* **13**, 1091–1097.

Couture, S., Houle, D. & Gagnon, C. 2012 Increases of dissolved organic carbon in temperate and boreal lakes in Quebec, Canada. *Environmental Science and Pollution Research* **19**, 361–371.

Delpla, I., Jung, A. V., Baures, E., Clement, M. & Thomas, O. 2009 Impacts of climate change on surface water quality in relation to drinking water production. *Environment International* **35**, 1225–1233.

Dhillon, G. S. & Inamdar, S. 2013 Extreme storms and changes in particulate and dissolved organic carbon in runoff: entering uncharted waters? *Geophysical Research Letters* **40**, 1322–1327.

Didan, K., Barreto-Munoz, A., Solano, R. & Huete, A. 2015 *MODIS Vegetation Index User's Guide (MOD13 Series), Version 3.00, June 2015 (Collection 6)*. Vegetation Index and Phenology Lab, The University of Arizona, pp. 1–32.

Dwyer, M. J. & Schmidt, G. 2006 The MODIS reprojection tool. In: *Earth Science Satellite Remote Sensing Vol. 2: Data, Computational Processing, and Tools* (Qu, J. J., Gao, W., Kafatos, M., Murphy, R. E. & Salomonson, V. V., eds). Tsinghua University Press, Beijing and Springer-Verlag GmbH Berlin Heidelberg, pp. 162–177.

El-Jabi, N., Caissie, D. & Turkkan, N. 2014 Water quality index assessment under climate change. *Journal of Water Resource and Protection* **6**, 533–542.

Erlandsson, M., Buffam, I., Fölster, J., Laudon, H., Temnerud, J., Weyhenmeyer, G. A. & Bishop, K. 2008 Thirty-five years of synchrony in the organic matter concentrations of Swedish rivers explained by variation in flow and sulphate. *Global Change Biology* **14**, 1191–1198.

Evans, C. D., Monteith, D. T. & Cooper, D. M. 2005 Long-term increases in surface water dissolved organic carbon: observations, possible causes and environmental impacts. *Environmental Pollution* **137**, 55–71.

Evans, C. D., Chapman, P. J., Clark, J. M., Monteith, D. T. & Cresser, M. S. 2006 Alternative explanations for rising dissolved organic carbon export from organic soils. *Global Change Biology* **12**, 2044–2053.

Fawcett, T. 2006 An introduction to ROC analysis. *Pattern Recognition Letters* **27**, 861–874.

Freeman, C., Fenner, N., Ostle, N. J., Kang, H., Dowrick, D. J., Reynolds, B., Lock, M. A., Sleep, D., Hughes, S. & Hudson, J. 2004 Export of dissolved organic carbon from peatlands under elevated carbon dioxide levels. *Nature* **430**, 195–198.

Frost, A., Ramchurn, A. & Smith, A. 2016 *The Bureau's Operational AWRA Landscape (AWRA-L) Model*. Bureau of Meteorology, Melbourne, p. 47.

Gavin, A. L., Nelson, S. J., Klemmer, A. J., Fernandez, I. J., Strock, K. E. & McDowell, W. H. 2018 Acidification and climate linkages to increased dissolved organic carbon in high-elevation lakes. *Water Resources Research* **54**, 5376–5393.

Granata, F., Papirio, S., Esposito, G., Gargano, R. & De Marinis, G. 2017 Machine learning algorithms for the forecasting of wastewater quality indicators. *Water* **9**, 105.

Grbić, R., Kurtagić, D. & Slišković, D. 2013 Stream water temperature prediction based on Gaussian process regression. *Expert Systems with Applications* **40**, 7407–7414.

Hansen, J., Kharecha, P., Sato, M., Masson-Delmotte, V., Ackerman, F., Beerling, D. J., Hearty, P. J., Hoegh-Guldberg, O., Hsu, S.-L., Parmesan, C., Rockstrom, J., Rohling, E. J., Sachs, J., Smith, P., Steffen, K., Van Susteren, L., von Schuckmann, K. & Zachos, J. C. 2013 Assessing 'dangerous climate change': required reduction of carbon emissions to protect young people, future generations and nature. *PLoS One* **8**, e81648.

Hejzlar, J., Dubrovský, M., Buchtele, J. & Růžička, M. 2003 The apparent and potential effects of climate change on the inferred concentration of dissolved organic matter in a temperate stream (the Malše River, South Bohemia). *Science of the Total Environment* **310**, 143–152.

Hinton, M. J., Schiff, S. L. & English, M. C. 1997 The significance of storms for the concentration and export of dissolved organic carbon from two Precambrian Shield catchments. *Biogeochemistry* **36**, 67–88.

Hudson, J., Dillon, P. & Somers, K. 2003 Long-term patterns in dissolved organic carbon in boreal lakes: the role of incident radiation, precipitation, air temperature, southern oscillation and acid deposition. *Hydrology and Earth System Sciences* **7**, 390–398.

Jiang, S., Shen, X. & Zheng, Z. 2019 Gaussian process-based hybrid model for predicting oxygen consumption in the converter steelmaking process. *Processes* **7**, 352.

Jönsson, P. & Eklundh, L. 2004 TIMESAT – a program for analyzing time-series of satellite sensor data. *Computers & Geosciences* **30**, 833–845.

Khalil, A., Almasri, M. N., McKee, M. & Kaluarachchi, J. J. 2005 Applicability of statistical learning algorithms in groundwater quality modeling. *Water Resources Research* **41**, W05010–05026.

Kim, Y. H., Im, J., Ha, H. K., Choi, J.-K. & Ha, S. 2014 Machine learning approaches to coastal water quality monitoring using GOCI satellite data. *GIScience & Remote Sensing* **51**, 158–174.

Kordestani, M. D., Naghibi, S. A., Hashemi, H., Ahmadi, K., Kalantar, B. & Pradhan, B. 2019 Groundwater potential mapping using a novel data-mining ensemble model. *Hydrogeology Journal* **27**, 211–224.

Lal, A. & Datta, B. 2018 *Genetic Programming and Gaussian Process Regression Models for Groundwater Salinity Prediction: Machine Learning for Sustainable Water Resources Management.*

Lary, D. J., Alavi, A. H., Gandomi, A. H. & Walker, A. L. 2016 Machine learning in geosciences and remote sensing. *Geoscience Frontiers* **7**, 3–10.

Loo, Y. Y., Billa, L. & Singh, A. 2015 Effect of climate change on seasonal monsoon in Asia and its impact on the variability of monsoon rainfall in Southeast Asia. *Geoscience Frontiers* **6**, 817–823.

Mohiuddin, A., Rajanayagam, C. & Kearney, C. 2014 Optimisation of non-ionic polymer to address production issues with high-color low-turbidity raw water. *Water: Journal of the Australian Water Association* **41**, 58–63.

Nadeau, C. & Bengio, Y. 2003 Inference for the generalization error. *Machine Learning* **52**, 239–281.

Najjar, R. G., Pyke, C. R., Adams, M. B., Breitburg, D., Hershner, C., Kemp, M., Howarth, R., Mulholland, M. R., Paolisso, M., Secor, D., Sellner, K., Wardrop, D. & Wood, R. 2010 Potential climate-change impacts on the Chesapeake Bay. *Estuarine, Coastal and Shelf Science* **86**, 1–20.

Office of Environment and Heritage 2017 *Upper Nepean State Conservation Area Draft Plan of Management*. Office of Environment and Heritage, Sydney, Australia.

O'Reilly, C. M, Sharma, S., Gray, S., Hampton, D. K., Read, S. E., Rowley, J. S., Schneider, R. J., Lenters, P., McIntyre, J. D., Kraemer, P. B., Weyhenmeyer, B. M., Straile, G. A., Dong, D., Adrian, B., Allan, R., Anneville, M. G., Arvola, O., Austin, L., Bailey, J., Baron, J. L., Brookes, J. S., de Eyto, J. D., Dokulil, E., Hamilton, M. T., Havens, D. P., Hetherington, K., Higgins, A. L., Hook, S. N., Izmest'eva, S., Joehnk, L. R., Kangur, K. D., Kasprzak, K., Kumagai, P., Kuusisto, M., Leshkevich, E., Livingstone, G., MacIntyre, D. M., May, S., Melack, L., Mueller-Navarra, J. M., Naumenko, D. C., Noges, M., Noges, P., North, T., Plisnier, R. P., Rigosi, P.-D., Rimmer, A., Rogora, A., Rudstam, M., Rusak, J. A., Salmaso, N., Samal, N. R., Schindler, D. E., Schladow, G., Schmid, M., Schmidt, S. R., Silow, E., Soylu, E., Teubner, K., Verburg, P., Voutilainen, A., Watkinson, A., Williamson, C. E. & Zhang, G. 2015 Rapid and highly variable warming of lake surface waters around the globe. *Geophysical Research Letters* **42**, 10773–10781.

Park, Y., Cho, K. H., Park, J., Cha, S. M. & Kim, J. H. 2015 Development of early-warning protocol for predicting chlorophyll-a concentration using machine learning models in freshwater and estuarine reservoirs, Korea. *Science of the Total Environment* **502**, 31–41.

Pärn, J. & Mander, Ü. 2012 Increased organic carbon concentrations in Estonian rivers in the period 1992–2007 as affected by deepening droughts. *Biogeochemistry* **108**, 351–358.

Parr, T. B., Inamdar, S. P. & Miller, M. J. 2019 Overlapping anthropogenic effects on hydrologic and seasonal trends in DOC in a surface water dependent water utility. *Water Research* **148**, 407–415.

Peel, M. C., Finlayson, B. L. & McMahon, T. A. 2007a Updated world map of the Köppen-Geiger climate classification. *Hydrology and Earth System Sciences Discussions* **4**, 439–473.

Peel, M. C., Finlayson, B. L. & McMahon, T. A. 2007b Updated world map of the Köppen-Geiger climate classification. *Hydrology and Earth System Sciences* **11**, 1633–1644.

Raupach, M., Briggs, P., Haverd, V., King, E., Paget, M. & Trudinger, C. 2009 *Australian Water Availability Project (AWAP): CSIRO Marine and Atmospheric Research Component: Final Report for Phase 3*. Centre for Australian Weather and Climate Research (Bureau of Meteorology and CSIRO), Melbourne, Australia, 67.

Read, A., Dowling, T., Gallant, J., Tickle, P. K. & Wilson, N. 2011 1Second SRTM Derived Hydrological Digital Elevation Model (DEM-H) version 1.0. Commonwealth of Australia (Geoscience Australia).

Rose, K. C., Winslow, L. A., Read, J. S. & Hansen, G. J. A. 2016 Climate-induced warming of lakes can be either amplified or suppressed by trends in water clarity. *Limnology and Oceanography Letters* **1**, 44–53.

Rostami, S., He, J. & Hassan, Q. K. 2018 Riverine water quality response to precipitation and its change. *Environments* **5**, 8.

Ruescas, A. B., Hieronymi, M., Koponen, S., Kallio, K. & Camps-Valls, G. 2017 Retrieval of coloured dissolved organic matter with machine learning methods. In: *2017 IEEE International Geoscience and Remote Sensing Symposium (IGARSS)*, pp. 2187–2190.

Ruescas, A. B., Hieronymi, M., Mateo-Garcia, G., Koponen, S., Kallio, K. & Camps-Valls, G. 2018 Machine learning regression approaches for colored dissolved organic matter (CDOM) retrieval with S2-MSI and S3-OLCI simulated data. *Remote Sensing* **10**, 1–20.

Savitzky, A. & Golay, M. J. E. 1964 Smoothing and differentiation of data by simplified least squares procedures. *Analytical Chemistry* **36**, 1627–1639.

Schoenheinz, D. & Grischek, T. 2011 Behavior of dissolved organic carbon during bank filtration under extreme climate conditions. In: *Riverbank Filtration for Water Security in Desert Countries* (Shamrukh, M., ed.). Springer Netherlands, Dordrecht, pp. 51–67.

Sixsmith, J., Thankappan, M., McIntyre, A., Tan, P. & Lymburner, L. 2015 *Dynamic Land Cover Dataset Version 2.1*. Geoscience Australia, Sydney, Australia.

Snauffer, A. M., Codden, C., Stubbins, A. & Mueller, A. V. 2018 High-frequency saltmarsh dissolved organic carbon estimates via machine learning. In: *AGU Fall Meeting Abstracts*, pp. EP51E-1875.

Sun, A. Y., Wang, D. & Xu, X. 2014 Monthly streamflow forecasting using Gaussian process regression. *Journal of Hydrology* **511**, 72–81.

Tranvik, L. J. & Jansson, M. 2002 Terrestrial export of organic carbon. *Nature* **415**, 861–862.

Vapnik, V. 2000 *The Nature of Statistical Learning Theory*. Springer, New York.

Whitworth, K. L., Baldwin, D. S. & Kerr, J. L. 2012 Drought, floods and water quality: drivers of a severe hypoxic blackwater event in a major river system (the southern Murray–Darling basin, Australia). *Journal of Hydrology* **450–451**, 190–198.

Winslow, L. A., Read, J. S., Hansen, G. J. A. & Hanson, P. C. 2015 Small lakes show muted climate change signal in deepwater temperatures. *Geophysical Research Letters* **42**, 355–361.

Winslow, L. A., Hansen, G. J. A., Read, J. S. & Notaro, M. 2017 Large-scale modeled contemporary and future water temperature estimates for 10774 Midwestern U.S. Lakes. *Scientific Data* **4**, 170053–170053.

Winterdahl, M., Bishop, K. & Erlandsson, M. 2014 *Acidification, Dissolved Organic Carbon (DOC) and Climate Change*, pp. 281–287.

Zhu, J., Hu, H., Tao, S., Chi, X., Li, P., Jiang, L., Ji, C., Zhu, J., Tang, Z., Pan, Y., Birdsey, R. A., He, X. & Fang, J. 2017 Carbon stocks and changes of dead organic matter in China's forests. *Nature Communications* **8**, 151.

First received 30 April 2020; accepted in revised form 25 July 2020. Available online 28 August 2020

Evolutionary Algorithms in Water Resources
doi: 10.2166/h2oj.2020.034

Modelling hydrological responses under climate change using machine learning algorithms – semi-arid river basin of peninsular India

G. Sireesha Naidu, M. Pratik and S. Rehana*

Lab for Spatial Informatics, International Institute of Information Technology, Hyderabad 500032, India

*Corresponding author. E-mail: rehana.s@iiit.ac.in

Abstract

Catchment scale conceptual hydrological models apply calibration parameters entirely based on observed historical data in the climate change impact assessment. The study used the most advanced machine learning algorithms based on Ensemble Regression and Random Forest models to develop dynamically calibrated factors which can form as a basis for the analysis of hydrological responses under climate change. The Random Forest algorithm was identified as a robust method to model the calibration factors with limited data for training and testing with precipitation, evapotranspiration and uncalibrated runoff based on various performance measures. The developed model was further used to study the runoff response under climate change variability of precipitation and temperatures. A statistical downscaling model based on K-means clustering, Classification and Regression Trees and Support Vector Regression was used to develop the precipitation and temperature projections based on MIROC GCM outputs with the RCP 4.5 scenario. The proposed modelling framework has been demonstrated on a semi-arid river basin of peninsular India, Krishna River Basin (KRB). The basin outlet runoff was predicted to decrease (13.26%) for future scenarios under climate change due to an increase in temperature (0.6 °C), compared to a precipitation increase (13.12%), resulting in an overall reduction in water availability over KRB.

Key words: Budyko hypothesis, climate change impacts, general circulation models, hydrological modelling, statistical downscaling

Highlights

- Catchment parameter estimation dynamically with precipitation, evapotranspiration and runoff using machine learning algorithms.
- Statistical downscaling model to relate climate and hydrological variables using machine learning algorithms.
- Assessment of hydrological responses under climate change of a semi-arid river basin of India.

INTRODUCTION

Hydrological models are promising tools to study various water resources engineering problems such as flood prediction and design, drought assessment, water quantity and quality assessment and hydrological responses under climate variability (Sood & Smakhtin 2015). The most important classification of hydrological models is empirical (data-driven), conceptual and physically based

Evolutionary Algorithms in Water Resources
doi: 10.2166/h2oj.2020.034

models (Aghakouchak & Habib 2010). Empirical models have evolved as data-driven using various regression (Zhang *et al.* 2020), machine learning algorithms such as Artificial Neural Network (ANN) (Hu *et al.* 2018), least square-Support Vector Regression (SVR) (Bharti *et al.* 2017), etc. Such empirical models are primarily based on observations without considering the various hydrological processes involved (Jaiswal *et al.* 2020), whereas the physically-based models work on the principles of conservation of mass, momentum and energy to consider various hydrological processes along with field measured parameters and other basin characteristics (Aghakouchak & Habib 2010). The conceptual models are based on the simplified mathematical conceptualization of a hydrological system by considering various components at catchment scales and use model parameters in simulating the discharges (Jaiswal *et al.* 2020). Physically-based models are complex and require fine spatio-temporal data, whereas purely data-driven empirical models are limited completely to observations. Compared to physical and empirical based hydrological models, the conceptual models have gained much interest in the research community due to their capability to include various hydrological processes (e.g. runoff, evapotranspiration, surface storage, etc.) and ease of implementation with available data (Ammann *et al.* 2020). In this context, various conceptual models have evolved in the past, such as HYMOD (Parra *et al.* 2018), HBV model (Vormoor *et al.* 2018), etc. Model parameters or calibration factors are most prominent in conceptual hydrological models affecting the accuracy of streamflow estimates (Arnold *et al.* 2012). In this context, auto-calibration approaches for hydrological models have progressed in recent years, for example, the Soil Water Assessment Tool – Calibration Uncertainty Programs (SWAT-CUP) (Ha *et al.* 2017), and the Shuffled Complex Evolution Algorithm (SCE-UA) which carries the limitation of multiple optimal parameters and corresponding accuracies (Pohlert *et al.* 2007). Calibration of hydrological model parameters using remotely sensed variables has also gained interest in recent years, based on satellite evapotranspiration data (Immerzeel & Droogers 2008; Rientjes *et al.* 2013). To this end, various catchment processes have been conceptualized in hydrological modelling by including model parameters whose values can be determined through calibration (Mianabadi *et al.* 2019). In this context, calibration factors were implemented for various processes such as closure of water balance (Jarsjö *et al.* 2008), radiation (Choudhury 1999) and vegetation dynamics (Donohue *et al.* 2007). Conceptual models developed by including such calibration factors on various processes have been used to study the hydrological responses under climate, land use and water changes as a function of various hydroclimatic variables and observed historical streamflows (e.g. Asokan *et al.* 2010).

Hydrological variables such as precipitation, evapotranspiration and streamflows have shown pronounced changes under climate signals and therefore the model parameters also should account for such changes (Srinivasa Raju & Nagesh Kumar 2018). Most of the conventional hydrological models apply such calibration parameters which were established based on observed historical streamflow data and other hydrological variables in the hydrological climate change impact assessment (e.g. Meenu *et al.* 2013). Such models integrate hydrological models and statistical downscaling models to study the hydrological responses based on the historically established model parameters (Hengade *et al.* 2018; Saraf & Regulwar 2018). Inclusion of time-invariant model parameters which are independent of the hydro-climatic variables and estimated solely based on the observed data are limited to capture the temporal variability of water-energy balance variables under climate change (Rehana *et al.* 2020a). The present study proposes to establish a relationship between the calibration factors and various hydroclimatic variables and uncalibrated runoff. Such an established functional form of calibration factors can be easily implementable for future scenarios under projected hydroclimatic variables. In this context, developing a calibration approach relating precipitation, evapotranspiration and uncalibrated streamflows can improve the model performance and understanding of hydrological processes (López *et al.* 2017). Therefore, estimating model parameters dynamically accounting for the expected hydrological variability under climate change can be promising. Furthermore, such approaches can be further extended with climate change projections to study the hydrological response of river basins.

The study proposed a modelling framework by relating the conceptual hydrological model parameter with various hydrological variables of precipitation, evapotranspiration and simulated runoff. Such established relationships will ease the estimation of model calibration factors for future scenarios with climate change projections obtained from statistical downscaling models. The major challenges in implementing such a framework are limited data and poor understanding of complex relationships between the calibration factors and various hydrological process-based variables. To overcome such limitations, the present study proposed an approach to model the parameters of conceptual models using various hydrological variables by adopting machine learning algorithms. The study emphasized the use of advanced machine learning algorithms for developing such complex relationships between model parameters and hydrological variables, which can form as a basis for studying the hydrological responses under climate variability. The proposed modelling framework has been demonstrated on a semi-arid river basin, Krishna River Basin (KRB), India.

MATERIALS AND METHODS

Basin description

Krishna River Basin (KRB) is the fifth largest river system in India and second largest river basin of peninsular India with a total catchment area of 258,948 km^2 (Figure 1). The KRB, 73°17′–81°9′E and 13°10′–19°22′N, has heavy precipitation over the Western Ghats with decreasing precipitation towards the upper and lower parts of the basin. The KRB originates in the Western Ghats of India and travels for about 1,400 km across the four states of Karnataka, Maharashtra, Andhra Pradesh and Telangana before joining the Bay of Bengal at Hamsaladeevi in Andhra Pradesh near the east coast of India. The major tributaries of the KRB include Ghataprabha, Malaprabha, Bhima, Tunga-Bhadra and Musi. About 44% of the KRB lies in Karnataka, with about 26, 15 and 15% of the basin in Maharashtra, Telangana and Andhra Pradesh respectively. The annual average precipitation over the basin is about 784 mm, with 90% of the precipitation occurring during the south-west monsoon from June to October (http://india-wris.nrsc.gov.in/wrpinfo/?title = Krishna). Most of the basin

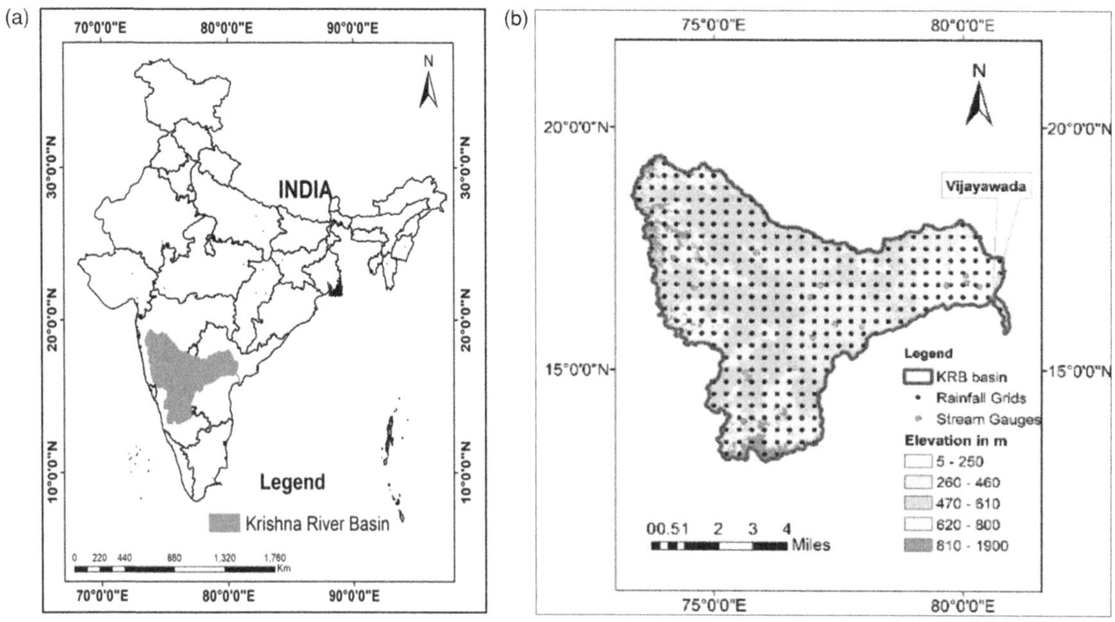

Figure 1 | Map of the Krishna River Basin (KRB). (a) Location of the catchment in India; (b) various hydro-meteorological stations and elevation map superimposed on the basin.

doi: 10.2166/h2oj.2020.034

has a prevailing semi-arid climate and the states covering the basin (Maharashtra, Karnataka and Telangana) are majorly drought-prone (http://india-wris.nrsc.gov.in). The major water use of the KRB is for irrigation with about 61.9 billion cubic meters per year (BCM/year) along with domestic and industrial water uses as 1.6 and 3.2 BCM/year (Rooijen *et al.* 2009).

Distributed hydrological model: PCRaster

In this study, hydrological modelling of the KRB was performed using the distributed hydrological model, PCRaster-water flow module using POLFLOW code (Wit 2001; Jarsjö *et al.* 2008) (http://pcraster.geo.uu.nl/pcraster/4.2.0/documentation/pcraster_manual/sphinx). The KRB was delineated and a local drainage direction (ldd) map was created using the topographical data (Figure 2). The basin was discretized with a rectangular grid (1 × 1 km) and for each grid various hydrological processes can be provided as input. The Digital Elevation Map (DEM) data used was at a resolution of 30 arc-seconds (approximately 1 km) from the Global 30 Arc-Second Elevation (GTOPO30) data set provided by the US Geological Survey. Spatially distributed data of daily precipitation (P) and temperature (T) for the KRB were from the India Meteorological Department (IMD). The gridded daily precipitation data at a spatial resolution of 0.25 × 0.25° (Pai *et al.* 2014) and temperature at a resolution of 1 × 1° (Srivastava *et al.* 2009) were cropped to the basin for a common time period from 1951 to 2014. The temperature, which was available at 1 × 1° resolution, was interpolated to 0.25 × 0.25° using the inverse distance weighting method. The discharge data was obtained from Krishna & Godavari Basin Organization (KGBO), Central Water Commission (CWC), Hyderabad, Government of India (www.kgbo-cwc.ap.nic.in) for about 25 discharge locations for the period 1966–2015 (Figure 1(b)).

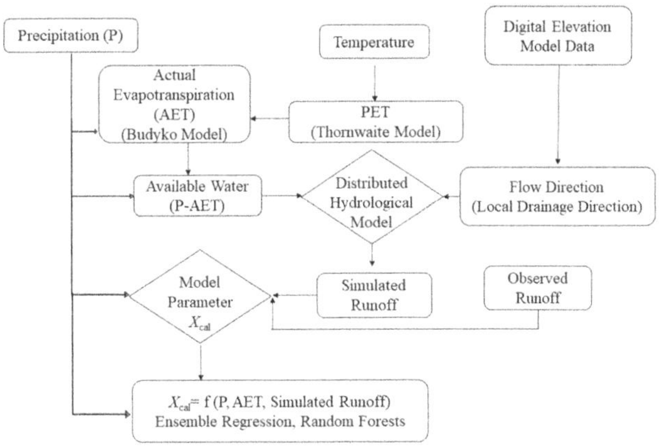

Figure 2 | Overview of the modelling approach to estimate catchment-scale calibration factors using machine learning algorithms.

The basic hydrological processes involved in the partitioning of precipitation (P) into evapotranspiration (AET), groundwater (GW), runoff (Q) and change of storage (ΔS) at catchment scale can be written as follows:

$$P + Q_{in} + GW_{in} - GW_{out} - Q_{out} - AET - \Delta S = 0 \quad (1)$$

For sufficiently long time-scales, the groundwater inflow (GW_{in}) and outflow (GW_{out}) volumes along with the change in storage volumes ΔS can be neglected. Such basic assumptions ultimately represent a water balance equation at river basin scale with P, AET and R as the major hydrological variables which can be related to estimating the precipitation surplus or Available Water (AW) and can be expressed as follows:

$$AW = P - AET \tag{2}$$

It can be noted that the study employed a conceptual catchment scale hydrological model on an annual time scale, neglecting the storage component of water in terms of groundwater in the water-balance equation. For long-term soil water storage changes the groundwater recharge influenced by geological features and human interaction water withdrawals are considered to be negligible (Wu *et al.* 2017). However, it can be noted that the storage component of water cannot be neglected at any temporal scale less than annual and the resulting phenomenon and the hydrological responses may vary due to the inclusion of groundwater storages being taken into account (Gunkel & Lange 2017).

If the basic hydrologic unit of a distributed hydrological model is considered as grid cell, i, the annual total precipitation (P_i) and AET (AET_i), the precipitation surplus ($AW_{cal,i}$) can be estimated as follows:

$$AW_{cal,i} = P_i - AET_i \tag{3}$$

where P_i and $AET_{clim,i}$ are at annual scale in mm/year. Due to the lack of direct and long-term AET data, most of the studies depend on either pan evaporation or empirical estimations of PET (Han *et al.* 2014). One of the widely applied methods to estimate AET flux in hydrology is estimating PET from the energy available for vaporization first, and then applying a limiting factor to account for the water availability (Anabalón & Sharma 2017). Such ET models based on P and PET works with readily available and modelled operational meteorological variables such as precipitation and temperature (Budyko 1974). In this context, Budyko's (1974) hypothesis has gained much interest in hydrology, which can relate to the most important hydrological variables of P, PET, AET along with model parameter 'ω' to represent climate variability, basin characteristics, vegetation and terrain, etc. Several studies used parametric formulations of the Budyko framework to estimate AET globally (e.g. Bai *et al.* 2020) and Indian case studies (e.g. Singh & Kumar 2015; Goyal & Khan 2017). A bottom-up probabilistic Budyko framework has been proposed by Singh & Kumar (2015) to estimate the vulnerability of available water of India under climate change. In another study by Sinha *et al.* (2018), the Budyko framework was used to study the anthropogenic stress and climatic variance under warming shifts across 55 catchments in peninsular India.

Such a basic parametric formulation of Budyko framework can be written as follows:

$$\frac{AET}{P} = 1 + \frac{PET}{P} - \left(1 + \left(\frac{PET}{P}\right)^{\omega}\right)^{(1/\omega)} \tag{4}$$

Several studies have modified the above formulation and the most widely applied one is based on one by Zhang *et al.* (2004) as a calibration-free formulation for the Budyko equation relating to P, PET and AET as follows:

$$AET_{climate} = \left[P\left(1 - exp\left(\frac{-PET}{P}\right)\right)PET\ tanh\left(\frac{P}{PET}\right)\right]^{0.5} \tag{5}$$

where PET can be estimated based on empirical formulations such as Penman (1950), Thornthwaite (1948) and Hargreaves & Allen (2003), etc. The present study considered the Thornthwaite model as one of the simplest models to estimate PET, which includes average air temperature and geographical location of the region of interest as input variables. The most dependable gridded observed meteorological data available for India is temperature among the other variables (e.g. wind speed, humidity,

doi: 10.2166/h2oj.2020.034

radiation, etc.) required to implement detailed PET estimates such as the Penman model. Therefore, the present study adopted the Thornthwaite model as the temperature-based PET model for the Indian context. The AET can be estimated on a monthly time scale with PET estimates. The AET estimated using Equation (5) is region-specific and ideally developed for given climatic and basin characteristics. Therefore, model parameters have to be introduced in Equation (5) to obtain runoff which is in agreement with that observed. As the AET is the most complicated component of the hydrological cycle, defined by various factors of climatic, vegetative, soil moisture and amount of available water, implementation of calibration factors on AET will be more promising. We have introduced the calibration factor on AET as the basic conceptual hydrological model considered was based on the closure of water balance involving the prominent hydrological variables of precipitation (P), AET and runoff (R), R = P-AET. A calibration factor (X_{Cal}) has been introduced on AET in the calibration of the observed runoff (R_{obs}) using the water balance equation, R_{obs} = (P- X_{Cal}AET). This way, a model parameter has been introduced on AET by accounting for the closure of water balance at the catchment outlet. Therefore, the AW with inclusion of calibration factor (X_{Cal}) can be written as follows:

$$AW_{actual,i} = P_i - X_{Cal}AET_{clim,i} \tag{6}$$

The discharge at the basin outlet ($R_{Cal,outlet}$) can be estimated by accumulating the flow at grid cell, i, ($RAW_{cal,i}$) and from all upstream grid cells ($RAW_{cal,\Omega}$) following the flow direction of the river and corresponding to the area of each grid cell (A_{Cell}) as follows:

$$R_{Cal,outlet} = (AW_{cal,i} + AW_{cal,\Omega}) * A_{Cell} \tag{7}$$

where $R_{Cal,outlet}$ is the uncalibrated total runoff from the basin outlet at steady state and it is generally not consistent with observed runoff, $R_{Obs,outlet}$:

$$R_{Obs,outlet} = (AW_{actual,i} + AW_{actual,\Omega}) * A_{Cell} \tag{8}$$

A comparison of observed and uncalibrated runoff estimations can be made to estimate the model parameter (X_{Cal}) based on Jarsjö *et al.* (2008) as follows:

$$\frac{R_{Cal,outlet}}{R_{Obs,outlet}} = \frac{P - AET_{clim}}{P - (X_{Cal} * AET_{clim})} \tag{9}$$

$$X_{Cal} = \frac{R_{Obs,outlet}}{R_{Cal,outlet}} + \left(1 - \frac{R_{Obs,outlet}}{R_{Cal,outlet}}\right) \frac{\sum P}{\sum AET_{clim}} \tag{10}$$

where $R_{Obs,outlet}$ and $R_{Cal,outlet}$ are the long-term annual average observed and simulated runoff at the basin outlet in m^3/s respectively and $\sum P$ and $\sum AET_{clim}$ are the long-term accumulated annual average observed precipitation and AET (Equation (5)) over the basin in mm/year. Such a basin averaged single calibration factor has been widely applied to study hydrological responses of soil, land cover and water use by Asokan *et al.* (2010) and Jarsjö *et al.* (2008). However, to study the hydrological responses under climate change, use of single time-invariant calibration factors may be limited to capture the variability of precipitation and AET. A dynamic model parameter which can capture the observed variability of precipitation and ET has been proposed, which can be further used to study the hydrological responses under climate change. Such a dynamic calibration factor can vary as a function of hydrological variables (P, AET and uncalibrated runoff) for each time step. To model the complex relation between the model parameters and hydrological variables of P, AET, and R, the present study proposed two machine learning algorithms, Ensemble Regression (ER) and Random Forest (RF).

Machine learning algorithms for modelling the calibration factors

The annual scale calibration factors (X_{Cal}) were estimated and considered as predictand variable and uncalibrated runoff (P-AET) resulting from the PCRaster model, along with P and AET which were considered as predictor variables in training and testing of machine learning models. The present study used two machine learning models, Ensemble Regression (ER) and Random Forest (RF) to predict the model parameters.

Ensemble regression (ER) model

Ensemble Regression (ER) methods are machine learning paradigms in which multiple methods which are often referred to as weak learners are trained to solve a problem and are combined to obtain better results (Friedman 2001). The ER model works on the hypothesis that a diverse set of models can make better predictions in comparison with an individual model. The ER models have gained interest in hydrological model assessments in recent years (Sajedi-Hosseini *et al.* 2018).

The input training dataset of 'N' points $\{X, Y\} = \{(x_i, y_i)\}_{i=1}^{N}$, where x_i is the set of predictors and y_i as the observed predictand value at the *i*th timestep, will be considered. Initially, all predictors are given equal weighting coefficients $\left(\text{equal importance} - \alpha_i = \frac{1}{N}\right)$ and an initial model ($F_0(x)$) will be developed to predict values of the form $y = F_0(x)$.

At every iteration, *m*, the residuals (r_{im}) will be calculated between the observed (y_i) and modelled predictand value ($F_{m-1}(x_i)$):

$$r_{im} = -\left[\frac{d\left(\sum_{i=1}^{N}[y_i - F_{m-1}(x_i)]^2\right)}{d(F_{m-1}(x_i))}\right] \quad \text{for i} = 1..\text{N} \tag{11}$$

A base-learner (h_m) will be fitted to these residuals using a loss function 'L' in the direction of the steepest gradient, i.e. weight, α_i of point '*i*' is increased corresponding to a higher value of residual using the training set $\{(x_i, r_{im})\}_{i=1}^{N}$. The model is then sequentially updated as follows:

$$F_m(x) = F_{m-1}(x) + argmin_h \sum_{i=1}^{n} L(y_i,\ F_{m-1}(x) + h_m(x_i, r_{im})\ ) \tag{12}$$

where the $argmin_h$ refers to minimization of base-learner error ($\mathrm{h_m}$). The least-squares loss function $L(y, F(x)) = \frac{1}{2}\ (y - F(x))^2$ is used to update the residuals.

Random forest (RF) model

In recent years, data mining algorithms such as Random Forest (RF) have been widely applied in various studies related to water resources engineering such as hydrological (Moore *et al.* 1991), soil moisture forecasting (Prasad *et al.* 2018) and water quality assessment (Bui *et al.* 2020). RFs are an ensemble of decision trees, which split the data space similar to the way humans and each node represents a predictor variable (Breiman 2001). The data splitting is generally done based on inequality condition, Gini index, which is a measure of the performance of the split and measures how diverse the data is until it reaches a terminal node (Breiman 2017). The RF tree size is determined based on the number of nodes, which serve as a basis to minimize the variance of each split. RFs can perform regression as well as classification using various decision trees based on bootstrap aggregation (bagging), which is a random data re-sampling methodology with replacement. The bootstrapping process was used to sample S data points with N variables by replacement. A decision tree was trained on each of these random samples. Bagging is the process of making overall predictions from average predictions of various bootstrapped samples trained on individual decision trees, making averaging predictions from each decision tree. The steps involved in designing the RF models are as follows:

1. Draw a bootstrap samples S* of size N from the training data
2. An RF tree T_b will be developed to the bootstrapped data recursively for each terminal node of the tree.
   - Selecting S variables at random from N variables
   - Choosing the best split-point among the S
   - Split the node into two daughter nodes
   - Continuing the above steps until the size of the minimum node n_{min} is reached.
3. The ensemble of trees can be estimated as $\{T_b\}_1^B$
4. Prediction at a new point, x, can be made using the regression formulation as

$$\hat{f}_{rf}^{B}(\mathrm{x}) = \frac{1}{B}\sum_{b=1}^{B} T_b(x) \tag{13}$$

The RF model has been identified as a robust model due to its ability to capture the complex relationships between features and labels efficiently, accounting for the uncorrelated decision trees in the prediction.

The trained and tested machine learning algorithms will be further used to predict the calibration factors which can be used to estimate the calibrated runoff at the catchment scale. To quantify the performance of models and to understand the dependence between calibration factors and hydrological variables, Root Mean Square Error (RMSE), Absolute Percentage Bias (APB), Mean Absolute Error (MAE), Nash–Sutcliffe Efficiency (NSE) (Nash & Sutcliffe 1970), Kling-Gupta Efficiency (KGE) (Knoben *et al.* 2019) and Likelihood were selected as performance measures.

Climate change projection model: statistical downscaling models (SDM)

Statistical downscaling models are the state-of-the-art climate change projections prediction models relating to large-scale climate variables with surface hydrological variables (e.g. precipitation) using statistical methods (Wilby & Dawson 2013). The present study employed a multisite statistical downscaling model to predict the climate change projections of precipitation and temperature (Figure 3). The basic formulation of the downscaling model includes data preprocessing to remove systematic bias in the modelled and actual climate observations (bias correction), data reduction method (principal component analysis, PCA), predictand variable states estimation (K-means clustering), fitting algorithm to relate predictors and predictand states (CART), and transfer function (Support Vector Regression). National Centers for Environmental Prediction/National Center for Atmospheric Research (NCEP/NCAR) climate data from 1948 to the present at a resolution of $2.5 \times 2.5°$ were considered as large-scale observed climate variables. The MIROC-ESM derived from the Atmosphere and Ocean Research Institute (The University of Tokyo), National Institute for Environmental Studies, and Japan Agency for Marine-Earth Science and Technology, Japan, was considered as the GCM output. Spatial resolution mismatch between NCEP and GCM data was resolved by applying an inverse distance weighting interpolation. The GCM predictor data undergoes bias correction based on quantile mapping method by comparing the cumulative distribution functions (CDFs) of NCEP and GCM predictor data for the

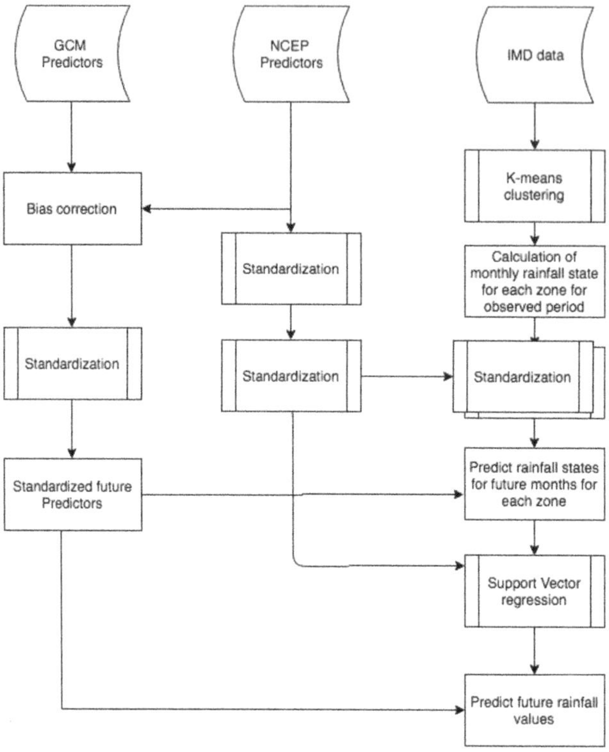

Figure 3 | Overview of the statistical downscaling model to predict projections of precipitation and temperatures over KRB.

historical and future scenarios to remove the systematic bias associated with the climate outputs (Li *et al.* 2010). Both bias-corrected GCM and reanalysis climate data undergoes standardization, which involves subtraction of mean and division by standard deviation estimated with all the data points of the time series. After data processing, statistically significant climate variables (predictors) were considered to predict the precipitation over various grid points of the basin. The predictor variables which have shown significant correlation coefficients with precipitation were identified as surface air temperature, wind speed, humidity, etc., which were also found as potential predictors in other studies over India (Rehana & Mujumdar 2012; Salvi *et al.* 2013). NCEP/NCAR reanalysis data for surface air temperature, mean sea level pressure, specific humidity at 500 mb pressure level, zonal and meridional wind velocity at the surface level is extracted for the latitudes in the range of 12.5–20°N and the longitudes in the range of 72.5–82.5°E, surrounding the entire Krishna Basin.

K-mean clustering

K-means clustering is an unsupervised machine learning algorithm that partitions n observations into K clusters for which each observation belongs to the cluster with the nearest mean. In this study, we use the K-means algorithm to achieve cross-correlation among the rain stations and group the months having similar rainfall. This technique reads the observed rainfall values for all grids in the basin in a month, clusters them, and provides a single representative value that is referred to as the state of rainfall for that particular month. Consider $X = x_1, x_2, x_3, \ldots, x_n$ as the set of data points and $V = \mu_1, \mu_2, \mu_3$ as the set of cluster means. For each data point, the nearest mean was identified and the point is assigned to the corresponding cluster. The objective of k-means algorithm is to iteratively minimize the squared error function given as follows:

$$J(V) = \sum_{i=1}^{c} \sum_{j=1}^{c_i} (\|x_i - v_j\|)^2 \tag{14}$$

doi: 10.2166/h2oj.2020.034

where $\|x_i - v_j\|$ represents the Euclidian distance between x_i and v_j, C is the number of clusters and $c_i$is the number of points in the cluster.

Classification and Regression Trees (CART)

Classification and Regression Trees is a decision tree learning technique used in prediction modelling. A decision tree is a machine learning model in which categorization of the data is carried out with each feature partitioning the data based on conditions or range of the features and works when the data has a finite set of values. CART helps in categorizing the rainfall into states by building a statistical relation between the continuous principal components extracted from predictor data and the rainfall states estimated using K-means clustering. The established relationship is assumed to be intact for the future predictors which are then taken as input for CART model and for which the future rainfall states are estimated. An advantage of CART over linear classification models is that they can capture non-parametric and nonlinear relationships as well as yield simple models. Cross-validation is carried out in order to ensure there is no risk of overfitting the data. For every iteration, calculate the Gini's impurity/diversity Index (E) (Loh 2011) of the data using the formula given in the equation as follows:

$$E = \sum_{i \neq j} P(w_i)P(w_j) = 1 - \sum_{i \neq j} P(w_i)^2 \tag{15}$$

where $P(w_j)$ represents the probability of data being in jth class for each value of the attribute. An attribute splitting will be selected which has minimum impurity index (E). It can be considered as 0 when all the patterns at node have the same class label. Continue until impurity (E) is less than a certain threshold (η) or on reaching the maximum number of iterations where $E < \eta$.

Support vector regression (SVR)

Individual regression models are built on separating the predictor and observed data based on the weather state category into individual datasets. The SVR model tries to fit the error within a certain threshold (ε), identifying a single separating hyperplane which maximizes the margin rather than solely minimizing the error, which helps to find the best model. Linear SVR was used as the regression technique to predict the precipitation and temperature without overfitting. The training data of $\{(x_1, y_1), \ldots\ldots.(x_n, y_n)\}$ with n patterns and function $f(x)$ will be identified with the consideration of the deviation from the observed target variables y_i (Lima *et al.* 2012). The input variables, X, will be mapped into a higher dimensional feature space using a nonlinear mapping function φ:

$$f(x; w) = \langle W, \varphi(x) \rangle + b \tag{16}$$

where $<, >$ denotes the inner product, and W and b are the regression coefficients, which can be estimated by minimizing the error between $f(x)$ and the observed values of y. SVR uses the $\in$-insensitive error to measure the error between $f(x)$ and the observed values of y, where $\in$ is the hyper-parameter:

$$|f(x; w) - y|_{\in} = \begin{cases} 0, & if |f(x; w) - y| < \in \\ |f(x; w) - y| - \in, & otherwise, \end{cases} \tag{17}$$

Using the training data of (x_i, y_i) the values of w and b are estimated by minimizing the objective function:

$$F = \frac{C}{N}\sum_{i=1}^{n} |f(x_i, w) - y_i|_{\in} + \frac{1}{2}||w||^2 \quad (18)$$

where C and $\in$ are the hyper-parameters. The minimization of the objective function, F, uses the Lagrange multiplier method, and the final regression equation with kernel function $K(X, X')$ can be in the form:

$$f(X) = \sum_{i} K(X, X_i) + b \quad (19)$$

RESULTS AND DISCUSSION

The study applied a PCRaster distributed hydrological model for 348 grids encompassing the KRB at $0.25 \times 0.25°$ resolution at the annual scale. The original water balance equation of (P-AET) was estimated for each grid from 1965 to 2014 to estimate the precipitation surplus. The precipitation surplus will be converted to runoff by using the area of each grid cell (Equation (7)). The basin outlet uncalibrated runoff will be estimated using the accuthresholdflux function of PCRaster using a local drainage network created based on DEM data. The observed and uncalibrated runoff along with precipitation and AET will be used to estimate the basin averaged model parameter, X_{cal} on an annual scale. The model parameters estimated based on the closure of water balance by comparing the observed and uncalibrated runoff will be considered as the predictand. The hydrological variables of P, AET and uncalibrated runoff (R_{cal}) are considered as the predictor variables in modelling of calibration factors using machine learning algorithms.

The calibration factors obtained based on the closure of water balance (Equation (10)) were considered as observed ones and we have used P, AET (Equation (5)) and uncalibrated runoff resulting from PCRaster (Equation (8)) as independent variables in the modelling of dynamic calibration factors using ER and RF models. The ER and RF models were trained with observed calibration factors obtained from Equation (10) as a dependent variable and P, AET and uncalibrated runoff obtained from PCRaster as independent variables for the 1966–2000 data. The trained ER and RF models have been tested with the independent data of P, AET and uncalibrated runoff for the period 2000–2014. The trained and tested ER and RF models were compared in terms of various performance measures, as given in Table 1. The error between the modelled and predicted values of calibration factors were found to be less for the RF model compared to ER. The RMSE and MAE values for the RF (ER) model was 0.04 (0.02) and 0.14 (0.09), respectively. Furthermore, the efficiency

Table 1 | Comparison of performance measures for RF and ER models in predicting the model parameters

Performance metric	Random Forest	Ensemble Regression
RMSE	0.041	0.142
APB	1.96	8.09
MAE	0.02	0.09
NSE	0.84	0.41
KGE	0.85	0.17
Likelihood	0.46	0.0001

doi: 10.2166/h2oj.2020.034

performance measures such as NSE and KGE have also shown that the RF model is robust to predict the calibration factors compared to the ER model with limited data points in training (34) and testing (15). Also, the RF model has been identified as a better fit compared to ER in terms of similarity measure (likelihood) and bias (APB), as given in Table 1 and Figure 4.

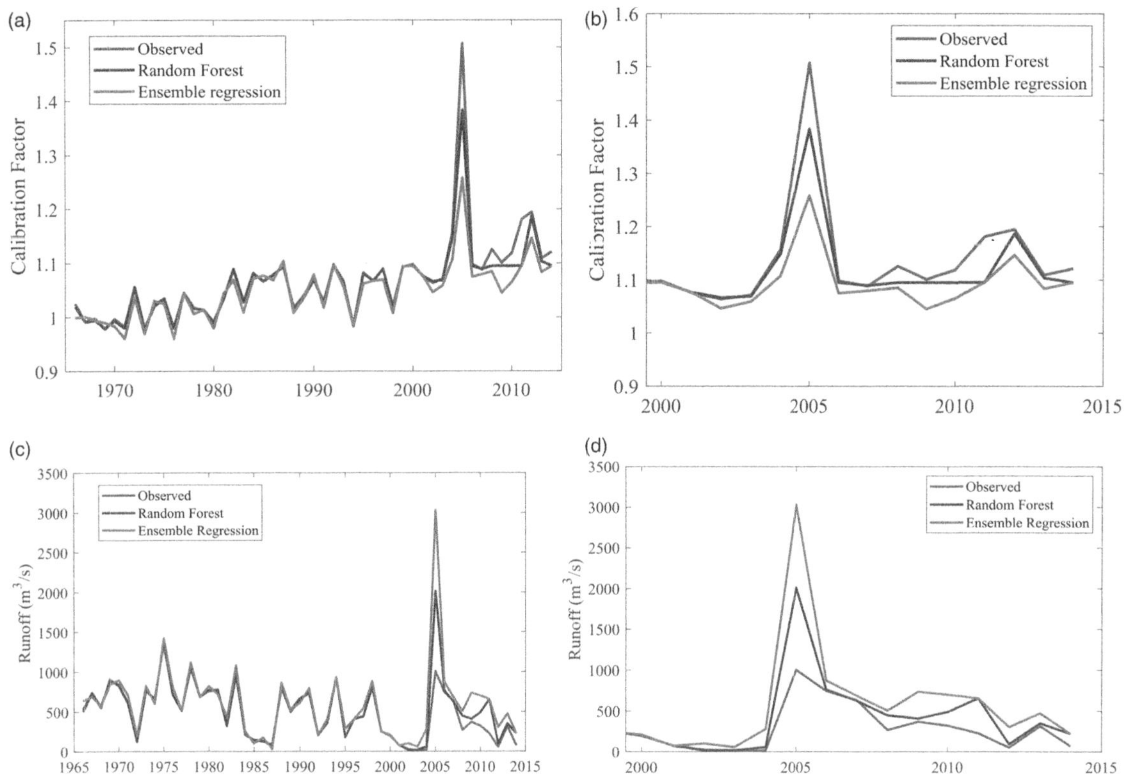

Figure 4 | Comparison of observed, Ensemble Regression and Random Forest modelled calibration factors obtained for: (a) training and (b) testing periods, simulated runoff for (c) training and (d) testing periods.

It can be noted that the calibration factors in Figure 4(a) and 4(b) are in the range of 0.9–1.5, which represents the multiplying factor to be applied on the AET. These are the range of values which have to be applied on the estimated AET based on Equation (5) to implement Equation (6) in the estimation of actual water availability in terms of runoff. The resulting runoff obtained after application of the calibration factor on AET will represent the calibrated runoff which is comparable with the observed runoff on the catchment scale. Figure 4(c) and 4(d) show the observed and simulated runoff values with ER and RF models for the training and testing periods. Due to the robustness of the RF model in modelling the calibration factors, the present study used the RF model to predict the calibration factors for the current and future scenarios. The trained and tested RF model has been used with projections of precipitation and temperature to estimate the projected runoff response over the basin. Figure 5 shows the comparison of NSE coefficients estimated between observed and NCEP and GCM simulations for the training period of 1951–1989 and the testing period of 1990–2005 of the statistical downscaling model for precipitation and temperature. The basin averaged NSE values estimated between observed and NCEP data for precipitation for the training and testing period were 0.65 and 0.44 respectively, whereas the NSE values estimated between observed and MIROC simulations of precipitation for the training and testing period were 0.57 and 0.32 respectively. Temperature predictions were more convincing compared to precipitation due to less variability in data. The basin averaged NSE values estimated between observed and NCEP data for

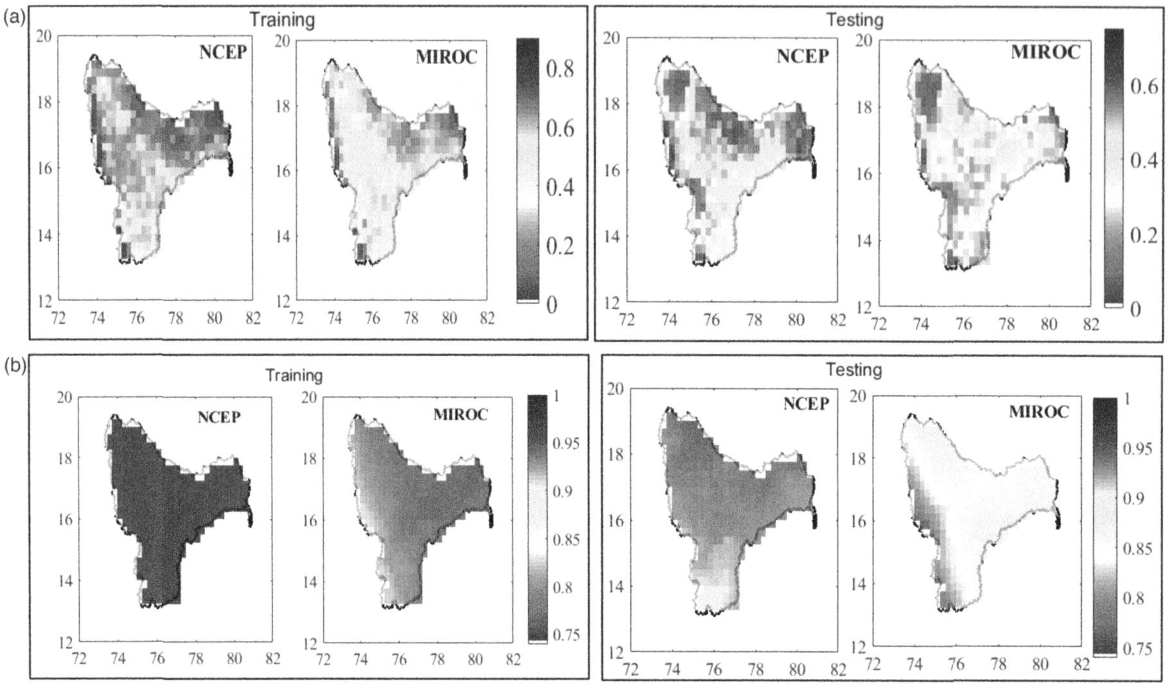

Figure 5 | Comparison of Nash–Sutcliffe coefficients estimated between observed and modelled with NCEP and GCM simulations for (a) precipitation and (b) temperature for the training and validation period.

temperature for the training and testing period were estimated as 0.98 and 0.93 respectively, whereas the NSE values estimated between observed and MIROC simulations of temperature for the training and testing period were 0.94 and 0.86 respectively. The observed basin averaged annual precipitation for the training and validation was estimated as 763.06 and 812.04 mm respectively, whereas the NCEP (MIROC) simulations for training and testing period were obtained as 663.68 (650.57) and 611.83 (648.26) mm, respectively. The observed mean annual temperature for the training (testing) period was noted as 26.26 (26.54) °C, with NCEP simulated as 26.25 (26.42) °C and MIROC simulated as 26.27 (26.20) °C. The future projections of precipitation and temperatures for the period of 2021–2080 were estimated with MIROC GCM under the Representative Concentration Pathway (RCP) 4.5 representing atmospheric radiation at 4.5 Wm^{-2} at the end of 2100. Both precipitation and temperatures were predicted to increase over the KRB for the period of 2021–2080. The annual average precipitation increase was predicted to be about 13.12% with a temperature increase of 0.6 °C for the period 2061–2080 compared to the observed period of 1951–1989 with MIROC GCM outputs. Overall, the precipitation and temperature are predicted to increase over the KRB, affecting the water-energy balance variables (Rehana *et al.* 2020a). Such increase of precipitation and temperature projections based on GCM outputs was found to be comparable with the earlier research findings over the KRB based on Regional Circulation Model (RCM) based Coordinated Regional Downscaling Experiment (CORDEX) projections (e.g. Rehana *et al.* 2020a, 2020b). Their study noted an approximate increase of 2.19% in precipitation and an approximate increase of 1.29 °C in temperature for the period 2021–2040 and 2041–2060 respectively compared to the observed period of 1966–2003 with the RCP 4.5 scenario with various CORDEX model outputs.

The projections of precipitation and temperature have been used as input to the hydrological modelling framework as shown in Figure 2 to estimate the runoff for the current and future scenarios. The annual AET for observed period of 1951–1989 and 1990–2005 was estimated as 340.22 and 495.96 mm respectively. The estimated AET with MIROC GCM projections of precipitation and temperature for the periods 2020–2040, 2041–2060 and 2061–2080 are 580.63, 613.75 and 648.49 mm respectively (Table 2). Overall, the AET has been predicted to increase under an increase

doi: 10.2166/h2oj.2020.034

Table 2 | Spatial average annual precipitation, temperature, PET, AET and runoff for current period (1951–1989 (training), 1990–2005 (testing)) and future period (2021–2040, 2041–2060, 2061–2080) for KRB

		Current		Future		
Hydrological variable	**Observed/simulated**	**Training**	**Validation**	**2021–2040**	**2041–2060**	**2061–2080**
Average annual precipitation (mm)	Observed	763.06	812.04	–	–	–
	NCEP	663.68	611.83	–	–	–
	MIROC	650.57	648.26	646.13	695.67	735.91
Annual temperature (°C)	Observed	26.26	26.54	–	–	–
	NCEP	26.25	26.42	–	–	–
	MIROC	26.27	26.20	26.95	26.79	26.85
Annual PET (mm)	Observed	1,772.31	1,831.13	–	–	–
	MIROC	1,717.49	1,697.60	1,909.87	1,859.31	1,868.61
Annual AET (mm)	Observed	340.22	495.96	–	–	–
	MIROC	575.09	592.16	580.63	613.75	648.49
Annual runoff at basin outlet (m^3/s)	Observed	586.56	362.63	–	–	–
	MIROC	687.60	575.68	498.83	500.39	527.79

of temperature and the consequent increase of PET over the basin. The decrease of basin outlet runoff for the periods 2021–2040, 2041–2060 and 2061–2080 compared to the observed period of 1951–1989 was noted as 87.73 (15%), 86.17 (14.69%), 58.77 (10.09%) m^3/s respectively over the KRB (Table 2). The increase of AET over the basin was found to be more pronounced compared to the increase of precipitation over the KRB using GCM climate change projections.

CONCLUSIONS

The present study proposed a modelling framework to study the hydrological response under climate change over the KRB. A conceptual hydrological model was adopted to study the runoff response under variability of P and AET. A time-variant calibration approach was developed to estimate the model parameters relating to precipitation, evapotranspiration and uncalibrated runoff using two machine learning algorithms, Ensemble Regression and Random Forest model. The Random Forest algorithm was identified as a robust method to model the calibration factors with limited data available for training and testing with precipitation, evapotranspiration and uncalibrated runoff. The developed model was further used to study the runoff response under climate change variability of precipitation and temperature. A statistical downscaling model based on K-means clustering, CART and SVR was developed to predict the future scenarios of precipitation and temperature. Both precipitation and temperature were predicted to increase by 13.12% and 0.6 °C respectively for the period 2061–2080 compared to the observed period of 1951–1989 with MIROC GCM outputs. The increase of temperature has predicted an increase of PET for the future scenarios. It can be noted that the AET estimates are defined by P and PET changes. The increase of temperature has resulted in an increase of PET and consequent increase of AET over KRB for the future scenarios. It can be noted that the projected increase of AET is more pronounced under an increase of PET compared to the increase of P, resulting in a decrease of runoff for the future scenarios over the KRB. The projected decrease of basin outlet runoff for the periods 2021–2040, 2041–2060 and 2061–2080 compared to the observed period 1951–1989 was noted as 15, 14.69 and 10.09% respectively. Overall, the basin outlet runoff was predicted to decrease (13.26%) for future scenarios under climate change due to an increase in temperature (0.6 °C) compared to the precipitation increase (13.12%) over the KRB. As the conceptual model applied is based on the basic hydrological variables of P and AET, the resulting runoff outlets were predicted to decrease under the projected increase of AET. Overall, severe water shortage was predicted under climate change due to the projected increase of AET compared to the precipitation over KRB.

The proposed modelling framework has considered one hydrological model with one particular GCM simulation under a particular concentration pathway scenario for demonstration purposes. It should be noted that hydrological response can be expected to vary with various hydrological models, calibration parameters, statistical downscaling models, various concentration path scenarios, and various machine learning algorithms. A detailed study may be evident under the possible range of uncertainties arising at each stage of the modelling framework, which can be a potential research problem.

The present study considered a calibration-free formulation of Budyko's hypothesis to estimate the annual AET on the catchment scale. The study intended to demonstrate how a hydrological calibration factor can be modelled based on various hydro-climatological variables and the same can be utilized for the climate change impact assessment. For this purpose, the catchment scale single calibration factor based conceptual hydrological model was adopted. However, the proposed methodology can be implemented with any such standard hydrological model, accounting for the number of calibration factors involved and the suitable data-driven model to be adopted. Furthermore, the present modelling framework can be extended by introducing multiple calibration factors accounting for the land use, soil, etc. (Jarsjö *et al.* 2008) in the hydrological modelling, which is a potential area of research. The main emphasis of the study was to develop a modelling framework to study the hydrological responses under climate change using a conceptual hydrological model with an integration of a calibration framework based on machine learning algorithms. The proposed modelling framework can be extendable for other river basins with sufficient data availability and validation by applying various machine learning algorithms. With the use of most readily available and operational hydrometeorological variables such as precipitation, temperature and runoff, the proposed model can be promising to study the catchment scale hydrological responses under climate change. However, validation of the modeled results is most critical in the implementation for other catchment studies. Specifically, the use of calibration-free Budyko formulation has to be validated with the water-balance or most dependable and state-of-the-art satellite based AET estimates. Furthermore, the study limited for annual water balances by neglecting the storage changes which can be further extended with the inclusion of appropriate storage components to study the water balances other than on annual scale. Due to the limitation over the availability of various meteorological variables in the estimation of PET, the present study adopted the Thornthwaite model, which can be extendable to the standard Penman model. Overall, the proposed methodology of inclusion of time-variant calibration factors in the conceptual hydrological model accounting for various hydrological variables is more reliable for catchment-scale water management under climate change.

ACKNOWLEDGEMENTS

The research work presented in the paper was funded by Science and Engineering Research Board (SERB), Department of Science and Technology, Government of India through Start-up Grant for Young Scientists (YSS) Project no. YSS/2015/002111 to Dr S. Rehana. We sincerely acknowledge the India Meteorological Department (IMD) for providing the gridded observations of precipitation and temperatures and details of the data are given in the website at https://iridl.ldeo.columbia.edu/SOURCES/.IMD/index.html?Set-Language = en. The authors sincerely thank the editors and anonymous reviewers for reviewing the manuscript and providing insightful comments.

DATA AVAILABILITY STATEMENT

Data cannot be made publicly available; readers should contact the corresponding author for details.

REFERENCES

Aghakouchak, A. & Habib, E. 2010 Application of a conceptual hydrologic model in teaching hydrologic processes. *International Journal of Engineering Education* **26** (4(S1)), 963–973. Retrieved from https://escholarship.org/uc/item/3sv066q5.

Ammann, L., Doppler, T., Stamm, C., Reichert, P. & Fenicia, F. 2020 Characterizing fast herbicide transport in a small agricultural catchment with conceptual models. *Journal of Hydrology* **586**, 124812. https://doi.org/10.1016/j.jhydrol.2020.124812.

Anabalón, A. & Sharma, A. 2017 On the divergence of potential and actual evapotranspiration trends: an assessment across alternate global datasets. *Earth's Future* **5** (9), 905–917. https://doi.org/10.1002/2016EF000499.

Arnold, J. G., Moriasi, D. N., Gassman, P. W., Abbaspour, K. C., White, M. J., Srinivasan, R., Santhi, C., Harmel, R. D., van Griensven, A., Van Liew, M. W., Kannan, N. & Jha, M. K. 2012 SWAT: model use, calibration, and validation. *Transactions of the ASABE* **55** (4), 1491–1508. https://doi.org/10.13031/2013.42256.

Asokan, S. M., Jarsjö, J. & Destouni, G. 2010 Vapor flux by evapotranspiration: effects of changes in climate, land use, and water use. *Journal of Geophysical Research: Atmospheres* **115** (D24). https://doi.org/10.1029/2010JD014417.

Bai, P., Liu, X., Zhang, D. & Liu, C. 2020 Estimation of the Budyko model parameter for small basins in China. *Hydrological Processes* **34** (1), 125–138. https://doi.org/10.1002/hyp.13577.

Bharti, B., Pandey, A., Tripathi, S. K. & Kumar, D. 2017 Modelling of runoff and sediment yield using ANN, LS-SVR, REPTree and M5 models. *Hydrology Research* **48** (6), 1489–1507. https://doi.org/10.2166/nh.2017.153.

Breiman, L. 2001 Random forests. *Machine Learning* **45** (1), 5–32. https://doi.org/10.1023/A:1010933404324.

Breiman, L. 2017 *Classification and Regression Trees*. Routledge, New York. https://doi.org/10.1201/9781315139470.

Budyko, M. 1974 *Climate and Life*. Academic Press, New York.

Bui, D. T., Khosravi, K., Tiefenbacher, J., Nguyen, H. & Kazakis, N. 2020 Improving prediction of water quality indices using novel hybrid machine-learning algorithms. *Science of The Total Environment* **721**, 137612. https://doi.org/10.1016/j.scitotenv.2020.137612.

Choudhury, B. J. 1999 Evaluation of an empirical equation for annual evaporation using field observations and results from a biophysical model. *Journal of Hydrology* **216** (1), 99–110. https://doi.org/10.1016/S0022-1694(98)00293-5.

Donohue, R. J., Roderick, M. L. & McVicar, T. R. 2007 On the importance of including vegetation dynamics in Budyko's hydrological model. *Hydrology and Earth System Sciences* **11**, 983–995. https://doi.org/10.5194/hess-11-983-2007.

Friedman, J. H. 2001 Greedy function approximation: a gradient boosting machine. *Annals of Statistics* **29** (5), 1189–1232. https://doi.org/10.1214/aos/1013203451.

Goyal, M. K. & Khan, M. 2017 Assessment of spatially explicit annual water-balance model for Sutlej River Basin in eastern Himalayas and Tungabhadra River Basin in peninsular India. *Hydrology Research* **48** (2), 542–558. https://doi.org/10.2166/nh.2016.053.

Gunkel, A. & Lange, J. 2017 Water scarcity, data scarcity and the Budyko curve – an application in the Lower Jordan River Basin. *Journal of Hydrology: Regional Studies* **12**, 136–149. https://doi.org/10.1016/j.ejrh.2017.04.004.

Ha, L. T., Bastiaanssen, W. G. M., van Griensven, A., van Dijk, A. I. J. M. & Senay, G. B. 2017 SWAT-CUP for calibration of spatially distributed hydrological processes and ecosystem services in a Vietnamese river basin using remote sensing. *Hydrology and Earth System Sciences Discussions* 1–35. https://doi.org/10.5194/hess-2017-251.

Han, S., Tian, F. & Hu, H. 2014 Positive or negative correlation between actual and potential evaporation? Evaluating using a nonlinear complementary relationship model. *Water Resources Research* **50** (2), 1322–1336. https://doi.org/10.1002/2013WR014151.

Hargreaves, G. H. & Allen, R. G. 2003 History and evaluation of Hargreaves evapotranspiration equation. *Journal of Irrigation and Drainage Engineering* **129** (1), 53–63. https://doi.org/10.1061/(ASCE)0733-9437(2003)129:1(53).

Hengade, N., Eldho, T. I. & Ghosh, S. 2018 Climate change impact assessment of a river basin using CMIP5 climate models and the VIC hydrological model. *Hydrological Sciences Journal* **63** (4), 596–614. https://doi.org/10.1080/02626667.2018.1441531.

Hu, C., Wu, Q., Li, H., Jian, S., Li, N. & Lou, Z. 2018 Deep learning with a long short-term memory networks approach for rainfall-runoff simulation. *Water* **10** (11), 1543. https://doi.org/10.3390/w10111543.

Immerzeel, W. W. & Droogers, P. 2008 Calibration of a distributed hydrological model based on satellite evapotranspiration. *Journal of Hydrology* **349** (3), 411–424. https://doi.org/10.1016/j.jhydrol.2007.11.017.

Jaiswal, R. K., Ali, S. & Bharti, B. 2020 Comparative evaluation of conceptual and physical rainfall–runoff models. *Applied Water Science* **10** (1), 48. https://doi.org/10.1007/s13201-019-1122-6.

Jarsjö, J., Shibuo, Y. & Destouni, G. 2008 Spatial distribution of unmonitored inland water discharges to the sea. *Journal of Hydrology* **348** (1–2), 59–72. https://doi.org/10.1016/j.jhydrol.2007.09.052.

Knoben, W. J. M., Freer, J. E. & Woods, R. A. 2019 Technical note: inherent benchmark or not? Comparing Nash–Sutcliffe and Kling–Gupta efficiency scores. *Hydrology and Earth System Sciences* **23** (10), 4323–4331. https://doi.org/10.5194/hess-23-4323-2019.

Li, H., Sheffield, J. & Wood, E. F. 2010 Bias correction of monthly precipitation and temperature fields from intergovernmental panel on climate change AR4 models using equidistant quantile matching. *Journal of Geophysical Research: Atmospheres* **115** (D10). https://doi.org/10.1029/2009JD012882.

Lima, A. R., Cannon, A. J. & Hsieh, W. W. 2012 Downscaling temperature and precipitation using support vector regression with evolutionary strategy. In: *The 2012 International Joint Conference on Neural Networks (IJCNN)*, Brisbane, Australia. pp. 1–8. https://doi.org/10.1109/IJCNN.2012.6252383.

Loh, W.-Y. 2011 Classification and regression trees. *WIRES Data Mining and Knowledge Discovery* **1** (1), 14–23. https://doi.org/10.1002/widm.8.

López, L., Sutanudjaja, P., Schellekens, E. H., Sterk, J., & Bierkens, G. & P, M. F. 2017 Calibration of a large-scale hydrological model using satellite-based soil moisture and evapotranspiration products. *Hydrology and Earth System Sciences* **21** (6), 3125–3144. https://doi.org/10.5194/hess-21-3125-2017.

Meenu, R., Rehana, S. & Mujumdar, P. P. 2013 Assessment of hydrologic impacts of climate change in Tunga–Bhadra river basin, India with HEC-HMS and SDSM. *Hydrological Processes* **27** (11), 1572–1589. https://doi.org/10.1002/hyp.9220.

Mianabadi, A., Coenders-Gerrits, M., Shirazi, P., Ghahraman, B. & Alizadeh, A. 2019 A global Budyko model to partition evaporation into interception and transpiration. *Hydrology and Earth System Sciences* **23**, 1–29. https://doi.org/10.5194/hess-23-4983-2019.

Moore, I. D., Grayson, R. B. & Ladson, A. R. 1991 Digital terrain modelling: a review of hydrological, geomorphological, and biological applications. *Hydrological Processes* **5** (1), 3–30. https://doi.org/10.1002/hyp.3360050103.

Nash, J. E. & Sutcliffe, J. V. 1970 River flow forecasting through conceptual models part I – a discussion of principles. *Journal of Hydrology* **10** (3), 282–290. https://doi.org/10.1016/0022-1694(70)90255-6.

Pai, D. S., Sridhar, L., Rajeevan, M. C., Sreejith, O. P., Satbhai, N. S. & Mukhopadhyay, B. 2014 Development of a new high spatial resolution (0.25° × 0.25°) long period (1901–2010) daily gridded rainfall data set over India and its comparison with existing data sets over the region. *Mausam* **65** (1), 1–18.

Parra, V., Fuentes-Aguilera, P. & Muñoz, E. 2018 Identifying advantages and drawbacks of two hydrological models based on a sensitivity analysis: a study in two Chilean watersheds. *Hydrological Sciences Journal* **63** (12), 1831–1843. https://doi.org/10.1080/02626667.2018.1538593.

Penman, H. L. 1950 The dependence of transpiration on weather and soil conditions. *Journal of Soil Science* **1** (1), 74–89. https://doi.org/10.1111/j.1365-2389.1950.tb00720.x.

Pohlert, T., Breuer, L., Huisman, J. A. & Frede, H.-G. 2007 Assessing the model performance of an integrated hydrological and biogeochemical model for discharge and nitrate load predictions. *Hydrology and Earth System Sciences* **11** (2), 997–1011. https://doi.org/10.5194/hess-11-997-2007.

Prasad, R., Deo, R. C., Li, Y. & Maraseni, T. 2018 Soil moisture forecasting by a hybrid machine learning technique: ELM integrated with ensemble empirical mode decomposition. *Geoderma* **330**, 136–161. https://doi.org/10.1016/j.geoderma.2018.05.035.

Rehana, S. & Mujumdar, P. P. 2012 Climate change induced risk in water quality control problems. *Journal of Hydrology* **444–445**, 63–77. https://doi.org/10.1016/j.jhydrol.2012.03.042.

Rehana, S., Sireesha Naidu, G., Monish, N. T. & Sowjanya, U. 2020a Modeling hydro-climatic changes of evapotranspiration over a semi-arid river basin of India. *Journal of Water and Climate Change*. jwc2020173. https://doi.org/10.2166/wcc.2020.173.

Rehana, S., Naidu, G. S. & Monish, N. T. 2020b Spatiotemporal variations of precipitation and temperatures under CORDEX climate change projections: a case study of Krishna River Basin, India. In: *Contemporary Environmental Issues and Challenges in Era of Climate Change* (Singh, P., Singh, R. P. & Srivastava, V. eds). Springer, Singapore, pp. 157–170. https://doi.org/10.1007/978-981-32-9595-7_8.

Rientjes, T. H. M., Muthuwatta, L. P., Bos, M. G., Booij, M. J. & Bhatti, H. A. 2013 Multi-variable calibration of a semi-distributed hydrological model using streamflow data and satellite-based evapotranspiration. *Journal of Hydrology* **505**, 276–290. https://doi.org/10.1016/j.jhydrol.2013.10.006.

Rooijen, D. J. V., Turral, H. & Biggs, T. W. 2009 Urban and industrial water use in the Krishna Basin. *India. Irrigation and Drainage* **58** (4), 406–428. https://doi.org/10.1002/ird.439.

Sajedi-Hosseini, F., Malekian, A., Choubin, B., Rahmati, O., Cipullo, S., Coulon, F. & Pradhan, B. 2018 A novel machine learning-based approach for the risk assessment of nitrate groundwater contamination. *Science of the Total Environment* **644**, 954–962. https://doi.org/10.1016/j.scitotenv.2018.07.054.

Salvi, K., Kannan, S. & Ghosh, S. 2013 High-resolution multisite daily rainfall projections in India with statistical downscaling for climate change impacts assessment. *Journal of Geophysical Research: Atmospheres* **118** (9), 3557–3578. https://doi.org/10.1002/jgrd.50280.

Saraf, V. R. & Regulwar, D. G. 2018 Impact of climate change on runoff generation in the Upper Godavari River Basin, India. *Journal of Hazardous, Toxic, and Radioactive Waste* **22** (4), 04018021. https://doi.org/10.1061/(ASCE)HZ.2153-5515.0000416.

Singh, R. & Kumar, R. 2015 Vulnerability of water availability in India due to climate change: a bottom-up probabilistic Budyko analysis. *Geophysical Research Letters* **42** (22), 9799–9807. https://doi.org/10.1002/2015GL066363.

Sinha, J., Sharma, A., Khan, M. & Goyal, M. K. 2018 Assessment of the impacts of climatic variability and anthropogenic stress on hydrologic resilience to warming shifts in Peninsular India. *Scientific Reports* **8** (1), 13833. https://doi.org/10.1038/s41598-018-32091-0.

Sood, A. & Smakhtin, V. 2015 Global hydrological models: a review. *Hydrological Sciences Journal* **60** (4), 549–565. https://doi.org/10.1080/02626667.2014.950580.

Srivastava, A. K., Rajeevan, M. & Kshirsagar, S. R. 2009 Development of a high resolution daily gridded temperature data set (1969–2005) for the Indian region. *Atmospheric Science Letters* **10** (4), 249–254. https://doi.org/10.1002/asl.232.

Srinivasa Raju, K. & Nagesh Kumar, D. 2018 Hydrological modeling. In: *Impact of Climate Change on Water Resources: With Modeling Techniques and Case Studies* (Srinivasa Raju, K. & Nagesh Kumar, D., eds). Springer, Singapore, pp. 137–167. https://doi.org/10.1007/978-981-10-6110-3_5.

Thornthwaite, C. W. 1948 An approach toward a rational classification of climate. *Geographical Review* **38** (1), 55–94. https://doi.org/10.2307/210739.

Vormoor, K., Heistermann, M., Bronstert, A. & Lawrence, D. 2018 Hydrological model parameter (in)stability – 'crash testing' the HBV model under contrasting flood seasonality conditions. *Hydrological Sciences Journal* **63** (7), 991–1007. https://doi.org/10.1080/02626667.2018.1466056.

Wilby, R. L. & Dawson, C. W. 2013 The statistical downScaling model: insights from one decade of application. *International Journal of Climatology* **33** (7), 1707–1719. https://doi.org/10.1002/joc.3544.

Wit, M. J. M. d. 2001 Nutrient fluxes at the river basin scale. I: the PolFlow model. *Hydrological Processes* **15** (5), 743–759. https://doi.org/10.1002/hyp.175.

Wu, J., Miao, C., Wang, Y., Duan, Q. & Zhang, X. 2017 Contribution analysis of the long-term changes in seasonal runoff on the Loess Plateau, China, using eight Budyko-based methods. *Journal of Hydrology* **545**, 263–275. https://doi.org/10.1016/j.jhydrol.2016.12.050.

Zhang, L., Hickel, K., Dawes, W. R., Chiew, F. H. S., Western, A. W. & Briggs, P. R. 2004 A rational function approach for estimating mean annual evapotranspiration. *Water Resources Research* **40** (2), W025021. https://doi.org/10.1029/2003WR002710.

Zhang, J., Zhang, Y., Song, J., Cheng, L., Kumar Paul, P., Gan, R., Shi, X., Luo, Z. & Zhao, P. 2020 Large-scale baseflow index prediction using hydrological modelling, linear and multilevel regression approaches. *Journal of Hydrology* **585**, 124780. https://doi.org/10.1016/j.jhydrol.2020.124780.

First received 30 April 2020; accepted in revised form 17 August 2020. Available online 11 November 2020

Evolutionary Algorithms in Water Resources
doi: 10.2166/h2oj.2020.123

Real-time monitoring of water level and storage dynamics of irrigation tank using IoT

Muthiah Krishnaveni[a,*], S. K. Praveen Kumar[a], E. Arul Muthusamy[b], J. Kowshick[a] and K. G. Arunya[a]

[a] Centre for Water Resources, Anna University, Chennai, India

[b] Electronics and Communication Engineering, Anna University, Chennai, India

*Corresponding author. E-mail: mkveni@annauniv.edu

Abstract

The internet of things (IoT), an emerging technological marvel, consists of a group of physical objects such as vehicles, machines and sensors to monitor and transfer data over the internet with much less human to machine interaction. It relies on a host of technologies like application programming interfaces (API), which in turn, help the devices to get connected with the internet. Efficient irrigation tank management requires a strong database on continuous water level dynamics for irrigation decision-making. Real-time tank water level monitoring is possible through an IoT device by integrating sensors and microcontroller that can send the water level data to the cloud. Google sheet is used to store the water level data that can be viewed using a web application as well as a mobile application. The contour map of the study tank is used to develop the stage (water level) vs volume curve. The volume of water present in the tank at any time can be arrived at for any tank water level using the above curve. The developed device can provide real-time continuous water level data with low cost and simple infrastructure, thus aiding tank water management.

Key words: internet of things, irrigation tank, real-time monitoring, sensor technology

Highlights

- The present study attempts to demonstrate the use of IoT technologies for the automated process of the real-time water level monitoring of an irrigation tank.
- An IoT device is developed by integrating sensors and microcontroller which can sense and send the real-time tank water level data to the cloud.
- The developed low cost IoT-based device will sense, analyse and provide the continuous data on water level and storage dynamics to the user departments and farmers for decision making. It can also be extended as a flood monitoring device to alert farmers in the command area of the irrigation tank.

Evolutionary Algorithms in Water Resources
doi: 10.2166/h2oj.2020.123

INTRODUCTION

India is well endowed with sufficient average rainfall of 1,215 mm per annum which is not distributed throughout the year. Eighty per cent of rainfall occurs during monsoon and 50% of this rainfall occurs during approximately 15 days of this monsoon period. India has the world's second largest irrigated area, accounting for about 64.57 Mha of which the area irrigated under tank irrigation is about 2.25 Mha. There are about 127,000 tanks in the southern region consisting of Andhra Pradesh, Tamil Nadu and Karnataka state. Tamil Nadu is a water-scarce state with annual per capita water availability of about 900 cm^3 as against India's average of 2,200 cm^3 (Anbumozhi *et al.* 2001). Most of the rivers in Tamil Nadu flow only during the monsoon season and it is not possible to construct more large-scale reservoirs due to the flat landscape (Arumugam *et al.* 2009). This has led to the construction of as many as 39,000 tanks as localized water harvesting structures. Most of the tank system is mainly used for irrigation purposes. Tanks are mostly managed by the local communities as a common property resource. Since tank irrigation systems are less capital-intensive and have a large geographical distribution, they are of special significance to a large number of small and marginal farmers (Gunnell & Krishnamurthy 2003).

Sustainable management of tank irrigation systems requires hydrological and meteorological data (Kortuem *et al.* 2010). The hydro-meteorological data are essential to estimate crop water requirement and to plan irrigation scheduling. Proper tank operation for irrigation release during normal, flood and drought periods essentially necessitates the water balance study of the tank on a daily basis. These data are rarely monitored and available in the case of tank systems. Thus, a cost-effective device using modern technology such as the internet of things (IoT) is necessary for the effective management of the tank irrigation systems of India.

IoT integrates the physical world with the virtual world by using the internet as the medium to communicate and exchange information (Mohammed Shahanas & Bagavathi Sivakumar 2016). It integrates several technologies that already exist, such as wireless sensor network (WSN), radio frequency (RF) identification, cloud computing, middleware systems and end-user applications. The capability of WSN to self-organize, self-configure, self-diagnose and self-heal has made it a good choice for smart agriculture and the food industry (Palanisami & Nanthakumaran 2009). It revolves around increased machine-to-machine communication and reduced human-to-human or human-to-computer interaction. It enables new forms of communication between people and things, and between things themselves. It is possible to monitor, transfer and store the data in real time. Thus, IoT helps various organizations and individuals to access information towards facilitating major decision-making processes. As such, the present study attempts to demonstrate the use of IoT technologies for the automated process of sensing the real-time data of water level of an irrigation tank. The developed IoT-based irrigation tank water level monitoring device will measure, analyse and provide a decision-supporting platform for the farmers.

The usefulness of IoT applications in various domains including water, medical, manufacturing, industrial, transportation, education, governance and mining is being explored. Several researchers have addressed the use of IoT in agriculture to enhance the different agricultural processes (Palanisami 2006). Zhu & Lin (2012) emphasized mostly the use of cloud enabled systems to show the relationship between the information cloud and IoT from the view point of agricultural data and its use cases. They argue that intelligent agriculture is one of the applications of IoT that has an extensive application and bright future (Patil *et al.* 2017). Khaled Reza *et al.* (2010) developed a device that could automatically control the water level in an overhead tank by switching on and off the pump using IoT (Govindasamy & Balasubramanian 1990). Mohammed Shahanas & Bagavathi Sivakumar (2016) framed an algorithm to monitor the water level dynamics in the overhead tank of a gated apartment community as an initiative to smart water management in smart cities of India (Shanmugam 2012).

doi: 10.2166/h2oj.2020.123

Conventionally, the water level in irrigation tanks is noted and recorded manually. This involves human error and this information will not be available to farmers and researchers. This paper illustrates the novel idea of real-time monitoring of water level dynamics of an irrigation tank through IoT. In addition to this, the humidity and temperature data are also measured using the relevant sensors. All these data can be shared with farmers, administrators and researchers digitally to assess the water balance and plan the optimal use of tank water.

STUDY AREA

The study tank chosen is the Melnelli big tank in Arcot taluk of Vellore district, Tamil Nadu which is located at latitude 12°43′45″N and longitude 79°28′10″E. The spatial spread of the catchment, water spread and the command areas of the irrigation tank are shown in Figure 1. The original storage capacity of the tank at full tank level (FTL) is 1.40 Mm^3. However, the current capacity of the tank

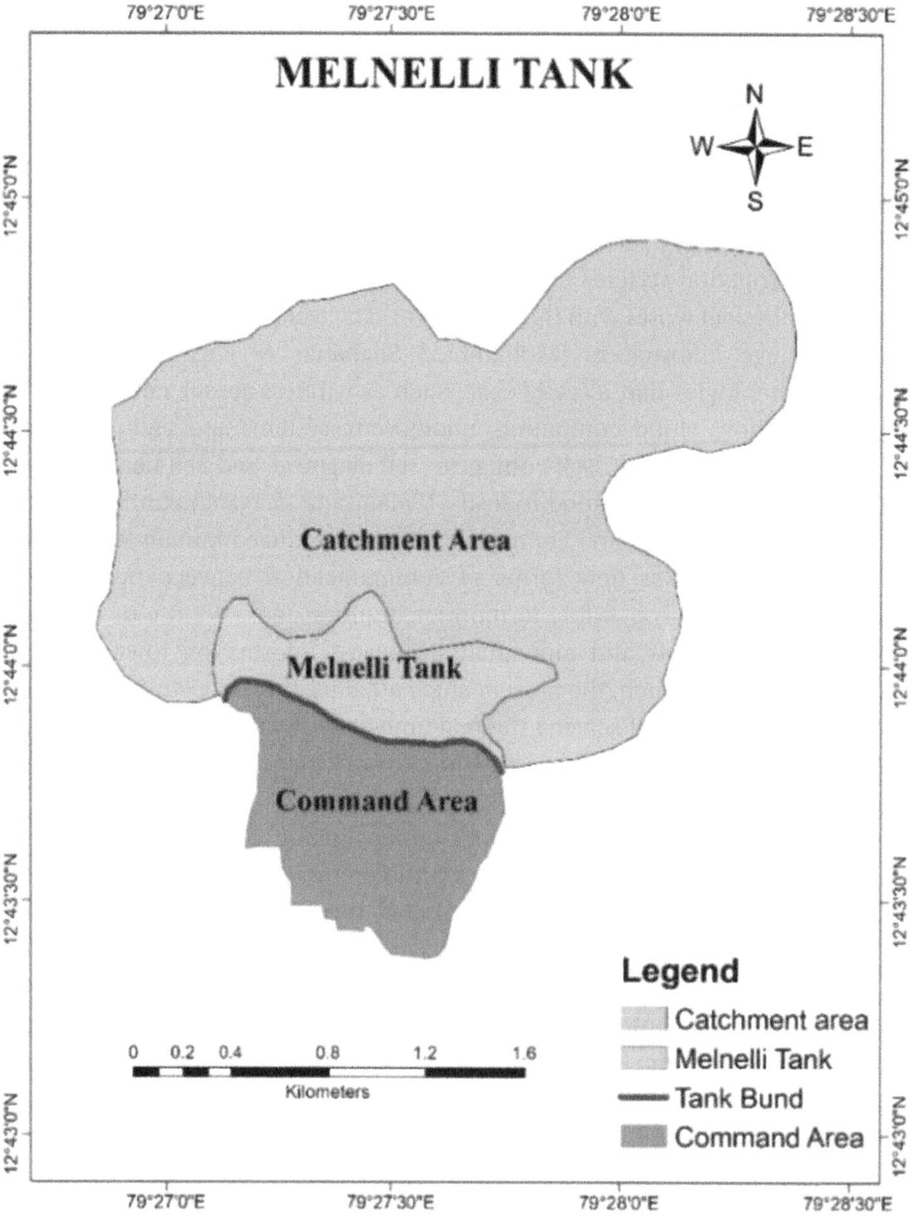

Figure 1 | Melnelli tank in Vellore district of Tamil Nadu.

Evolutionary Algorithms in Water Resources
doi: 10.2166/h2oj.2020.123

is 1.28 Mm^3. The reduction in capacity is mainly due to siltation in the tank bed. The FTL of the tank is +120,460 m above the mean sea level (msl) and the deepest point in the tank is +117,000 m above msl. The tank command area is 56.67 hectares.

Stage vs volume curve of the irrigation tank

Elevation data for different points in the tank bed were collected from the Public Works Department of Tamil Nadu. A contour map of 0.7 m interval was created with AutoCAD (Figure 2) using the elevation data. This map was imported into GIS environment using ArcGIS and a digital elevation model (DEM) was created. The elevation vs storage graph for the study tank was developed and plotted as shown in Figure 3. The elevation represents the water level of the tank also. By observing

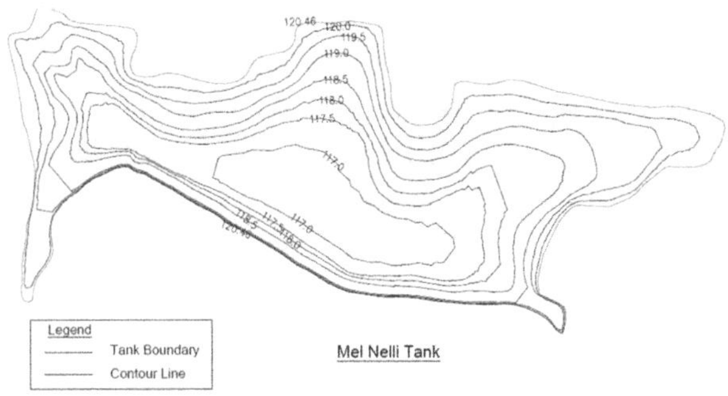

Figure 2 | Contour map of Melnelli tank in Vellore district of Tamil Nadu.

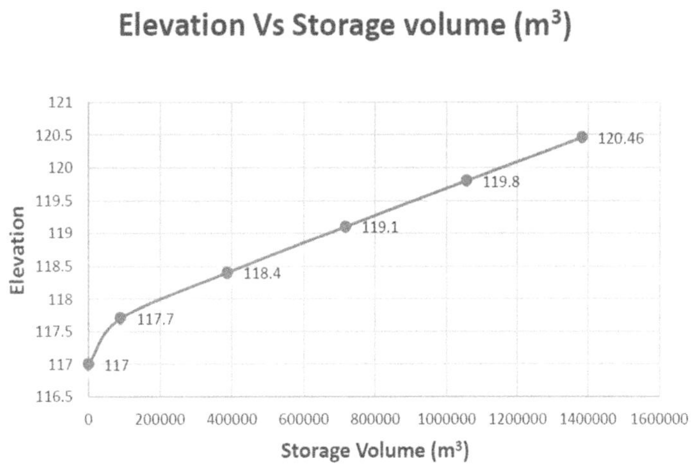

Figure 3 | Elevation vs storage volume curve of Melnelli tank.

the water level in the tank in real time, one can obtain the volume of water stored in the tank at any time using the above graph. Hence, the water resources' status of the tank can be obtained by the continuous monitoring of water level in the irrigation tank.

SENSOR CALIBRATION OF WATER LEVEL MEASUREMENT

The most widely used sensors in the field of water level monitoring are ultrasonic sensor and LIDAR sensor (Trinklein 1990; Wang *et al.* 2006). The success of this project depends upon accuracy and reliability of the data from the device. Working-wise, LIDAR and ultrasonic sensors are quite similar, but in LIDAR laser waves are used and large depths of up to 40 m can be measured. LIDAR measurements are not affected by temperature variations but the ultrasonic sensors do vary with an increase or decrease in temperature. The one and only factor affecting the sound waves is temperature, which causes some errors in the readings. It can be rectified with the help of the following formula (Fan 2013):

$$V(m/s) = 331(m/s) + 0.6T(C) \tag{1}$$

where V is velocity in m/s, T is temperature in Celsius. Hence, by knowing the temperature, the velocity of waves is corrected. Now the corrected velocity is substituted in the following formula to find the distance between the sensor and the water level:

$$Distance(m) = (duration/1000)(ms)*(V/2)(m/s) \tag{2}$$

In this study, the ultrasonic sensor and LIDAR sensor were tested and compared for better depth measurement accuracy. The best of the two sensors is identified through calibration testing for measuring the water level. The calibration of LIDAR and ultrasonic sensors was made using an evaporation pan in the field laboratory of the Centre for Water Resources, Anna University (Figure 4). The results are compared with the actual depth of water level in the evaporation pan and are shown in Table 1.

Figure 4 | Sensors' calibration using evaporation pan.

Table 1 | Comparison of sensors

Actual depth in cm	LIDAR value	Ultrasonic sensor value	Ultrasonic sensor after temperature correction
50	47–53	49–49.5	49.8–50.3
100	97–103	98–98.5	99.7–100.3
150	147–153	147.5–148	149.7–150.3

It can be seen that the ultrasonic sensor gives greater accuracy than the LIDAR. After temperature correction, accuracy of the sensor goes up to ± 0.3 cm whereas LIDAR has an error of ± 3 cm, hence the ultrasonic sensor HR-SC04 module is chosen in this study. Ultrasonic sensor and DHT 11 module were connected with the NodeMCU controller.

METHODOLOGY AND MATERIALS

The main aim of this study is the remote monitoring of temperature, humidity and water level dynamics of the tank to estimate the volume of water available for irrigation. The conceptual flow chart explaining

doi: 10.2166/h2oj.2020.123

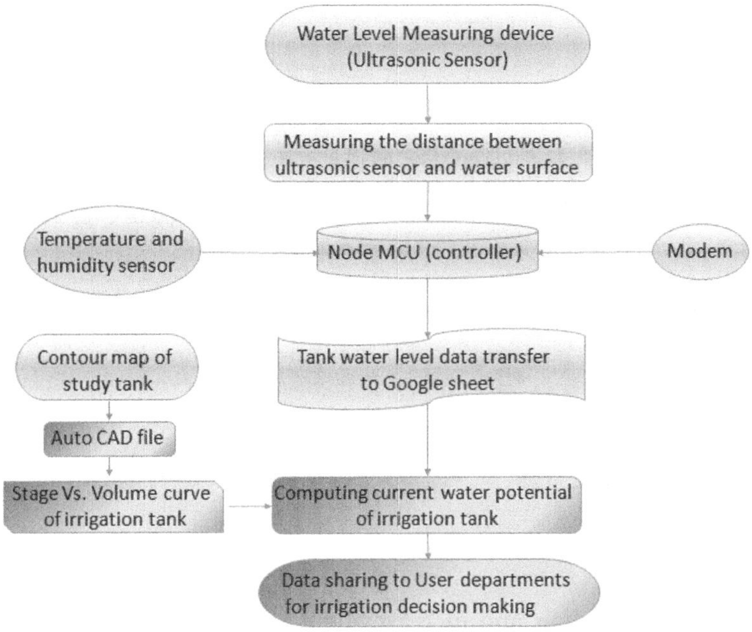

Figure 5 | Methodology for remote monitoring of tank water level dynamics.

the various processes involved in the real-time monitoring of tank water level is given in Figure 5. The IoT-based device consists of NodeMcu, sensors and power supply networks, which are explained in the following sections. NodeMcu is used to control the overall system automatically that reduces the design and control complexity. NodeMcu takes input from the sensor unit which senses the water level, temperature and humidity. The collected data from the sensor are sent to the cloud network by means of WiFi for further data analysis. The actual water level present in the tank, temperature and humidity are displayed on the Google sheets which are stored in cloud storage of Google. Anybody can use this portal to get information about the water available in the lake.

Sensor network

NodeMCU is an integrated chip which provides internet connectivity in a relatively small package based on ESP8266-12E WiFi module. It can be programmed directly through a USB port using Arduino IDE. NodeMCU can serve as the WiFi access point and station as well as the microcontroller. It consumes 5 V. The schematic representation of the sensor network is shown in Figure 6. The ultrasonic sensor and temperature-humidity sensor are connected to the digital pin of NodeMcu. A basic ultrasonic sensor consists of one or more ultrasonic transmitters, a receiver and a control module.

The ultrasonic sensor emits high frequency ultrasonic waves of about 40 KHz. These ultrasonic waves reflect back if there is a change in density of ultrasonic pulse travel medium. The reflected waves are received by the receiver in the sensor and it takes 2t time to receive the reflection. The depth between the sensor and water level (D) will be equal to the product of half of the duration (t/2) and velocity of the ultrasonic wave (v). The DHT11 temperature and humidity sensor module is a composite sensor containing a calibrated digital signal output of the temperature and humidity. This sensor can sense the relative humidity ranging from 20 to 90% with an accuracy of $\pm 5\%$ relative humidity and can sense the temperature ranging from 0 to 50 °C with an accuracy of ± 2 °C.

Power supply through solar panel

Power requirement for the total circuit is supplied by the solar panel which is made with high efficiency silicone cells. The solar cells are coated with blue nitride anti-reflection coating which

doi: 10.2166/h2oj.2020.123

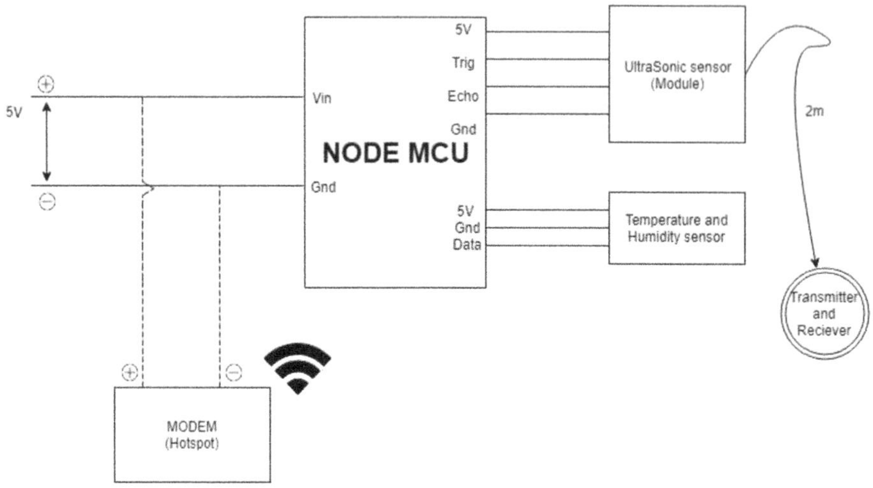

Figure 6 | Sensor network.

provides a uniform blue hue colour and increases light absorption in all weather conditions. Each panel is equipped with low-iron patterned tempered glass which provides up to 60% more strength as compared to conventional photovoltaic glass. A schematic representation of the power supply network is shown in Figure 7.

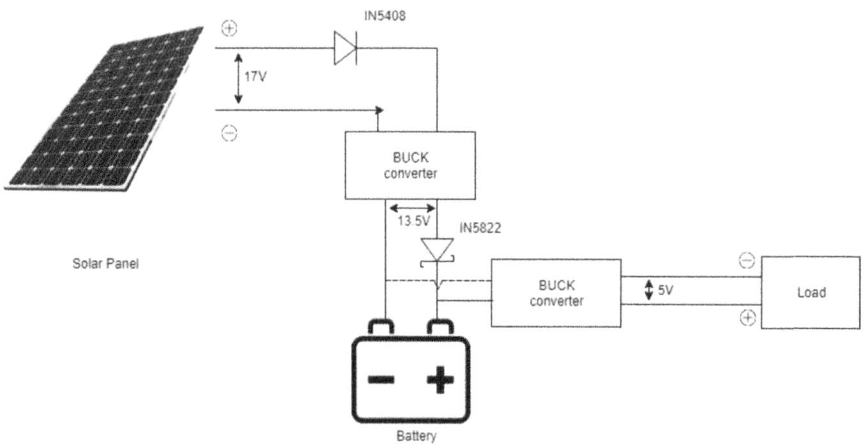

Figure 7 | Power supply.

The solar panel is capable of producing up to 17 V but the sensor network needs only 5 V DC power supply. A buck converter (step-down converter) is a DC-to-DC power converter which steps down voltage (while stepping up current) from its input (supply) to its output (load). It is a class of switched-mode power supply (SMPS) typically containing at least two semiconductors (a diode and a transistor, although modern buck converters frequently replace the diode with a second transistor used for synchronous rectification) and at least one energy storage element, a capacitor, inductor, or the two in combination. To reduce voltage ripple, filters made of capacitors (sometimes in combination with inductors) are normally added to such a converter's output (load-side filter) and input (supply-side filter). The buck converter is used when the DC output voltage needs to be lower than the DC input voltage. Efficiency of the buck converters are higher than 90%, so that it can be used to convert a main supply voltage of nearly 17–5 V DC.

Evolutionary Algorithms in Water Resources
doi: 10.2166/h2oj.2020.123

Water level monitoring in irrigation tank

The IoT device is installed in the irrigation sluice which is located in the deepest level of the tank bed (+117 m above msl). The experiment was conducted during November 2017. The water level, temperature and depth are sensed and sent to the Google cloud storage through the WiFi module. The sensor provides 500 depth readings per second. A Google script program is written in Arduino IDE to record the values of depth, temperature and humidity for each second. The program performs the temperature correction on depth and pushes the corrected data to cloud storage through the built-in Wifi module of NodeMCU. Google cloud storage is utilized to store and retrieve the sensor data. A detailed program is written in Google script to retrieve the data through Google sheet and to plot the graph showing temporal variation in tank water level as shown in Figure 8. The ID indicates the time in seconds. The table shows the values of temperature, humidity and depth of water level (cm) for each second. Permission to access these data on Google sheet can be given to the concerned authorities, such as Public Works Department for decision-making. Using the stage vs volume plot, the authorities can infer the volume of water available in the tank. This device enables monitoring and studying the water level dynamics of the tank through mobile or PC remotely.

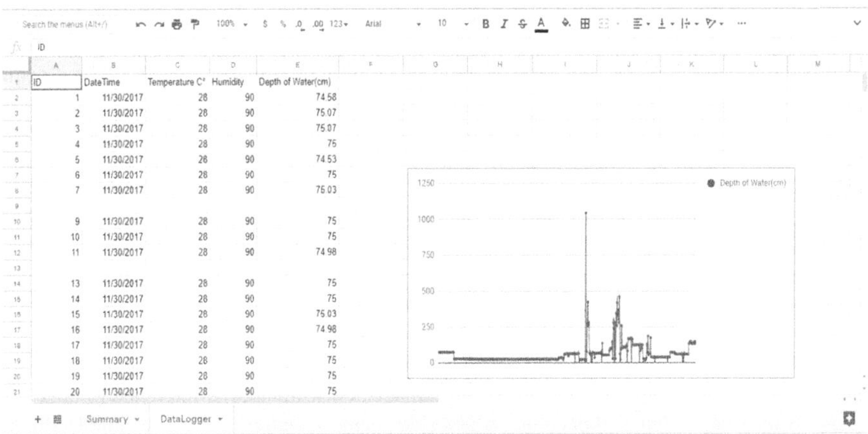

ID	DateTime	Temperature C°	Humidity	Depth of Water(cm)
1	11/30/2017	28	90	74.58
2	11/30/2017	28	90	75.07
3	11/30/2017	28	90	75.07
4	11/30/2017	28	90	75
5	11/30/2017	28	90	74.53
6	11/30/2017	28	90	75
7	11/30/2017	28	90	75.03
9	11/30/2017	28	90	75
10	11/30/2017	28	90	75
11	11/30/2017	28	90	74.98
13	11/30/2017	28	90	75
14	11/30/2017	28	90	75
15	11/30/2017	28	90	75.03
16	11/30/2017	28	90	74.98
17	11/30/2017	28	90	75
18	11/30/2017	28	90	75
19	11/30/2017	28	90	75
20	11/30/2017	28	90	75

Figure 8 | Water level dynamics monitoring through web application.

CONCLUSION

IoT has been gradually bringing a sea of technological changes in our daily lives, which in turn, helps to make our lives simpler and more comfortable though various applications. A low cost, eco-friendly and simple IoT device is developed in this study for the remote monitoring of tank water level dynamics for the water administrators and the farming community who depend mainly on the tank system as their source of irrigation. This study found that the ultrasonic sensor can give accurate results when compared to that of the LIDAR sensor. The overall accuracy of the ultrasonic sensor is ± 2.5 cm. Apart from water level, the temperature and humidity are also monitored by the device. Data are available through mobile and web portals that make them very easy to access and to understand. Continuous data on water availability will be very useful for operating tank sluices, selecting crops, changing cropping pattern and other irrigation decision-making. It can also be used as a flood monitoring device which is indirectly used to warn people in the command area of the irrigation tank. The developed device can be extended to all the 39,000 tanks and 116 reservoirs of Tamil Nadu state. The IoT technologies can be applied for further areas of water resources like flood monitoring, dam seepage monitoring, overhead tanks for domestic water supply, automated evaporimeters in meteorological stations and water quality monitoring.

ACKNOWLEDGEMENT

We would like to thank the Centre for Technology Development and Transfer (CTDT), Anna University for funding this innovative project study. Our sincere thanks to Public Works Department, Palar river basin division, Vellore district for providing the necessary data.

DATA AVAILABILITY STATEMENT

Data cannot be made publicly available; readers should contact the corresponding author for details.

REFERENCES

Anbumozhi, V., Matsumoto, K. & Yamaji, E. 2001 Towards improved performance of irrigation tanks in semi-arid regions of India: modernization. *Irrigation and Drainage Systems* **15** (4), 293–309.

Arumugam, N., Mohan, S. & Ramaprasad, R. 2009 Sustainable development and management of tank irrigation systems in south India. *Water International* **22** (2), 90–97.

Fan, T.-K. 2013 Smart agriculture based on cloud computing and IOT. *Journal of Convergence Information Technology* **8** (2), 210–216.

Govindasamy, R. & Balasubramanian, R. 1990 Tank irrigation in India: problems and prospects. *International Journal of Water Resources Development* **6** (3), 211–217.

Gunnell, Y. & Krishnamurthy, A. 2003 Past and present status of runoff harvesting systems in dryland peninsular India: a critical review. *AMBIO* **32** (4), 320–324.

Khaled Reza, S. M., Shah Ahsanuzzaman Md, T. & Mohsin Reza, S. M. 2010 Microcontroller based automated water level sensing and controlling: Design and implimentation issue. In: *Proceedings of the World Congress on Engineering and Computer Science*, October 20–22, San Francisco, CA, USA.

Kortuem, G., Kawsar, F., Sundramoorthy, V. & Fitton, D. 2010 Smart objects as building blocks for the internet of things. *IEEE Internet Computing* **14** (1), 44–51.

Mohammed Shahanas, K. & Bagavathi Sivakumar, P. 2016 Framework for a smart water management system in the context of smart city initiatives in India. *Procedia Computer Science* **92**, 142–147.

Palanisami, K. 2006 Sustainable management of tank irrigation systems in India. *Journal of Developments in Sustainable Agriculture* **1**, 34–38.

Palanisami, K. & Nanthakumaran, A. 2009 Water resources management with special reference to tank irrigation with groundwater use. In: *Second National Workshop on Strategic Issues in Indian Irrigation*, New Delhi, India, pp. 305–333.

Patil, K. N., Tushar, K., Swaranjali, M. S. & Shreya, M. A. 2017 IOT based water level monitoring system for lake. *International Research Journal of Engineering and Technology* **4** (2), 1425–1427.

Shanmugam, K. R. 2012 *Monitorable Indicators and Perfomances: Tamilnadu*, 17th edn. Madras School of Economics, Chennai, India.

Trinklein, F. E. 1990 *Modern Physics*. Holt Rinehart and Winston, New York, USA.

Wang, N., Zhang, N. & Wang, M. 2006 Wireless sensors in agriculture and food industry – recent development and future perspective. *Computers and Electronics in Agriculture* **50** (1), 1–14.

Zhu, X. & Lin, Y. 2012 ZigBee implementation in intelligent agriculture based on internet of things. In: *2nd International Conference on Electronic & Mechanical Engineering and Information Technology*, Paris, France, pp. 1842–1846.

First received 28 March 2020; accepted in revised form 3 August 2020. Available online 30 September 2020

Evolutionary Algorithms in Water Resources
doi: 10.2166/h2oj.2020.117

Development of algorithms for evaluating performance of flood simulation models with satellite-derived flood

Tushar Surwase[a,*], P. Manjusree[a], Sachin Prakash[b] and Saikiran Kuntla[c]

[a] Special User Flood and Landslide Monitoring Division (SFLMD)/NRSC/ISRO, Hyderabad, 500037, India

[b] Application Software and Computing Infrastructure Division (ASCID)/NRSC/ISRO, Hyderabad, 500037, India

[c] Department of Civil Engineering, Indian Institute of Technology, Delhi (IIT-D), Delhi, 110016, India

*Corresponding author. E-mail: tusharsurwase@ymail.com

Abstract

Flood inundation simulation models are widely used for simulating severe events of flood, generating hazard maps, risk assessment, and to identify flood vulnerable locations. It is important to assess the degree of accuracy of flood model results as these results may be one of the triggering parameters considered in developing flood hazard maps, flood mitigation policies, and land using planning where multi-criteria analysis is approached. In the present study, an algorithm is developed in order to know the performance of flood models by validating it with flood footprints extracted from synthetic aperture radar (SAR) images using multi-segmentation and Otsu's thresholding technique. Evaluation of the performance of the model is based on two best fit criteria called F1 and F2. For this, HEC-RAS model is used for simulating the severe event of flood witnessed in Mahanadi River in Odisha stretching between Tikarpara and Mundali during September 2008. Three simulations were made by considering three different Manning's roughness for river and floodplain. The model gives appreciable results and best fit $F1 = 0.85$ and $F2 = 0.74$ was found for Manning's roughness 0.020.

Key words: evaluation, HEC-RAS, model performance, multi-segmentation and Otsu's thresholding

Highlights

- The study showcases development of algorithms for extraction of flood disaster footprints without any manual intervention which can help disaster respondents to respond to the event in near real time.
- It also showcases development of algorithms for evaluation performance of flood models which help flood modellers to evaluate the used flood model performance in validation with satellite-based observed flood.

INTRODUCTION

Flooding is a global phenomenon that causes human and property loss as mentioned by Teng *et al.* (2017). A flood inundation hydraulic model predicts flood extent and it may be an important source of information for flood risk assessment studies and in determining flood-vulnerable areas (Liu *et al.* 2018; Shastry & Durand 2019). Flood models also play an important role in flood preparedness and flood risk reduction as stated strongly by Chen *et al.* (2019). Flood hydrodynamic models give the result in the form of depth, water surface elevation and velocities with respect to time. Researchers and hydraulic modelers simulate flood for a river basin to study its impact assessment and to design hydraulic structures. They generally validate their simulated flood extent results (total area) with observed flood extent (ground truth or satellite observed) total area. Validating by considering only total area simulated versus total area observed cannot be a correct practice because simulated extent may be more than observed extent at some places and at some places the reverse may be true. Thus, results compared in terms of total area may give appreciable results, however, there will be ambiguity in the validation. Hence, simply evaluating total area flood simulated versus total area flood observed cannot give a correct picture of hydraulic/flood model performance. The simulated results must be assessed pixel-wise with satellite observed extent in order to evaluate flood model performance. There are many statistical approaches for evaluating flood model performance, one of which was developed by SENAGUA (2014) to compare the water surface elevation result from the model to the ground scenario. Bates & De Roo (2000) evaluated the model performance based on flood extent. Horrit *et al.* (2007) and Di Baldassarre *et al.* (2009) worked on evaluating flood extent in terms of measures of fit F1 and F2 which was proposed for flood studies by Moya-Quirogaa *et al.* (2016). The present study uses Horrit and Di Baldassarre techniques, which were further modified to develop an algorithm in Python for evaluating the performance of flood model (F1 and F2) automatically by inputting two rasters, i.e., flood observed and flood simulated. This technique is selected as it evaluates flood extent in terms of measures of fit, which is limited to the scope of the present study. The significance of the developed algorithm over the existing technique is that the algorithm minimizes the number of GIS operations and computational efforts on two rasters (flood simulated and flood observed) in order to evaluate fitting values F1 and F2 and estimate cell values, which are correctly predicted, overpredicted, and underpredicted by the model. To test the algorithm, 2D flood simulation model is set up for part of Mahanadi River in Odisha in the HEC-RAS model. HEC-RAS (Hydrologic Engineering Centre – River Analysis System) is a freely available and popular hydraulic model developed by the United States Army Corps of Engineers (USACE). The flood inundation can be simulated by using 1D (one-dimensional) or 2D (two-dimensional) models. HEC-RAS version prior to 5.0.1 was limited to simulating 1D flow, which can be suitable to analyze the behavior of flow in open channels (natural or artificial) where water is flowing below danger level, and thus can be considered to be 1D flow as the water is flowing only in the longitudinal direction. When water levels exceed threat (danger) level and water approaches floodplains, there is the possibility of water moving in two directions, thus 2D flow modeling becomes essential for simulating floods with greater accuracy. Taking this into account, USACE in February 2016 developed version 5.0.1, which has the capability to simulate 2D flow. HEC-RAS is open source hydraulic software which performs better and achieves high accuracy in inundation modeling, Surwase *et al.* (2019) concluded in their flood inundation simulation study. The study uses HEC-RAS 2D model version 5.0.1 February 2016 release to simulate the floods for the September 2008 Mahanadi river flood. The simulations are performed by providing the daily river discharge values as an input which were recorded at the time of the flood event and obtained from Central Water Commission. The model shows the simulation results in the form of flood extent with depth and velocity. The results were further validated with RADARSAT SAR satellite image. SAR has proven to be an effective tool for water surface detection as stated by Tholey *et al.* (2015), Hoque *et al.* (2011), Henry *et al.* (2006), Kuenzer *et al.* (2013), for estimation of flood depth studied by Iervolino *et al.* (2015), and flooding beneath vegetation canopies in certain conditions observed by Richards *et al.* (1987), Hess *et al.* (1990), Townsend (2002a, 2002b), Kasischke & Bourgeau-Chavez (1997). Commonly used SAR-based flood extent mapping approaches include simple visual interpretation (Oberstadler *et al.* 1997), supervised classification (De Roo *et al.* 1999; Townsend 2002a, 2002b), image texture algorithms (Schumann *et al.* 2005), histogram thresholding (Schumann *et al.* 2010), various multitemporal change detection methods (Bazi *et al.* 2005), and active contour models (Bates *et al.* 1997; Matgen *et al.* 2007). Among these, Cao *et al.*

Evolutionary Algorithms in Water Resources
doi: 10.2166/h2oj.2020.117

(2019) mentioned that histogram thresholding is a common and simple approach which can be used to separate flood and non-flood areas in a large area (swathe) due to the specular backscattering characteristics of active radar pulses on plain water surfaces and the resultant low signal return.

Visual inspection and manual editing can help in training regions of interest (ROIs) and reducing false positives/negatives, demonstrated by Pulvirenti *et al.* (2010). As manual editing is time-consuming and manpower-driven, it cannot be applied to rapid response to flood disasters, especially during back-to-back flood events (Shen *et al.* 2019). The common technique used is the thresholding technique which is a pixel-based operation and considered as the most efficient way for rapid mapping. The specular reflection properties of non-obstructed water have driven numerous endeavors to decide a limit, beneath which, pixels are recognized as water, which was demonstrated by Yamada (2001), Giustarini *et al.* (2013), Hirose *et al.* (2001), and Matgen *et al.* (2011). A single threshold may not hold well in large-area water bodies (Tan *et al.* 2004) or for the whole swathe of a SAR image since it suffers from the heterogeneity of the environment, caused by wind-roughening and satellite system parameters (Martinis *et al.* 2009). Martinis & Rieke (2015) demonstrated the temporal heterogeneity of the backscattering of permanent water bodies, implying the temporal variability of the threshold. To address the spatial variability, Martinis *et al.* (2009) applied a split-based approach (SBA) together with an object-oriented (OO) segmentation method as described by Baatz (1999). Matgen *et al.* (2011) introduced a histogram segmentation method, and Giustarini *et al.* (2013) generalized the calibration process. Pulvirenti *et al.* (2011) developed an image segmentation method consisting of dilution and erosion operators to remove isolated groups of water pixels and small holes in water bodies, which are believed to be caused by speckle. Giustarini *et al.* (2013), Matgen *et al.* (2011), and Lu *et al.* (2014) employed the region growing algorithm (RGA) to extend inundation areas from detected water pixels. The RGA starts from seeding pixels and then keeps absorbing homogenous pixels from neighbors until no more homogenous pixels exist in neighboring areas. Martinis *et al.* (2009) applied SBA, as described by Bovolo & Bruzzone (2007), to determine the global threshold for binary (water and non-water) classification. In SBA, a SAR image is first divided into splits (sub-tiles) to determine their individual thresholds using the Kittler and Illingworth (KI) method, global minimum, and quality index described by Kittler & Illingworth (1986). Then, only global threshold can be obtained by qualified splits showing sufficient water and non-water pixels.

In this study, automation of the delineation of the flood is carried out using a multi-segmentation technique combined with Otsu's threshold technique and executed with Python scripts. The post-processing component includes removal of stray pixels and applying a water mask to delineate the final flood layer. The GDAL module in Python was used to read the pre-processed image/intermediate product and water mask, along with their projection system. Later, the image is processed for classification with numpy and scikit-image modules. The final output is exported using GDAL module with the same input projection system.

STUDY AREA

The flood simulation was carried out for a stretch of the Mahanadi River in Odisha state, India. The study area is from Tikarpara to Mundali and lies between latitudes 20° 07′58.8″ N and 20° 43′8.4″ N and longitudes 84°44′42″ E and 85° 46′4.8″ E. Mahanadi is the major river flowing through Odisha state and in the study area. A location map of the study area is shown in Figure 1.

There are two gauge stations, Tikarapara and Mundali, where discharge is measured. The study area is located at the downstream side of Hirakud dam, which intercepts a catchment area of 48,700 km^2 that is responsible for floods in Mahanadi deltaic area (Parhi *et al.* 2012). Odisha witnessed peak floods in September 2008 affecting millions of people, damaging agricultural crops and infrastructure.

doi: 10.2166/h2oj.2020.117

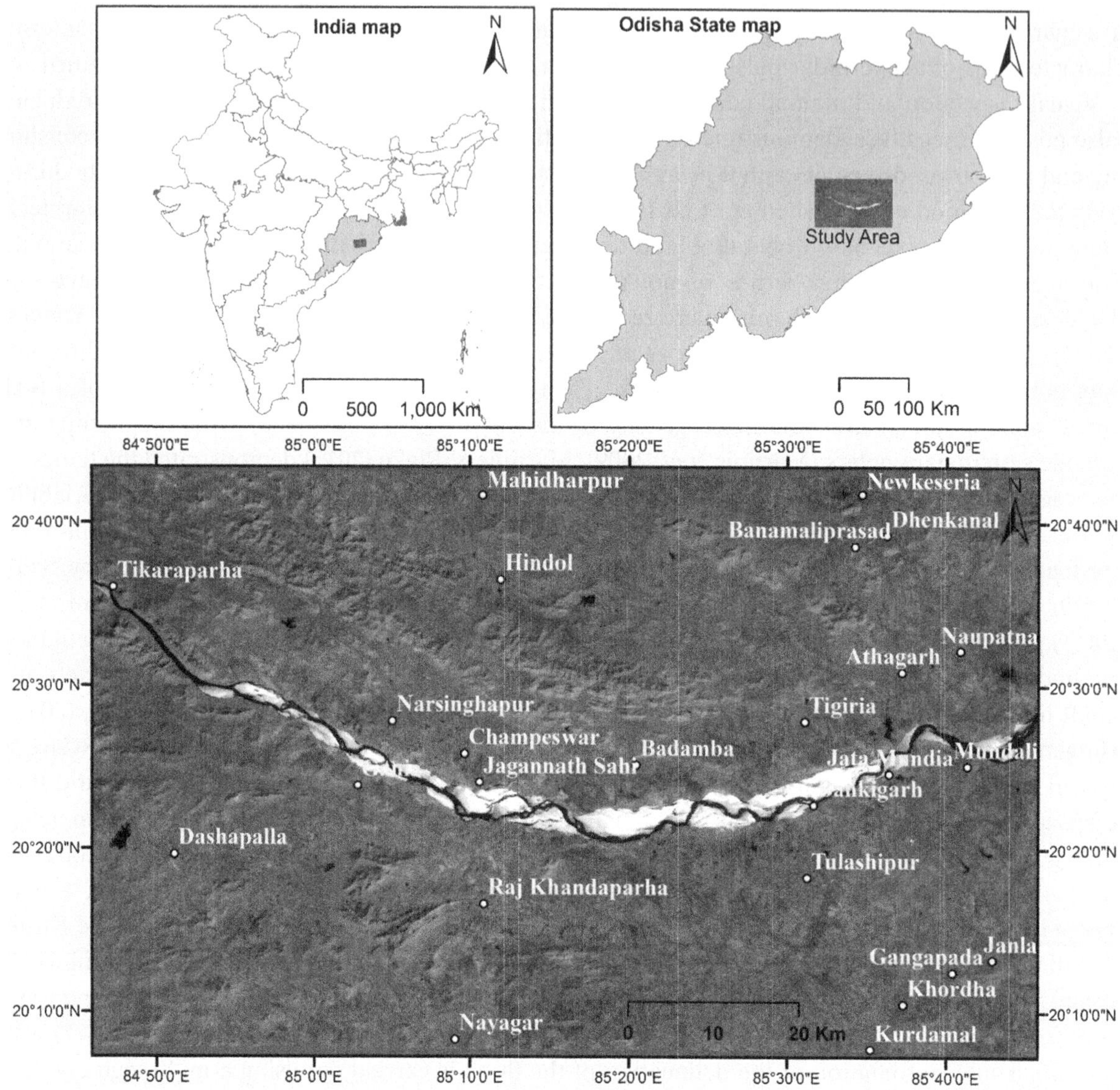

Figure 1 | Study area showing location map and major settlements (IRS-LISS-III image).

Hence, flood inundation simulation for the September 2008 flood event is carried out for the study area.

DATA USED

Flood inundation simulation was carried out by using Carto-DEM 10 meter spatial resolution. Land-use land cover (LULC) at 1:250,000 scale derived from Advanced Wide Field Sensor (AWiFS) image acquired in 2016–2017, was used for assigning Manning's roughness coefficient of the terrain. Discharge data collected from Central Water Commission (CWC) of Tikarpara gauge station were used as inflow boundary condition of the setup model in HEC-RAS. The study simulated the flood event between 2 September 2008 and 30 September 2008. Radarsat satellite SAR image was used to validate the simulated results.

doi: 10.2166/h2oj.2020.117

METHODS

The methodology is split into three categories, namely, flood simulation, development of an algorithm for extraction of flood footprints from SAR image, and development of an algorithm for evaluating the performance of flood models.

2D flood simulation

Flood is simulated by using HEC-RAS version 5.0.1 for the flood event between 5 September and 30 September, 2008. The study area is defined by 2D closed polygon. The 2D polygon is an area where flood inundation is expected to occur. The polygon is digitized by connecting higher elevation like hills and mountains where flood inundation will not extend beyond the same features. After defining the boundary of a polygon, the computational square cells are generated. The cells are in a square grid called a staggered grid and at the boundary of the 2D polygon the cells are in a non-staggered grid which is a composition of polygons having three to a maximum of eight sides. The spacing of the square grid was kept at an interval 10 m × 10 m in x and y directions which was kept equal to spatial resolution of the digital elevation model (DEM) used. Further, two boundary conditions were assigned to the 2D polygon. The first one is the inflow boundary condition which represents observed discharge in the form of a hydrograph at Tikarpara gauge station. The observed inflow hydrograph in between the simulation period is shown in Figure 2. The width of the river at the gauge station is 760 m which covers 76 cells, and the inflow boundary condition was applied to these 76 cells. The second boundary condition is the outflow boundary condition normal depth assigned at Mundali station. Considering topographic conditions from DEM, normal depth slope 0.001 was assigned. Three simulations were carried out for three different roughness coefficients for river banks (floodplains), with roughness coefficient for the channel being kept the same for all three simulations (Table 1).

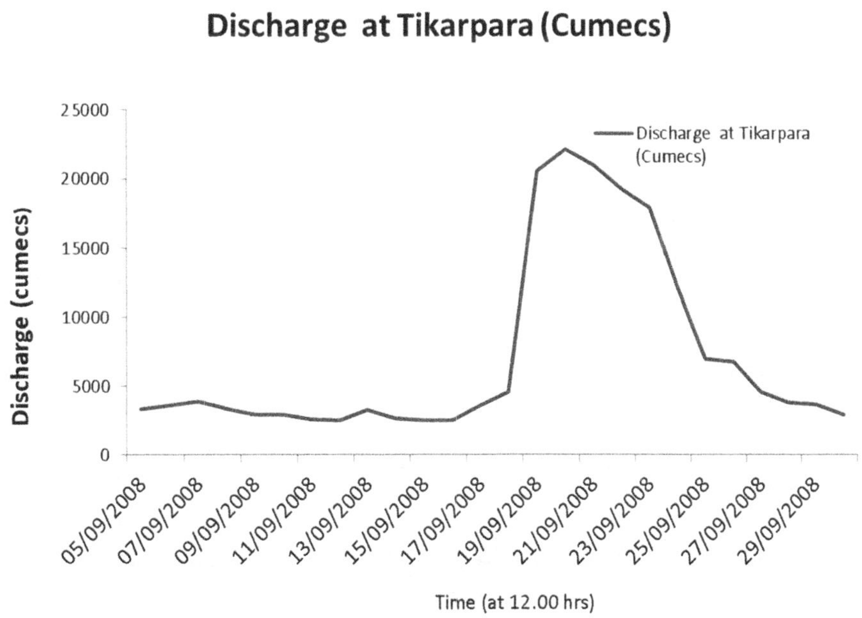

Figure 2 | Observed inflow hydrograph at Tikarpara gauge station.

doi: 10.2166/h2oj.2020.117

Table 1 | Roughness coefficient for three simulations

Simulation	Roughness coefficient for channel (n_c)	Roughness coefficient for floodplains (n_p)
1	0.032	0.020
2	0.032	0.030
3	0.032	0.040

The simulation is carried out using full momentum Equations (1) and (2):

$$\frac{\partial p}{\partial t}+\frac{\partial}{\partial x}\frac{(p^2)}{(h)}+\frac{\partial}{\partial y}\frac{(pq)}{(h)}=-n^2pg\frac{\sqrt{(p^2+q^2}}{h^2}-gh\frac{\partial s}{\partial x}+pf+\frac{\partial}{\rho\partial x}(h\tau_{xx})+\frac{\partial}{\rho\partial y}(h\tau_{xy}) \quad (1)$$

$$\frac{\partial q}{\partial t}+\frac{\partial}{\partial x}\frac{(q^2)}{(h)}+\frac{\partial}{\partial y}\frac{(pq)}{(h)}=-n^2qg\frac{\sqrt{(p^2+q^2}}{h^2}-gh\frac{\partial s}{\partial y}+qf+\frac{\partial}{\rho\partial y}(h\tau_{yy})+\frac{\partial}{\rho\partial x}(h\tau_{xy}) \quad (2)$$

where, h is the water depth in meters, p and q are specific flow in x and y directions (m^2/sec), s is the surface elevation in meters, n is the Manning's resistance, g is gravitational acceleration in m/sec^2, ρ (rho) is the density of water in kg/m^3, f is the Coriolis force (/sec) and τ_{xx}, τ_{xy}, and τ_{yy} are the components of effective shear stress. Computational time step was evaluated based on the Courant–Friedrichs–Lewy condition given in Equation (3):

$$cr=\frac{C\,\Delta t}{\Delta x} \quad (3)$$

where, cr is Courant's number, c is celerity in m/sec; Δt is a computational time step in seconds, Δx is grid cell size in meters. The celerity 'c' was calculated considering the maximum depth of water observed at Tikarpara gauge station. Thus, Equation (3) can be returned as:

$$\frac{\sqrt{gh}}{\Delta x}\frac{\Delta t}{1}=\leq 1 \quad (4)$$

Since $c=\sqrt{gh}$, g is gravitational acceleration (m/sec^2) and h is the maximum depth of water observed at Tikarpara gauge station.

Development of algorithm for evaluating performance of flood simulation models

The developed algorithm is based on measure of fit of two raster datasets, i.e., simulated raster R1 and satellite-derived observed flood raster R2. Both of these rasters are of the order of one bit format. In order to find the correctly predicted cells (A), overpredicted cells (B), and underpredicted cells (C), the process starts with raster calculator $2\times R1-R2$. If the value of the cell is 1 then the corresponding cell values are correctly predicted. If the value of the cell is 2, then the corresponding cell values are overpredicted and if the value of the cell is -1, then the corresponding cell values are underpredicted. The evaluation is carried out using the formula shown by Equations (5) and (6). The performance evaluation algorithm was scripted/coded in ArcPy, which is available in the scripting module within Arc-Map (ArcGIS software module) version 10.6.1. The script requires a spatial analyst toolbox license and extension to execute the code. The map algebra function under the spatial analyst tool was used to extract the cells that are correctly predicted, underpredicted, and overpredicted. Code developed for performance evaluation of the flood model is shown in Figure 3. Figure 4 depicts the simulated layer, observed flood extent, correctly predicted, overpredicted, and underpredicted cells. The adopted methodology for evaluation of flood models is shown in Figure 5. The best fit

Evolutionary Algorithms in Water Resources
doi: 10.2166/h2oj.2020.117

```
# Import arcpy module
import arcpy
from arcpy.sa import *
# Check out any necessary licenses
arcpy.CheckOutExtension("spatial")
#Modelled inundation raster
ras1 = "E:\Mahanadi_project_tushar\srtm30\p2\\depth.tif"
#Actual satellite inundation raster
ras2 = "E:\\Mahanadi_project_tushar\\fld1\\fld707aug.img"
#Temporary calculated raster
out_file = "E:\\Mahanadi_project_tushar\\fld1\\test1121"
#Check for pixel size compatibility for raster calculator
result = 2*Raster(ras1) - Raster(ras2)
result.save(out_file)
rows = arcpy.SearchCursor(out_file)
for row in rows:
    if row.getValue("VALUE") == -1: #Underestimated
        C = row.getValue("COUNT")
        print "count of underesti cells is %d" %C
    elif row.getValue("VALUE") == 2:  #Overestimated
        B = row.getValue("COUNT")
        print "count of overesti cells is %d" %B
    elif row.getValue("VALUE") == 1: #Correctly Predicted
        A = row.getValue("COUNT")
        print "count of correctly predi cells is %d" %A
rows_sat = arcpy.SearchCursor(ras2)
for row in rows_sat:
    if row.getValue("Value") == 1:
        tot_ras2 = row.getValue("Count")
        #print "count of total cells in satellite image
##C = tot_ras2 - A
##print "count of underesti cells is %d" %C
F1 = (float(A))/(float(A) + float(B) + float(C))
F2 = (float(A) - float(B))/(float(A) + float(B) + float(C))
print "The value of F1 is %f" %F1
print "The value of F2 is %f" %F2
```

Figure 3 | Code developed for performance evaluation of flood model.

value F1 and F2 is given by Equations (5) and (6), respectively.

$$F1 = A/(A + B + C) \tag{5}$$

$$F2 = (A - B)/(A + B + C) \tag{6}$$

The value of F1 ranges from 0 to 1; F2 value ranges from −1 to 1. If the value of F1 and F2 is closer to 1, it means that the model performance is better, and if F1 and F2 values are exactly 1, it means all the cells are correctly predicted and there is not a single cell which is under- or overpredicted.

Development of algorithm for extraction of flood footprints from SAR image

SAR images should be pre-processed before carrying out the analysis. The pre-processing steps include applying orbit file, calibration, speckle filtering, terrain correction, linear to decibel conversion, adding elevation band, and deriving land/water mask using GPT tool with the help of Python scripts. After pre-processing, the image is divided into segments, each of size 100 × 100 pixels. Each segment is further divided into four sub-segments of size 50 × 50 pixels. The size of segments and sub-segments was fixed after testing on more than 50 images of different areas. The procedure followed by Martinis *et al.* (2015) is used for the selection of segments automatically. The mean

doi: 10.2166/h2oj.2020.117

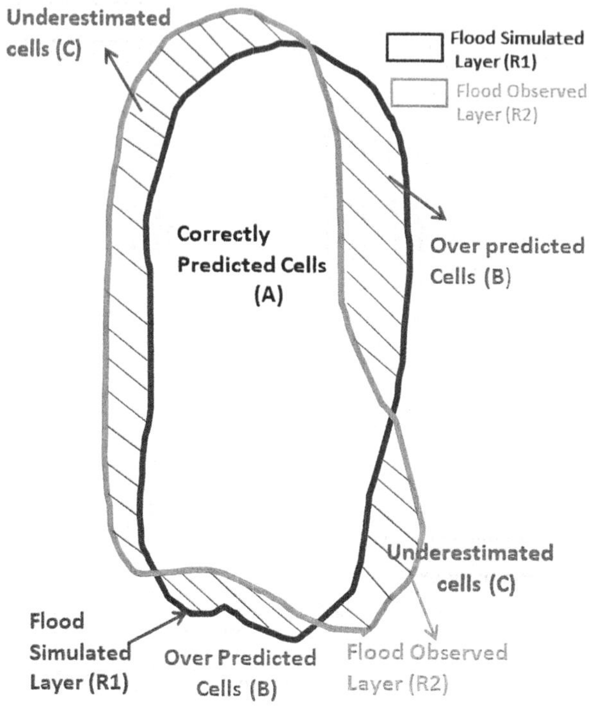

Figure 4 | Schematic diagram of flood simulated extent and observed flood extent.

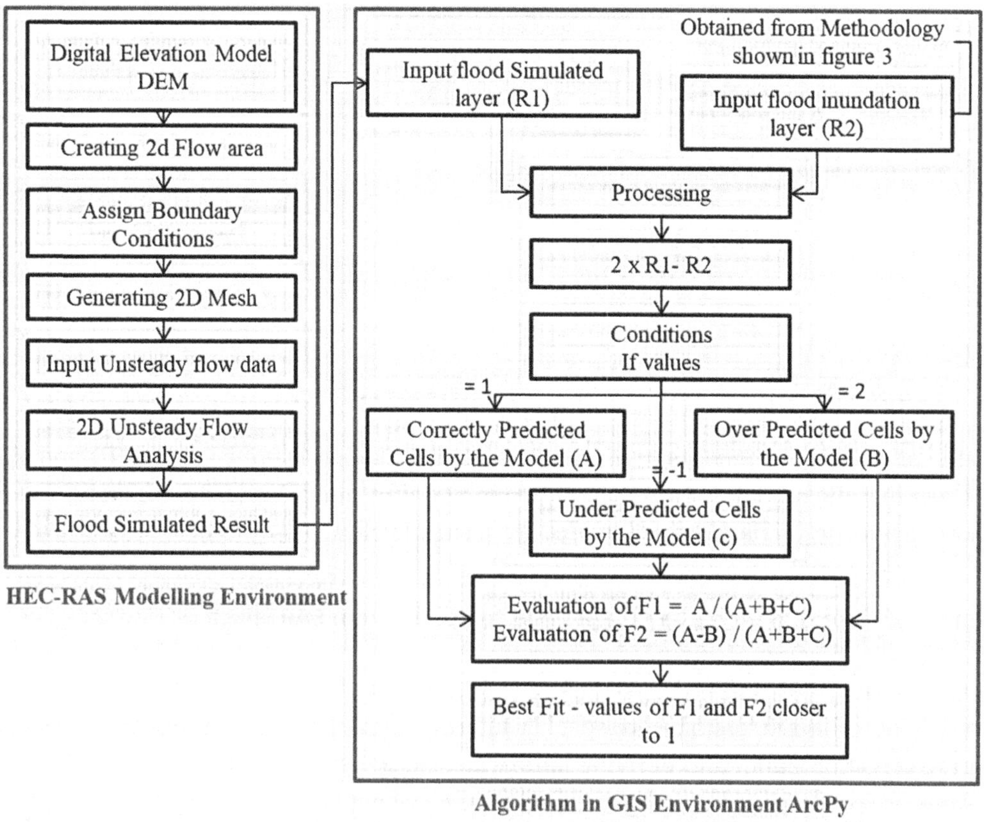

Figure 5 | Methodology flow chart for evaluation of flood models.

doi: 10.2166/h2oj.2020.117

intensity values are calculated in each sub-segment and using the respective mean values of four sub-segments, the standard deviation of the segment is calculated. Two conditions were checked, namely, first, the standard deviation of the segment must be high (>95% quantile). This criterion helps in finding the segments where there is a greater probability of distribution of more than one class within it and the number of pixels of each class is of good ratio. Second, the mean intensity value of the segment must be less than the mean intensity value of the whole image. This ensures that the selected segments have a good number of water pixels. If the number of segments which are satisfying the above-mentioned conditions is less than five then the size of segments and sub-segments is reduced to half and the quantile for the choice of tile is reduced to (90%) make sure that the segments also are chosen having information of a comparatively low extent of water surfaces or with smaller spread water bodies. Further, the top five segments in terms of standard deviation are selected for further analysis, since that is sufficient for threshold computation (Twele *et al.* 2016).

Otsu's thresholding technique is used to automatically perform clustering-based image threshold (Sezgin & Sankur 2004). The algorithmic rule assumes that the SAR image contains two categories of pixels which follow a bi-modal bar-chart, termed as foreground and background pixels, it then evaluates the optimum threshold separating the two categories so that their combined spread (intra-class variance, defined as a weighted sum of variances of the two classes) is minimal, or equivalently, so that their inter-class variance is maximal by (Otsu 1979). This technique has been employed by Vala & Baxi (2013) as it is effective and found to give reliable results to discover the threshold for each segment which was selected in the previous stage. The arithmetic mean of the thresholds of the five selected segments is finally generated and it is used as the global threshold for applying to the whole image to get a preliminary classified image. The preliminary classified image is refined using SRTM 1Sec HGT DEM data, applying majority filter and water mask. The overclassified water pockets are eliminated using SRTM 1Sec HGT DEM data obtained in the pre-processing stage from the terrain-corrected input image and areas where elevation ought to be above 15 meters are eliminated and classified as water. This ultimately reduces the commission errors which are usually caused because of shadows in the hills. Majority filter is also known as mode/modal filter. It is used to replace the central pixel value within a pre-mentioned size kernel with the pixel value which is reiterated a number of times in the same kernel. This operation reduces the stray pixels on the preliminary classified image and produces a smoothened image. SRTM water mask of the study area is used to exclude the permanent water bodies so as to generate a flood layer. Figure 6 shows the methodology adopted

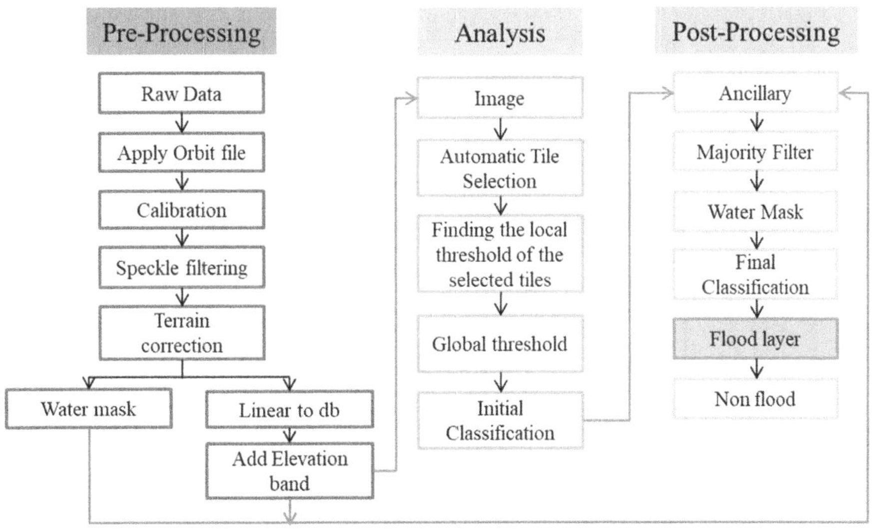

Figure 6 | Methodology flow chart for extraction of flood footprints from SAR image.

for extracting flood footprints from the SAR image. Table 2 shows the technical specifications of hardware and software used for automation.

Table 2 | Technical specifications of hardware and software used for automation

Programming language used	Python
Platform	Anaconda 3.7.1
System configuration	Intel i5 8th generation, 16 GB RAM
Modules used	GDAL, Numpy, Scikit-image, tkinter, matplotlib and Graph Processing Tool (GPT) by ESA's SNAP
Processing time	14 minutes

RESULTS AND DISCUSSION

HEC-RAS-simulated output results deliver temporal flood extent with depth, velocity of water, and surface water elevation. The simulated output results are stored in tiff file format with floating point type. The simulated output results can be generated for the desired date and time from the simulated period. For the present study, the simulated result for the desired date and time of 23 September, 2008 at 0600 hours was exported from the model to validate with RADARSAT satellite SAR data acquired on the same date and time. Delineated flood footprints from RADARSAT SAR image using multi-segmentation and Otsu's thresholding technique are shown in Figure 7. The flood simulated result for roughness coefficient 0.02 in floodplains is shown in Figure 8. Similarly, flood-simulated results for flood roughness coefficient 0.030 and 0.040 are shown in Figures 9 and 10,

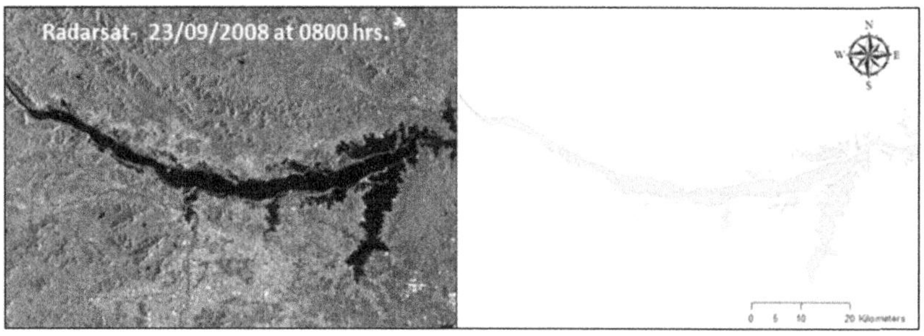

Figure 7 | RADARSAT SAR image (left) and its delineated flood layer (right).

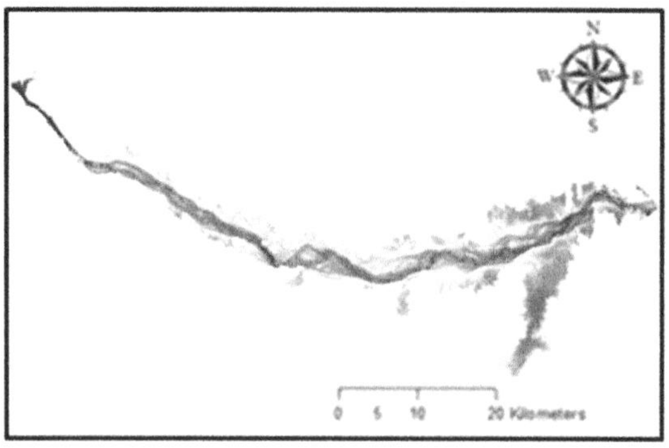

Figure 8 | Flood simulated by HEC-RAS on 23 September 2016 at 0,600 hours with Manning's roughness coefficient 0.020.

doi: 10.2166/h2oj.2020.117

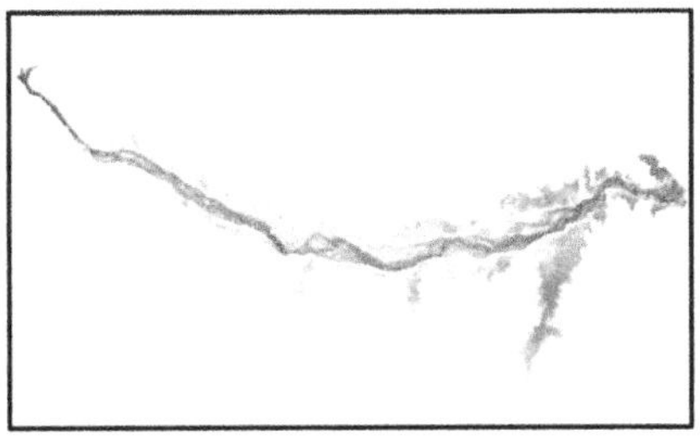

Figure 9 | Flood simulated results for roughness coefficient 0.030 for floodplains.

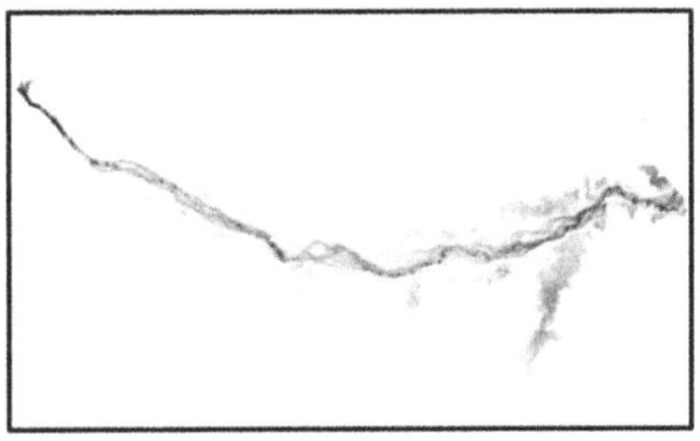

Figure 10 | Flood simulated results for roughness coefficient 0.040 for floodplains.

respectively. The F1 and F2 measures of fit have been evaluated for all three simulations using the algorithm and are shown in Tables 3 and 4. From the results, it is observed that F1 and F2 values are closer to 1 for the roughness coefficient 0.020 for floodplains when compared to others. Further, the inundated areas for simulated and observed is calculated by multiplying the number of pixels with a single pixel area (100 m^2) as 10 m is the spatial resolution. It is observed that the total inundated area for this coefficient is matching 94% with the simulated. as shown in Table 4.

Table 3 | Model performance for different roughness coefficients

Sr. no.	Manning's roughness coefficient n_c	n_p	Correctly predicted flood cells 'A'	Over predicted flood cell 'B'	Under predicted flood cell 'C'	F1	F2
1	0.032	0.020	1,553,668	206,750	52,092	0.85	0.74
2	0.032	0.030	1,554,075	519,470	51,685	0.73	0.50
3	0.032	0.040	1,554,381	745,570	48,379	0.67	0.35

Table 4 | Flood inundated area observed vs. simulated

Sr. no.	Manning's Roughness coefficient n_c	n_p	Flood observed (FO) hectares	Flood simulated (FS) hectares	FO/FS
1	0.032	0.020	17,057.6	18,125.10	0.94
2	0.032	0.030	17,057.6	21,252.30	0.80
3	0.032	0.040	17,057.6	23,483.30	0.72

CONCLUSION

The present study uses HEC-RAS model for simulating the severe flood event of Mahanadi River in Odisha. Three simulations were run using different Manning's coefficient for floodplains.

Satellite-based flood footprints were extracted by multi-segmentation and Otsu's thresholding technique which will be helpful for disaster responders to prepare flood hazard maps in near real time without any manual intervention. The development of an algorithm for performance evaluation of the HEC-RAS model for three different simulations in validation with extracted flood prints from RADARSAT shows appreciable results for Manning's roughness coefficient 0.020 for floodplains and 0.032 for channel as measures of fit F1 and F2 are 0.85 and 0.74, respectively, which are closest to 1 among the other two simulation values. From the results it is inferred that the HEC-RAS model is sensitive towards the Manning's roughness coefficient. The developed algorithm for performance evaluation can be used for any flood model (MIKE, Tuflow, ANUGA, etc.) which has the capability to export the flood inundation extent result in GIS formats.

DATA AVAILABILITY STATEMENT

All relevant data are included in the paper or its Supplementary Information.

REFERENCES

Baatz, M. 1999 Object-oriented and multi-scale image analysis in semantic networks. In: *Proceedings of the 2nd International Symposium on Operationalization of Remote Sensing*, 16–20 August 1999, Enschede, The Netherlands.

Bates, P. D. & De Roo, A. P. J. 2000 A simple raster-based model for flood inundation simulation. *J. Hydrol.* **236** (1–2), 54–77. https://doi.org/10.1016/S0022-1694(00)00278-X.

Bates, P. D., Horrit, M. S., Smith, C. N. & Mason, D. C. 1997 Integrating remote sensing observations of flood hydrology and hydraulic modelling. *Hydrological Processes* **11**, 1777–1795.

Bazi, Y., Bruzzone, L. & Melgani, F. 2005 An unsupervised approach based on the generalized Gaussian model to automatic change detection in multitemporal SAR images. *IEEE Trans. Geosci. Remote Sens.* **43**, 874–887.

Bovolo, F. & Bruzzone, L. 2007 A split-based approach to unsupervised change detection in large-size multitemporal images: application to tsunami-damage assessment. *IEEE Trans. Geosci. Remote Sens.* **45** (6), 1658–1670. doi:10.1109/tgrs.2007.895835.

Cao, H., Zhang, H., Wang, C. & Zhang, B. 2019 Operational flood detection using Sentinel-1 SAR data over large areas. *Water* **11**, 786. doi:10.3390/w11040786.

Chen, Z., Luo, J., Chen, N., Xu, R. & Shen, G. 2019 RFim: a real-time inundation extent model for large floodplains based on remote sensing big data and water level observations. *Remote Sens.* **11**, 1585.

De Roo, A., Van Der Knijff, J., Horritt, M., Schmuck, G. & De Jong, S. 1999 Assessing flood damages of the 1997 Oder flood and the 1995 Meuse flood. In: *Proceedings of the Second International ITC Symposium on Operationalization of Remote Sensing*, 16–20 August 1999, Enschede, The Netherlands.

Di Baldassarre, G., Schumann, G. & Bates, P. D. 2009 A technique for the calibration of hydraulic models using uncertain satellite observations of flood extent. *J. Hydrol.* **367**, 276–282. doi:10.1016/j.jhydrol.2009.01.020.

Giustarini, L., Hostache, R., Matgen, P., Schumann, G. J., Bates, P. D. & Mason, D. C. 2013 A change detection approach to flood mapping in urban areas using terrasar-x. *IEEE Trans. Geosci. Remote Sens.* **51**, 2417–2430.

Henry, J. B., Chastanet, P., Fellah, K. & Desnos, Y. L. 2006 Envisat multi-polarized ASAR data for flood mapping. *Int. J. Remote Sens.* **27**, 1921–1929.

Hess, L. L., Melack, J. M., Simonett, D. S. & Sieber, A. J. 1990 Radar detection of flooding beneath the forest canopy: a review. *Int. J. Remote Sens.* **11**, 1313–1325.

Hirose, K., Maruyama, Y., Do Van, Q., Tsukada, M. & Shiokawa, Y. 2001 Visualization of flood monitoring in the lower reaches of the Mekong River. In: *Proceedings of the 22nd Asian Conference on Remote Sensing*, 5–9 November 2001, Singapore.

Hoque, R., Nakayama, D., Matsuyama, H. & Matsumoto, J. 2011 Flood monitoring, mapping and assessing capabilities using RADARSAT remote sensing, GIS and ground data for Bangladesh. *Nat. Hazards* **57**, 525–548.

Horrit, M., Di Baldasarre, G., Bates, P. & Brath, A. 2007 Comparing the performance of a 2-D finite element and a 2-D finite volume model of floodplain inundation using airborne SAR imagery. *Hydrol. Processes* **21**, 2745–2759.

Iervolino, P., Guida, R., Iodice, A. & Riccio, D. 2015 Flooding water depth estimation with high-resolution SAR. *IEEE Trans. Geosci. Remote Sens.* **53**, 2295–2307.

Kasischke, E. S. & Bourgeau-Chavez, L. L. 1997 Monitoring South Florida wetlands using ERS-1 SAR imagery. *Photogramm. Eng. Remote Sens.* **63**, 281–291.

Kittler, J. & Illingworth, J. 1986 Minimum error thresholding. *Pattern Recognition* **19** (1), 41–47. https://doi.org/10.1016/0031-3203(86)90030-0.

Kuenzer, C., Guo, H., Huth, J., Leinenkugel, P., Li, X. & Dech, S. 2013 Flood mapping and flood dynamics of the Mekong Delta: ENVISAT-ASAR-WSM based time series analyses. *Remote Sens.* **5**, 687–715.

Liu, Z., Merwade, V. & Jafarzadegan, K. 2018 Investigating the role of model structure and surface roughness in generating flood inundation extents using 1D and 2D hydraulic models. *J. Flood Risk Manage.* e12347. 10.1111/jfr3.12347.

Lu, J., Giustarini, L., Xiong, B., Zhao, L., Jiang, Y. & Kuang, G. 2014 Automated flood detection with improved robustness and efficiency using multi-temporal SAR data. *Remote Sens. Lett.* **5**, 240–248.

Martinis, S. & Rieke, C. 2015 Backscatter analysis using multi-temporal and multi-frequency SAR data in the context of flood mapping at River Saale, Germany. *Remote Sens.* **7**, 7732–7752.

Martinis, S., Twele, A. & Voigt, S. 2009 Towards operational near-real time flood detection using a split-based automatic thresholding procedure on high resolution TerraSAR-X data. *Nat. Hazards Earth Syst. Sci.* **9**, 303–314. https://doi.org/10.5194/nhess-9-303-2009.

Martinis, S., Kersten, J. & Twele, A. 2015 A fully automated TerraSAR-X based flood service. *ISPRS J. Photogramm. Remote Sens.* **4**, 203–212. https://doi.org/10.1016/j.isprsjprs.2014.07.014.

Matgen, P., Schumann, G., Henry, J. B., Hoffmann, L. & Pfister, L. 2007 Integration of SAR-derived river inundation areas, high-precision topographic data and a river flow model toward near real-time flood management. *Int. J. Appl. Earth Obs. Geoinf.* **9**, 247–263.

Matgen, P., Hostache, R., Schumann, G., Pfister, L., Hoffmann, L. & Savenije, H. 2011 Towards an automated SAR-based flood monitoring system: lessons learned from two case studies. *Phys. Chem. Earth Parts A/B/C* **36**, 241–252.

Moya Quiroga, V., Kure, S., Udo, K. & Mano, A. 2016 Application of 2D numerical simulation for the analysis of the February 2014 Bolivian Amazonia flood: application of the new HEC-RAS version 5. *Ribagua* **3** (1), 25–33. doi:10.1016/j.riba.2015.12.001.

Oberstadler, R., Hönsch, H. & Huth, D. 1997 Assessment of the mapping capabilities of ERS-1 SAR data for flood mapping: a case study in Germany. *Hydrol. Processes* **11**, 1415–1425.

Otsu, N. 1979 A threshold selection method from gray-level histograms. *IEEE Transactions on Systems, Man and Cybernetics* **9** (1), 62–66. doi:10.1109/TSMC.1979.4310076.

Parhi, P. K., Sankhua, R. N. & Roy, G. P. 2012 Calibration of channel roughness for Mahanadi River, (India) using HEC-RAS model. *J. Water Resour. Prot.* **4**, 847–850.

Pulvirenti, L., Pierdicca, N. & Chini, M. 2010 Analysis of COSMO-SkyMed observations of the 2008 flood in Myanmar. *Ital. J. Remote Sens.* **42**, 79–90.

Pulvirenti, L., Chini, M., Pierdicca, N., Guerriero, L. & Ferrazzoli, P. 2011 Flood monitoring using multi-temporal COSMO-SkyMed data: image segmentation and signature interpretation. *Remote Sens. Environ.* **115**, 990–1002. doi:10.1016/j.rse.2010.12.002.

Richards, J. A., Woodgate, P. W. & Skidmore, A. K. 1987 An explanation of enhanced radar backscattering from flooded forests. *Int. J. Remote Sens.* **8**, 1093–1100.

Schumann, G., Henry, J., Hoffmann, L., Pfister, L., Pappenberger, F. & Matgen, P. 2005 Demonstrating the high potential of remote sensing in hydraulic modelling and flood risk management. In: *Proceedings of the Annual Conference of the Remote Sensing and Photogrammetry Society with the NERC Earth Observation Conference*, 6–9 September 2005, Portsmouth, UK.

Schumann, G., Baldassarre, G. D., Alsdorf, D. & Bates, P. D. 2010 Near real-time flood wave approximation on large rivers from space: application to the River Po, Italy. *Water Resour. Res.* **46**, 1–8.

SENAGUA (SecretaríaNacionaldel Agua) 2014 *Vulnerability Assessment to Flood Events of the Santa Bárbara River.* Vol. 2. *Hydraulic Study*, ENAGUA, Quito, p. 108. (in Spanish).

Sezgin, M. & Sankur, B. 2004 Survey over image thresholding techniques and quantitative performance evaluation. *J. Electron. Imaging* **13**, 146–168. doi:10.1117/1.1631315.

Shastry, A. & Durand, M. 2019 Utilizing flood inundation observations to obtain floodplain topography in data-scarce regions. *Front. Earth Sci.* **6**, 243. doi:10.3389/feart.2018.00243.

Shen, X., Wang, D., Mao, K., Anagnostou, E. & Hong, Y. 2019 Inundation extent mapping by synthetic aperture radar: a review. *Remote Sens.* **11**, 879.

Surwase, T., SrinivasaRao, G., Manjusree, P., Begum, A., Nagamani, P. V. & JaiSankar, G. 2019 Flood inundation simulation of Mahanadi River, Odisha during September 2008 by using HEC-RAS 2D model. In: *Proceedings of International Conference on Remote Sensing for Disaster Management. Springer Series in Geomechanics and Geoengineering* (Rao, P., Rao, K. & Kubo, S., eds). Springer, Cham, Switzerland.

Tan, Q., Bi, S., Hu, J. & Liu, Z. 2004 Measuring lake water level using multi-source remote sensing images combined with hydrological statistical data. In: *Proceedings of the IGARSS'04, 2004 IEEE International Conference on Geoscience and Remote Sensing Symposium*, 20–24 September 2004, Anchorage, AK, USA, pp. 4885–4888.

Teng, J., Jakeman, A. J., Vaze, J., Croke, B., Dutta, D. & Kim, S. 2017 Flood inundation modelling: a review of methods, recent advances and uncertainty analysis. *Environ. Modell. Software* **90**, 201–216. doi:10.1016/j.envsoft.2017.01.006.

Tholey, N., Clandillon, S. & De Fraipont, P. 2015 The contribution of spaceborne SAR and optical data in monitoring flood events: examples in northern and southern France. *Hydrol. Processes* **11**, 1409–1413.

Townsend, P. 2002a Relationships between forest structure and the detection of flood inundation in forested wetlands using C-band SAR. *Int. J. Remote Sens.* **23**, 443–460.

Townsend, P. A. 2002b Estimating forest structure in wetlands using multitemporal SAR. *Remote Sens. Environ.* **79**, 288–304.

Twele, A., Cao, W., Plank, S. & Martinis, S. 2016 Sentinel-1-based flood mapping: a fully automated processing chain. *Int. J. Remote Sens.* **37** (13), 2990–3004. https://doi.org/10.1080/01431161.2016.1192304.

Vala, H. J. & Baxi, A. 2013 A review on Otsu image segmentation algorithm. *Int. J. Advanc. Res. Comput. Eng. Technol.* **2** (2), 387–389.

Yamada, Y. 2001 Detection of flood-inundated area and relation between the area and micro-geomorphology using SAR and GIS. In: *Proceedings of the IGARSS'01, IEEE 2001 International Conference on Geoscience and Remote Sensing Symposium*, 9–13 July 2001, Sydney, NSW, Australia, pp. 3282–3284.

First received 21 February 2020; accepted in revised form 3 June 2020. Available online 22 July 2020

Evolutionary Algorithms in Water Resources
doi: 10.2166/h2oj.2020.005

Modelling runoff in an arid watershed through integrated support vector machine

Sandeep Samantaray and Dillip K. Ghose*

Department of Civil Engineering, NIT Silchar, Assam, India

*Corresponding author. E-mail: dillipghose_2002@yahoo.co.in

Abstract

Modelling of runoff is a significant practice in water resources engineering. Therefore, discovering consistent and advanced methods for prediction of runoff is crucial for hydrologic processes. Here, a narrative integrated intelligence model attached with PSR (phase space reconstruction) is anticipated to estimate runoff for five watersheds of Balangir, Odisha, India. Monthly monsoon precipitation, temperature, humidity data of five watersheds over 28 years (1990–2017) are employed and validated. Here, the proposed model is an integration of support vector machine (SVM) with firefly algorithm (FFA) and PSR. Various indices such as NSE (Nash–Sutcliffe), RMSE (root mean square error) and WI (Willmott's index) are used to find the performance of the model. The developed PSR-SVM-FFA model demonstrates pre-eminent WI value ranging from 0.97 to 0.98 while the SVM and SVM-FFA models encompass 0.92 to 0.93 and 0.94 to 0.95, respectively. Also, an assessment of data from the suggested model is schemed and validated. The proposed PSR-SVM-FFA model gives better accuracy results and error limiting up to 2–3%.

Key words: hydrology, PSR-SVM-FFA, runoff, SVM, SVM-FFA, watershed

Highlights

- The most relevant techniques are used for performing the output of rainfall-runoff process at a single outlet. The advances gained from a modelling point of view are considered and discussed.
- The models developed using three techniques are support vector machine (SVM), support vector machine with firefly algorithm (SVM-FFA) and support vector machine and firefly algorithm with Phase space reconstruction (PSR-SVM-FFA) in the study.
- The study is different from previous studies by including and comparing two new techniques, SVM-FFA and PSR-SVM-FFA, while past research has used hybridization of SVM for predicting runoff in a watershed.
- Interaction of all techniques along with five different models is presented.
- The model proposed reasonably predicts the runoff at a single outlet and improves the method of measuring runoff without the detailed study for all sub-outlets of the watershed.

INTRODUCTION

Improvement in water resources is essential for balancing water demand around the globe. Due to the complex nature and uncertainty in water parameters, traditional methods like statistical method,

Dickens formula (1865), Ryves formula (1984), and Khosla formula (1960) are insufficient to understand the cycle of rainfall-runoff (Das 2009). Innovative computational algorithms and their application to water resources engineering are necessary to understand the behaviour of the water cycle in the context of climate change. Computational algorithms in view of modern techniques like FFA, SVM and PSR are important for engineering applications. Attempts have also been made to apply these techniques to water resources engineering (Tayfur 2014).

Tabari *et al.* (2012) employed the SVM, ANFIS (adaptive neuro-fuzzy inference system), MLR (multiple linear regression) method utilizing six climate data variables as inputs to estimate evapotranspiration at Nozheh station, Iran. Outcomes suggested that SVM and ANFIS models perform better with greater precision than those of regression and climate-based models. Baydaroglu & Kocak (2014) applied support vector regression (SVR) to envisage water loss using solar radiation, relative humidity, wind speed, evaporation and temperature as input. SVR helps in predicting evaporation effectively since it has a good generalization ability with greater determination coefficients of 83 and 97% for univariate and multivariate time series entrenching. Raghavendra & Deka (2014) in their review found that SVM is more suitable for prediction than other techniques pertinent to hydrology. Cha *et al.* (2014) projected the SVM-FFA model for more precise forecast of malaria incidences in the Jodhpur and Bikaner area. Performance of the projected hybrid model was compared with ANN, autoregressive moving average as well as SVM and results indicated that the proposed integrated model provides forecasts with better accuracy as compared to traditional methods. The FNN (feed forward neural network) technique was used for monthly stream flow forecasting for a period of 53 years in the United States by Vignesh *et al.* (2015). Obtained outcomes are valuable in identifying suitable model intricacy at discrete stations, arrangements across regions and sub-regions, interpolating and extrapolating data, and classifying catchments. SVM-FFA and SVM-wavelet were used to evaluate precipitation trends in 29 meteorological stations in Serbia from 1946 to 2012 by Gocic *et al.* (2016). Estimation and prediction outcomes of hybrid models were contrasted and results showed that the SVM-wavelet approach gives improved prediction accuracy and ability to generalize. Moghaddam *et al.* (2016) implemented SVM-FFA for predicting the exhaustion lifetime of polyethylene terephthalate (PET) reformed asphalt mixture taking PET percentages, stress levels and environmental temperatures as inputs. Prediction results by SVM-FFA were compared against ANN and genetic programming (GP) and it was found that the hybrid model performs better. Al-Shammari *et al.* (2016) projected a hybrid method integrating SVM with FFA for predicting daily dew point temperature (T_{dew}) for an Iranian city and performance of the hybrid model is compared to SVM, ANN and GP. They observed that SVM-FFA is certainly very efficient in predicting T_{dew} with better accuracy and consistency. Integrated PSR–ANN and ANN techniques were utilized to predict daily river flow on river gauging stations in the USA by Delafrouz *et al.* (2017). The outcomes of PSR–ANN, ANN and gene expression programming models were compared and it was found that the projected hybrid model gives the best prediction accuracy for daily river flow. Mehr *et al.* (2018) employed integration of FFA and SVM for rainfall forecasting on a monthly basis at Tabriz and Urmia stations, Iran. The effectiveness of the hybrid model was cross-checked with SVR and GP-based forecasting models and it was found that the SVM-FFA model performs well with promising accuracy in terms of rainfall forecasting. Ghose & Samantaray (2019) employed LRNN (layer recurrent NN), RBFN (radial basis function NN) and FFBPN (feed forward back propagation NN) to evaluate runoff as a loss function of evapotranspiration, temperature and precipitation. Outcomes revealed that performance of the LRNN model is best as compared to FFBPN and RBFN for predicting runoff in the watershed which helps in planning, designing and managing hydraulic structures in the neighbourhood of the watershed. Zaini *et al.* (2018) used an SVM model and also a hybridized SVM-PSO (particle swarm optimization) model to scheme daily stream flow at Upper Bertam watershed, Malaysia. Hybridized models SVM-PSO1 and SVM-PSO2 perform best in comparison to SVM1 and SVM2 at forecasting river flow 1–7 days ahead. Tao *et al.* (2018) used SVM and SVM-FFA to anticipate rainfall

doi: 10.2166/h2oj.2020.005

at Chhattisgarh. The proposed hybrid model significantly improved forecasting accuracy and can also be used for monthly rainfall forecasting in provincial areas of India. Gandomi *et al.* (2011) utilized FFA to solve mixed variable structural optimization problems and also the implications of FFA are thoroughly analysed in comparison with PSO, GA, simulated annealing for future research. Ji & Sun (2013) presented a method for multitasking multiple classes of SVM on the basis of minimizing regularization functions and found it to be best among other multitask learning methods. Shamshirband *et al.* (2015) proposed a hybrid SVM-FFA model for estimating monthly mean horizontal global solar radiation (HGSR). Performance of the hybrid model is evaluated by comparing it with ANN, GP and ARMA (autoregressive moving average) models and results revealed that the projected hybrid model performs best.

Several researchers have found different individual algorithms for predicting runoff. Some have also developed optimization methods and least integrated methods to predict runoff. Thus, an attempt has been made to think about the hybrid model via integral algorithms. The objective of this research is to compare results of the SVM-FFA, PSR-SVM-FFA model with the SVM and empirical models to tune up the magnitude of runoff for controlling flood in the region. Basically, it is applicable for efficient planning and management of water resources such as flood control and management of a watershed.

STUDY AREA

Balangir district is located in the west part of Odisha, India, with an area of 5,165 km^2, as shown in Figure 1. It lies between the latitudes of 20°59′00″ and longitudes of 83°32′22″. It has an average elevation of 115 metres (377 feet). The watershed is located towards the mid-north edge of Balangir district. Five gauging stations, Loisinga, Balangir, Tushura, Saintala and Patnagarh, are considered in the present research.

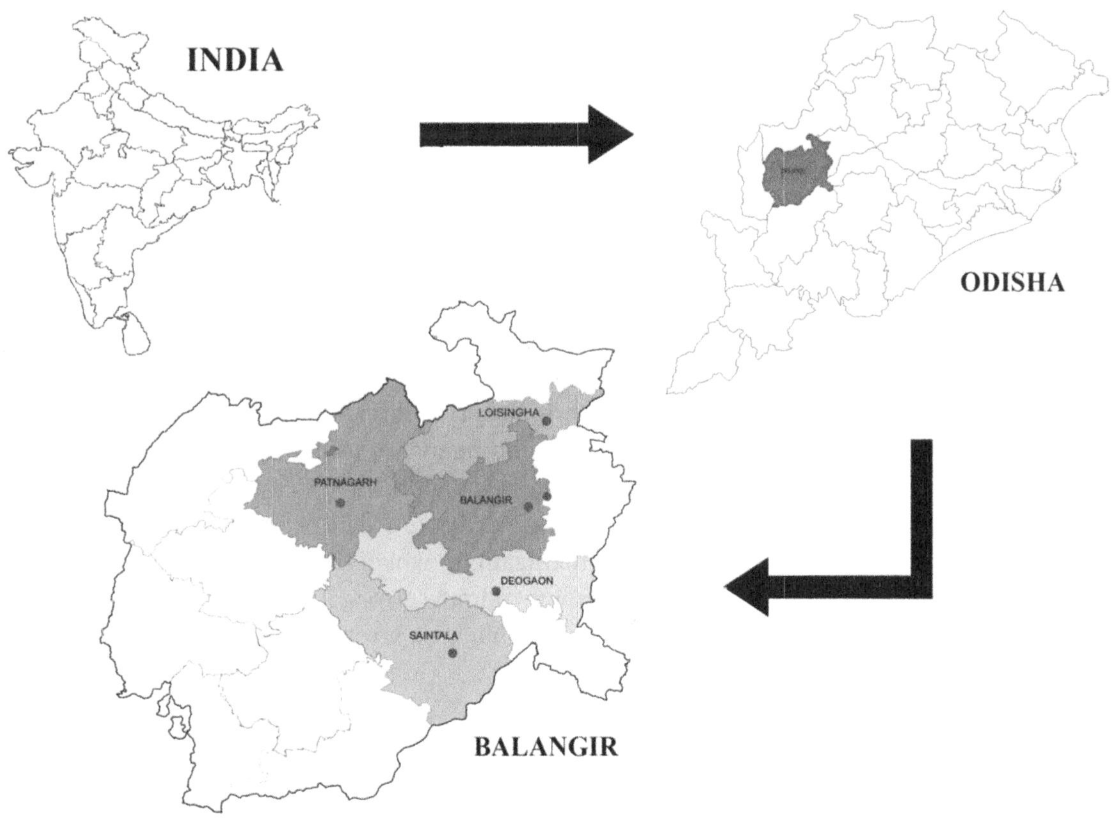

Figure 1 | Schematic diagram of the proposed watershed.

Physical and statistical characteristics of the gauging stations are described in Table 1. Here, prediction of runoff is assessed by taking the dataset from 1990 to 2017. Mean monthly precipitation and temperature data for the monsoon months (May to October) from the period 1990–2017, spanning over 28 years, are obtained from IMD (India Meteorological Department) Bhubaneswar. Runoff data are collected from the soil conservation office, Bolangir. Here, the runoff data are computed using Khosla's empirical equation to understand the coherence between developed empirical data and observed data collected from that department.

Table 1 | Physical and statistical characteristics of five stations in Balangir, Odisha, India

Station name	Drainage area (km^2)	Latitude	Longitude	Elevation (m)	Mean rainfall (mm)	Min.–Max. rainfall (mm)
Loisinga	317.6	20.86N	83.50E	162	91.22	7.83–158.8
Balangir	386.98	20.7N	83.5E	115	87.82	7.93–157.29
Deogaon	432.01	20.71N	83.49E	148	93.69	7.28–155.26
Saintala	454.43	20.26N	83.20E	182	87.795	7.45–159.56
Patnagarh	591.45	20.72N	83.15E	243	87.23	7.89–151.34

METHODOLOGY

Khosla formula

Khosla considered precipitation, discharge and temperature data for different catchments in India and the USA to arrive at a pragmatic connection as follows in Equation (1). It is circuitously anchored in the water-balance perception and mean monthly temperature to replicate losses due to evapotranspiration:

$$R_m = P_m - L_m \quad (1)$$
$$L_m = 0.48\ T_m, \text{ for } T_m > 4.5\,^{\circ}\mathrm{C}$$

where R_m is monthly runoff, P_m is monthly rainfall, L_m is monthly losses and T_m is mean monthly temperature of the catchment.

Support vector machine model

A technique with statistical learning theory and promising empirical results for recognition of pattern with high dimension data to avoid error with decision boundary using training in terms of support vector is known as a support vector machine (Cortes & Vapnik 1995).

A graphical description of the SVM model is shown in Figure 2. The principles of SVM include statistical learning known as structural risk minimization, which affords a higher boundary for generalization error of a classifier (R) concerning training error (Re), model complexity (h) and number of training examples (N) to evaluate probability of error and is defined as (Chen *et al.* 2018; Thanh & Kappas 2018; Wang *et al.* 2018a):

$$R \leq R_e + \varphi\left(\frac{h}{N}, \frac{\log \eta}{N}\right) \quad (2)$$

where φ is monotonic increasing function of the model complexity.

doi: 10.2166/h2oj.2020.005

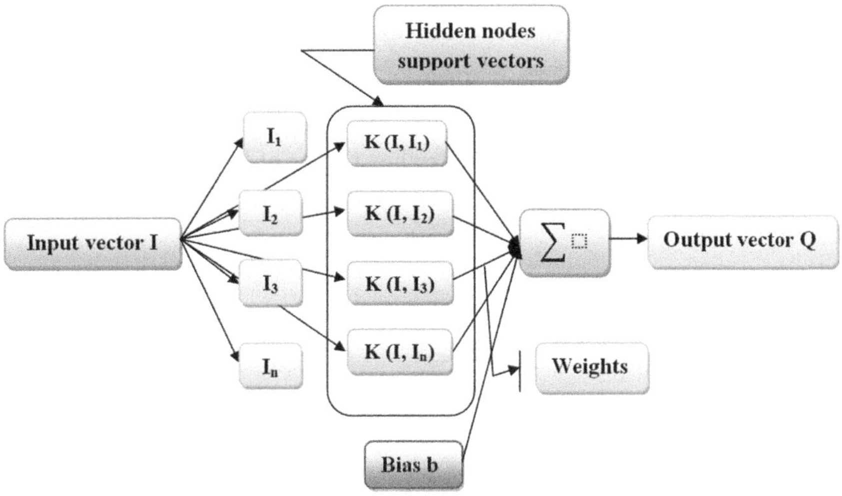

Figure 2 | Graphical description of SVM model.

This is enviable when designing a linear classifier which maximizes margins of their decision boundaries to minimize error. In this regard, SVM is an effective linear classifier which searches a hyper plane with the largest margin called the maximal marginal classifier. The decision boundary of the linear classifier is:

$$w \times x + b = 0 \tag{3}$$

The training phase of SVM involves estimation of constraints weight (w), bias (b) of decision boundary from the training data set. Parameters are preferred with the below conditions:

$$\begin{aligned} & w \times x_i + b \geq 1, \textit{ if } y_i = 1 \\ & w \times x_i + b \leq -1, \textit{ if } y_i = -1 \end{aligned} \tag{4}$$

The objective function is the Lagrangian multiplier and denoted as:

$$L_p = \frac{1}{2}\|w\|^2 - \sum_{i=1}^{N} \lambda_i\{y_i(w \times x_i + b) - 1\} \tag{5}$$

The linear programming is to be utilized through Karush–Kuhn–Tucker (KKT) conditions and the KKT condition is:

$$\begin{aligned} & \lambda_i \geq 0 \\ & \lambda_i\{y_i(w \times x_i + b) - 1\} = 0 \end{aligned} \tag{6}$$

Integration of SVM with FFA model fitness

The FFA is a natural algorithm based on the blinking manners of fireflies (Yang 2009; Abd-Elazim & Ali 2018; Lieu *et al.* 2018; Wang *et al.* 2018b, 2018c). Since the complex nature of fireflies induces global communication among swarming particles, effective multi-objective optimization is necessary to recognize the multifaceted algorithm of FFA. In this algorithm, fireflies use flashing light to search for mates, to attract potential prey for protection from their predators. In order to achieve efficient optimal solutions, flashing light intensity induces fireflies to a more attractive position associated with objective function of the problem.

doi: 10.2166/h2oj.2020.005

There are some components in FFA, such as distance, attractiveness and movement.

Euclidian distance

Euclidian distance is the distance between two fireflies i and j at locations u_i and u_j, defined as Euclidean distance (r_{ij}) applying Equation (7), where $u_{i,k}$ is the kth factor of spatial coordinate u_i of ith firefly and d is dimension number:

$$r_{ij} = \|u_i - u_j\| = \sqrt{\sum_{k=1}^{d} (u_{i,k} - u_{j,k})^2} \tag{7}$$

Attractiveness

Attractiveness task of a firefly is revealed in Equation (8):

$$\beta_r = \beta_0 \times exp(-\gamma r^m),\ m \geq 1 \tag{8}$$

Movement

Progress of firefly i is towards an extra striking firefly j and is specified by Equation (9):

$$\text{updated } u_i = u_i + \beta_0 \times \exp(\gamma r_{ij}^2) \times (u_j - u_i) + \alpha\left(r - \frac{1}{2}\right) \tag{9}$$

where u_i is the current position of firefly, $\beta_0 \times \exp(\gamma r_{ij}^2) \times (u_j - u_i)$ is active for allowing a firefly's attractiveness to light intensity and $\alpha\left(r - \frac{1}{2}\right)$ is applied for random movement of a firefly.

The settings of firefly algorithm (FA) parameters: γ (light absorption coefficient) $= 1.0$, α (randomization parameter) $= 0.3$, β_0 (attractiveness value) $= 1.0$ and random number generator consistently disseminated in space $[0, 1] = 0.2$. A flowchart of the SVM-FFA model is shown in Figure 3.

Objective function

$F(Y),\ Y = (Y_1, Y_2, Y_3, \ldots\ldots\ldots, Y_k)^Z$
Fireflies' initialization, $Y_j(j = 1,2,3,4,\ldots\ldots\ldots m)$
Concentration of light L_j *at* Y_j, $f(Y_j)$
Light preoccupation coefficient β
While $z <$ *maximum generation do*
        For $j = 1{:}m$ all m fireflies do
            For $i = 1{:}i$ all m fireflies do
                If $L_i > L_j$ then
  Transfer firefly j towards i in k dimension
                End if
  Attractiveness diverges with distance r through $exp[-\beta r]$
  Calculate high result and update intensity of light
                End for
            End for

doi: 10.2166/h2oj.2020.005

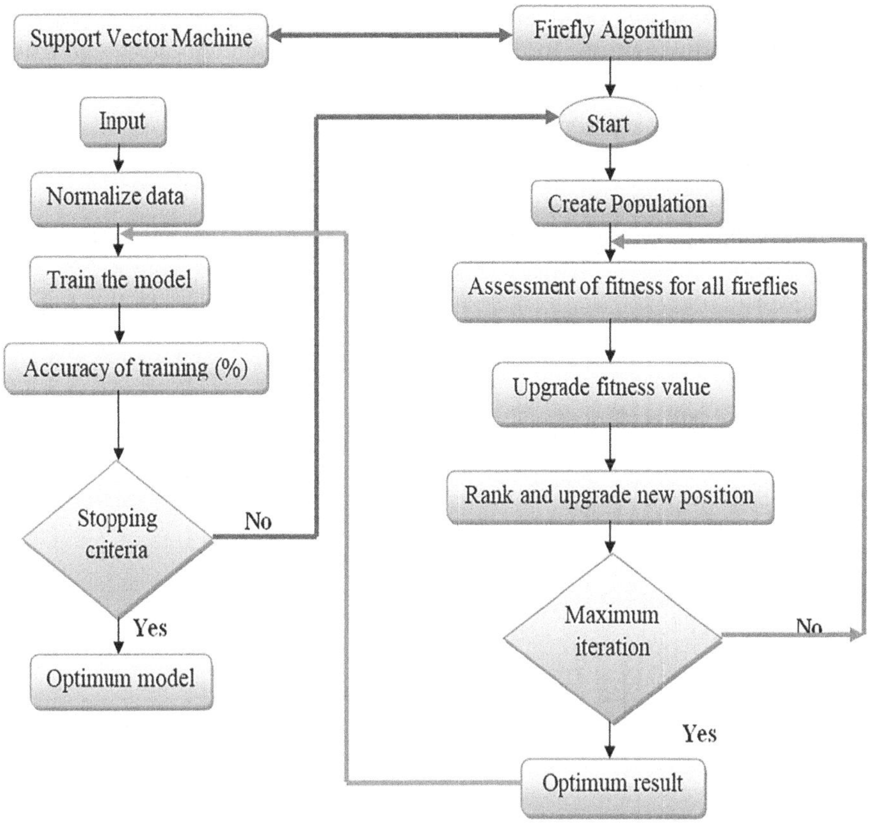

Figure 3 | Descriptive flowchart of SVM-FFA model.

Evaluate best value
End while
Final result and conception

Integration of PSR coupled with SVM-FFA model

Illustration of hybrid extravagance model is demonstrated in Figure 4. The models involve the subsequent modelling process (Dutta *et al.* 2018; Fan *et al.* 2018; Hajiloo *et al.* 2018; Sun & Wang 2018).

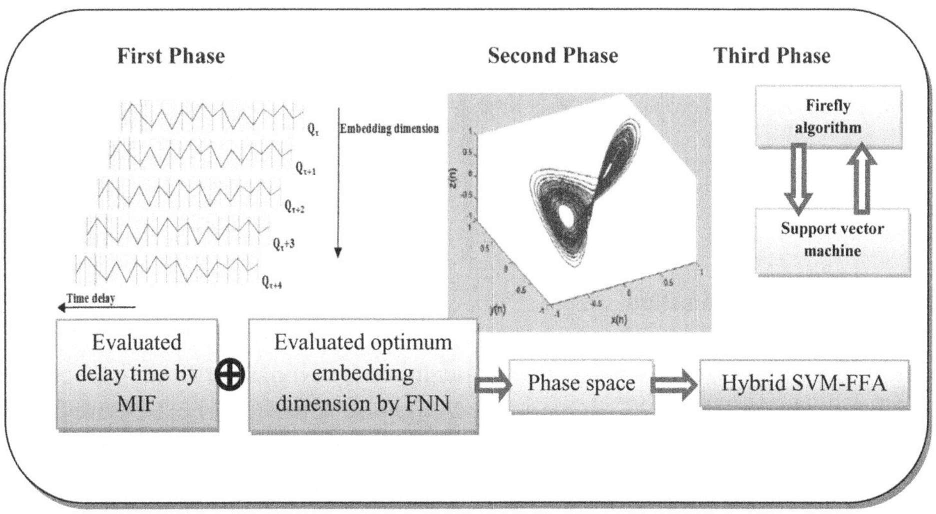

Figure 4 | Graphical representation of hybrid PSR-SVM-FFA.

(i) First phase, optimal τ and m is established to accomplish MIF and FFN approaches for embattled index phase space reconstruction. In this manner, phase space will converge from input aspects and abstract critical dynamics of chaotic time series.
(ii) Second phase, input matrix is assembled rooted in several dimensions $X_t, X_{t-\tau}, X_{t-2\tau}, \ldots, X_{t-(m-1)\tau}$ by considering optimal τ and m, for targeted value X_{t+1}. According to the PSR technique, an input matrix is proposed (Sivakumar *et al.* 2001).
(iii) Third phase, the hybrid SVM-FFA model is fed by the produced matrix using phase space signal.

Objective function

For univariate time signal $X_1 = (X_{1,1}, X_{1,2}, \ldots\ldots\ldots\ldots, X_{1,N})^T$, PSR is

$$X^{r}_{1i} = (x_{1,i}, x_{1,i+\tau_1}, \ldots\ldots x_{1,i+(m_1-1)\tau_1}),\ i = 1, 2, \ldots\ldots\ldots\ldots\ldots N - (m_1 - 1)\tau_1$$

where, N is time series, τ_1 is lag and m_1 is embedding dimension for X_1.

For multivariate time series,

$$X = [X_1, X_2, X_3, \ldots\ldots\ldots X_s] = \begin{bmatrix} x_{1,1} \ldots\ldots\ldots x_{s,i} \\ \cdot \\ \cdot \\ \cdot \\ \cdot \\ x_{1,N} \ldots\ldots\ldots\ldots x_{S,N} \end{bmatrix}$$

PSR computes the reconstruction for each time series:

$$X^{r}_{1} = (X^{r}_{1,i}, X^{r}_{2,i}, X^{r}_{3,i} \ldots\ldots\ldots\ldots X^{r}_{s,i})\ i = 1, 2, 3, \ldots\ldots\ldots\ldots N - (max\{m_i\} - 1)(max\{\tau i\})$$

s: measurement number.

Delay estimation

For delay, PSR allows average mutual information (AMI).
For reconstruction, time delay is first local lowest AMI.

AMI is computed as: $AMI(T) = \sum_{i=1}^{N} p(x_i, x_{i+T}) log_2 \left[\frac{p(x_i, x_{i+T})}{p(x_i)p(x_{i+T})} \right]$

T = 1 [for max. lag].

Embedding dimension (ED) estimation

ED for PSR is appraised utilizing the false nearest neighbour (FNN) algorithm. For I point at d dimension, points X^r_i and adjacent point $X^{r^*}_i$ in reconstructed phase space $\{X^r_i\}$, i = 1:N, are false neighbours if $\sqrt{\frac{R^2_i(d+1) - R^2_1(d)}{R^2_1(d)}}$ > distance threshold where, $R^2_1(d) = \|X^r_i - X^{r*}_i\|^2$ is distance metric, d is smallest value; P_{fnn} < *per cent false neighbours* and P_{fnn} is ratio of FNN points to total PSR quantity.

doi: 10.2166/h2oj.2020.005

Processing and preparation of data

Humidity, temperature and monthly average rainfall are composed of data from the meteorological department of India for a season of monsoon months (May to October) for 1990–2017. Data from

Table 2 | Input constraints for SVM, SVM-FFA model

Input	Model name (SVM)	Model name (SVM-FFA)
Q_{t-1}	SVM1	SVMI-FFA
Q_{t-1}, Q_{t-2}	SVM2	SVM2-FFA
Q_{t-1}, Q_{t-2}, Q_{t-3}	SVM3	SVM3-FFA
Q_{t-1}, Q_{t-2}, Q_{t-3}, Q_{t-4}	SVM4	SVM4-FFA
Q_{t-1}, Q_{t-2}, Q_{t-3}, Q_{t-4}, Q_{t-5}	SVM5	SVM5-FFA

where, Q_{t-1} is one-month lag runoff, Q_{t-2} is two-month lag runoff, Q_{t-3} is three-month lag runoff, Q_{t-4} is four-month lag runoff and Q_{t-5} is five-month lag runoff.

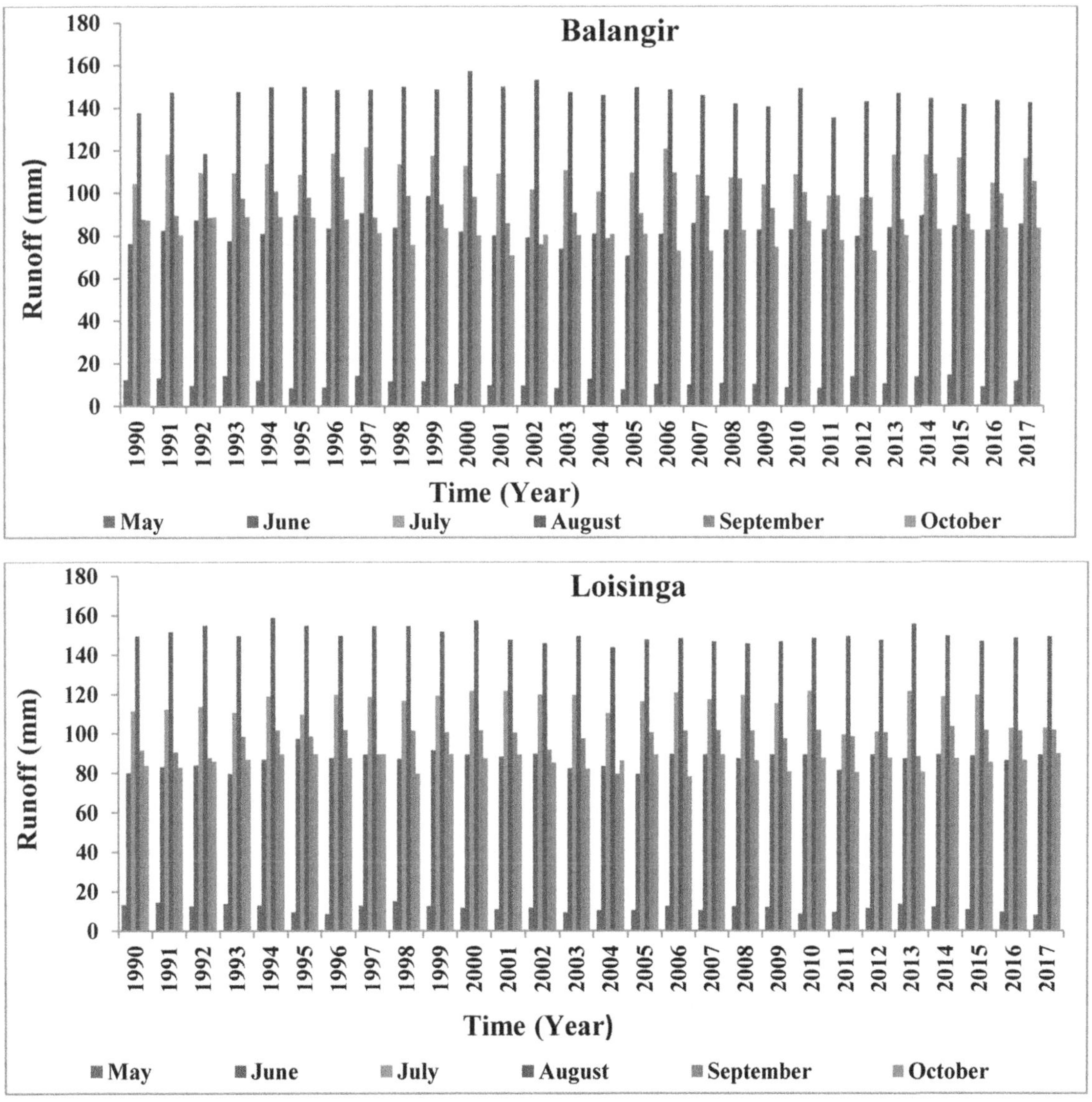

Figure 5 | Estimated mean monthly runoff of monsoon season for proposed watershed. *(Continued.)*

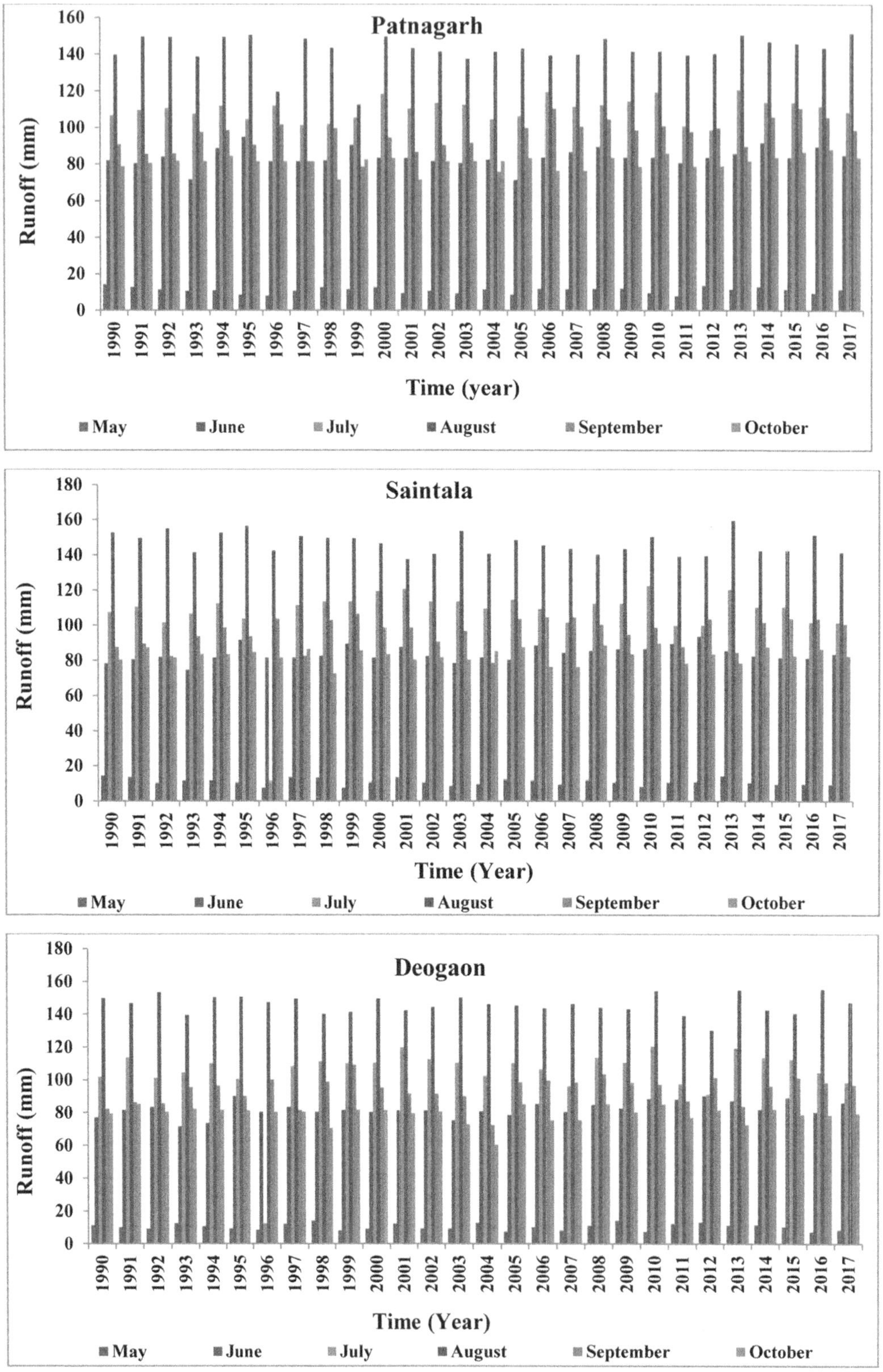

Figure 5 | Continued.

1990 to 2009 are employed for training and from 2010 to 2017 are considered for testing the model. Daily data are transformed into monthly data, which finally helps in training and testing the model. Subsequent runoff arrangements are employed as input (Table 2).

doi: 10.2166/h2oj.2020.005

Model performance evaluation

The indicators NSE, RMSE and WI are used to assess the performance of model efficiency. The formulae can be articulated as:

$$NSE = 1 - \left[\frac{\sum_{i=1}^{N}(O_i - P_i)^2}{\sum_{i=1}^{N}(O_i - \overline{O_i})^2}\right] - \infty < NS \leq 1 \tag{10}$$

$$RMSE = \sqrt{\frac{1}{N}\sum_{i=1}^{N}(P_i - O_i)^2} \tag{11}$$

$$WI = 1 - \left[\frac{\sum_{i=1}^{N}(O_i - P_i)^2}{\sum_{i=1}^{N}(|P_i - \overline{O_i}| + |O_i - \overline{O_i}|)^2}\right] \tag{12}$$

where, O_i and P_i are observed and predicted i^{th} runoff and $\overline{O_i}$ is mean of observed runoff.

RESULTS AND DISCUSSION

Runoff is estimated by using the Khosla formula for five watersheds in the monsoon period (May–October) for 1990–2017. A graphical representation of mean monthly runoff is presented in Figure 5.

The performances of the SVM model with different inputs for the five proposed watersheds are presented in Table 3. Three evaluating parameters, NSE, RMSE and WI, are estimated for both testing and training phases as explained below.

Table 3 | Results of SVM model for runoff prediction

		Training period			Testing period		
Positions	**Model**	**NSE**	**RMSE**	**WI**	**NSE**	**RMSE**	**WI**
Balangir	SVMI	0.513	91.93	0.841	0.486	88.93	0.849
	SVM2	0.548	87.17	0.857	0.509	85.48	0.864
	SVM3	0.597	84.86	0.874	0.552	81.39	0.881
	SVM4	0.642	80.05	0.869	0.601	77.91	0.886
	SVM5	0.681	76.98	0.935	**0.653**	**73.92**	**0.942**
Loisinga	SVMI	0.497	93.78	0.826	0.451	91.95	0.845
	SVM2	0.518	89.42	0.831	0.486	88.12	0.857
	SVM3	0.561	85.18	0.862	0.537	83.85	0.884
	SVM4	0.598	77.04	0.886	0.552	75.93	0.905
	SVM5	0.636	72.73	0.919	**0.601**	**70.21**	**0.925**
Patnagarh	SVMI	0.505	92.75	0.818	0.478	90.68	0.832
	SVM2	0.531	88.51	0.835	0.507	85.08	0.856
	SVM3	0.574	84.94	0.862	0.541	80.59	0.881
	SVM4	0.605	81.62	0.881	0.576	78.38	0.907
	SVM5	0.648	76.83	0.927	**0.613**	**73.41**	**0.931**
Saintala	SVMI	0.527	92.85	0.795	0.485	90.03	0.821
	SVM2	0.551	88.22	0.829	0.517	86.62	0.857
	SVM3	0.605	83.94	0.857	0.554	81.79	0.882
	SVM4	0.663	79.05	0.874	0.617	75.91	0.914
	SVM5	0.692	73.82	0.938	**0.658**	**69.73**	**0.947**
Deogaon	SVMI	0.518	92.65	0.805	0.496	90.52	0.831
	SVM2	0.547	88.91	0.826	0.501	86.01	0.852
	SVM3	0.582	85.57	0.853	0.549	82.59	0.886
	SVM4	0.621	81.48	0.882	0.584	78.71	0.918
	SVM5	0.683	77.73	0.945	**0.648**	**74.05**	**0.953**

Bold font indicates best value of model performance.

Among the five models, the SVM5 model shows the best result with WI 0.935 and 0.942 for training and testing period, respectively, while Q_{t-1}, Q_{t-2},Q_{t-3},Q_{t-4}, Q_{t-5} are considered as input parameters for Balangir watershed. For the remaining watersheds, model SVM5 gives best WI value for both phases out of five simulations. In the case of the Loisinga watershed, the pre-eminent WI values for training and testing periods are 0.919 and 0.925 for model SVM5. For the training phase at Patnagarh, Saintala and Deogaon watersheds, principal WI values are 0.927, 0.938 and 0.945, respectively. The result for the SVM-FFA model based on NSE, RMSE and WI for testing and training periods is given in Table 4.

Table 4 | Outcomes of SVM-FFA model for runoff prediction

		Training period			Testing period		
Positions	**Model**	**NSE**	**RMSE**	**WI**	**NSE**	**RMSE**	**WI**
Balangir	SVMI-FFA	0.694	89.71	0.846	0.673	86.48	0.859
	SVM2-FFA	0.717	86.04	0.862	0.698	82.71	0.864
	SVM3-FFA	0.759	80.54	0.899	0.736	79.03	0.903
	SVM4-FFA	0.782	76.82	0.926	0.758	75.66	0.941
	SVM5-FFA	0.815	74.34	0.965	**0.791**	**70.72**	**0.971**
Loisinga	SVMI-FFA	0.681	90.55	0.835	0.662	90.55	0.841
	SVM2-FFA	0.708	86.81	0.847	0.689	86.27	0.865
	SVM3-FFA	0.746	82.59	0.892	0.716	80.91	0.901
	SVM4-FFA	0.772	76.68	0.926	0.751	73.78	0.937
	SVM5-FFA	0.807	70.39	0.958	**0.783**	**68.49**	**0.961**
Patnagarh	SVMI-FFA	0.679	91.84	0.818	0.649	88.81	0.825
	SVM2-FFA	0.692	87.96	0.858	0.662	83.96	0.864
	SVM3-FFA	0.737	83.68	0.885	0.709	78.78	0.899
	SVM4-FFA	0.769	78.17	0.929	0.726	75.93	0.937
	SVM5-FFA	0.791	74.09	0.953	**0.754**	**70.07**	**0.962**
Saintala	SVMI-FFA	0.688	90.38	0.825	0.648	88.81	0.836
	SVM2-FFA	0.711	86.17	0.859	0.672	84.35	0.868
	SVM3-FFA	0.745	81.04	0.907	0.705	79.16	0.913
	SVM4-FFA	0.781	77.91	0.924	0.751	73.49	0.935
	SVM5-FFA	0.813	70.55	0.955	**0.793**	**67.88**	**0.966**
Deogaon	SVMI-FFA	0.693	90.01	0.848	0.654	89.28	0.855
	SVM2-FFA	0.713	85.63	0.869	0.689	84.96	0.885
	SVM3-FFA	0.759	83.86	0.914	0.723	80.81	0.929
	SVM4-FFA	0.792	79.47	0.939	0.751	76.67	0,946
	SVM5-FFA	0.828	74.28	0.961	**0.779**	**70.11**	**0.974**

Here five different models are used to estimate NSE, RMSE and WI value for five projected watersheds. When Q_{t-1}, Q_{t-2}, Q_{t-3}, Q_{t-4}, Q_{t-5} is used as input scenario, WI value provides the best result. Considering Balangir, the best values for WI are 0.965 and 0.971 for the training and testing periods, respectively. For the training phase, the WI values for Loisinga, Patnagarh, Saintala, Deogaon watershed are 0.958, 0.953, 0.955, 0.961, respectively. Similarly, the WI values of the testing phase are 0.968, 0.962, 0.966 and 0.974 for Loisinga, Patnagarh, Saintala and Deogaon watershed, respectively. Results of the PSR-SVM-FFA model for both testing and training phases are presented in Table 5.

Table 5 | Optimal parameters of τ, m for proposed watersheds and results of PSR-SVM-FFA model for runoff prediction

		Training period			Testing period		
Positions	**Optimal parameters of τ, m for PSR**	**NSE**	**RMSE**	**WI**	**NSE**	**RMSE**	**WI**
Balangir	$(m=14, \tau=5)$	0.905	43.74	0.984	0.874	42.27	0.985
Loisinga	$(m=14, \tau=5)$	0.938	39.18	0.970	0.908	37.96	0.967
Patnagarh	$(m=14, \tau=5)$	0.912	40.38	0.971	0.893	39.08	0.974
Saintala	$(m=14, \tau=5)$	0.897	37.66	0.972	0.869	36.11	0.976
Deogaon	$(m=14, \tau=5)$	0.923	41.82	0.982	0.897	40.35	0.988

doi: 10.2166/h2oj.2020.005

Assessment of results for recommended model

A comparison between the PSR-SVM-FFA, SVM-FFA and SVM models for testing of all projected watersheds is presented in Figures 6–8. The pre-eminent WI values for SVM, SVM-FFA and PSR-SVM-FFA models are 0.935, 0.965 and 0.984, correspondingly, for Balangir watershed. Similarly, for Loisinga station, the paramount WI value is 0.919, 0.958 and 0.9702 for SVM, SVM-FFA and

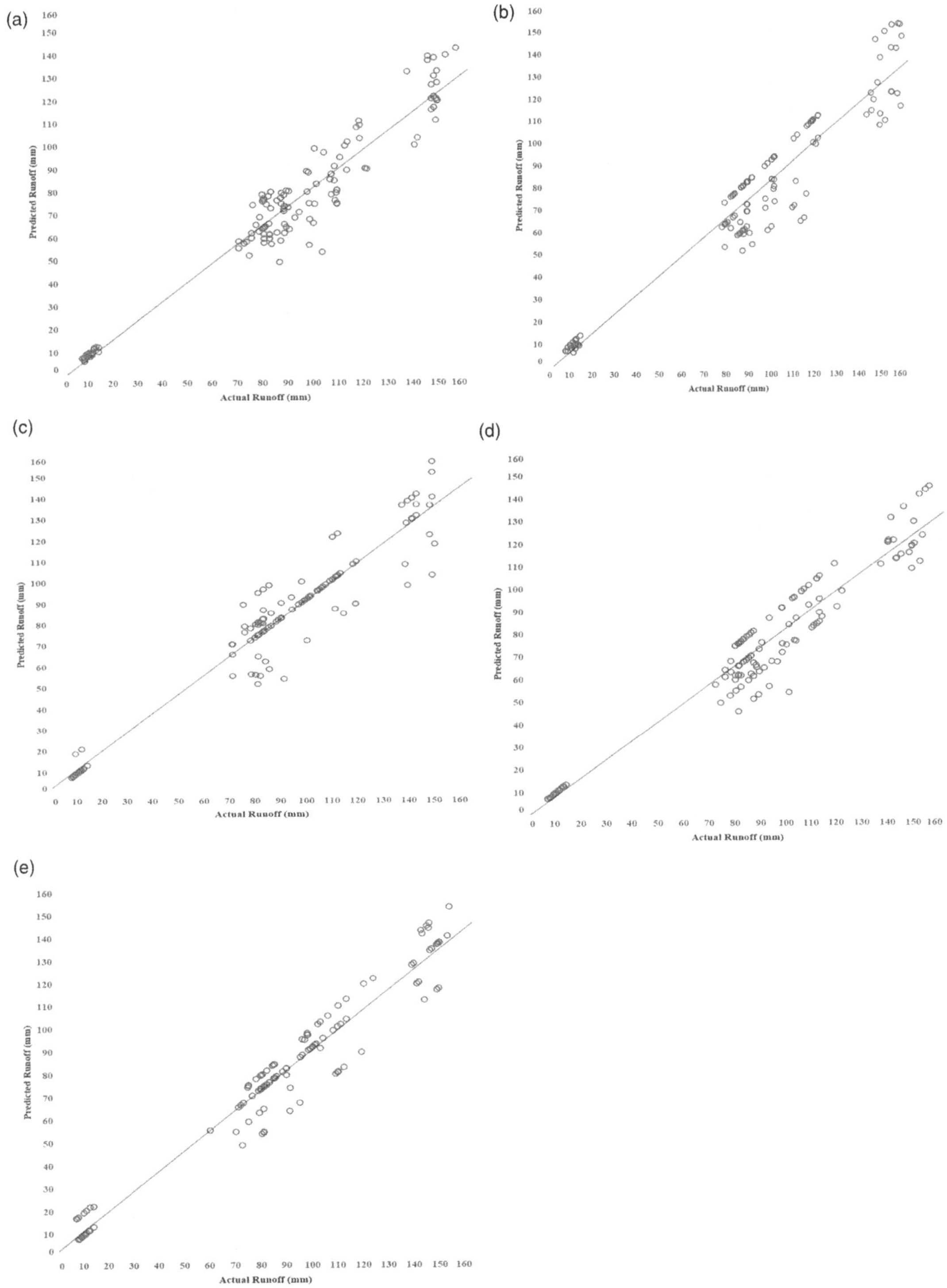

Figure 6 | Best fit lines by SVM model at (a) Balangir, (b) Loisinga, (c) Patnagarh, (d) Saintala and (e) Deogaon watersheds during training phase.

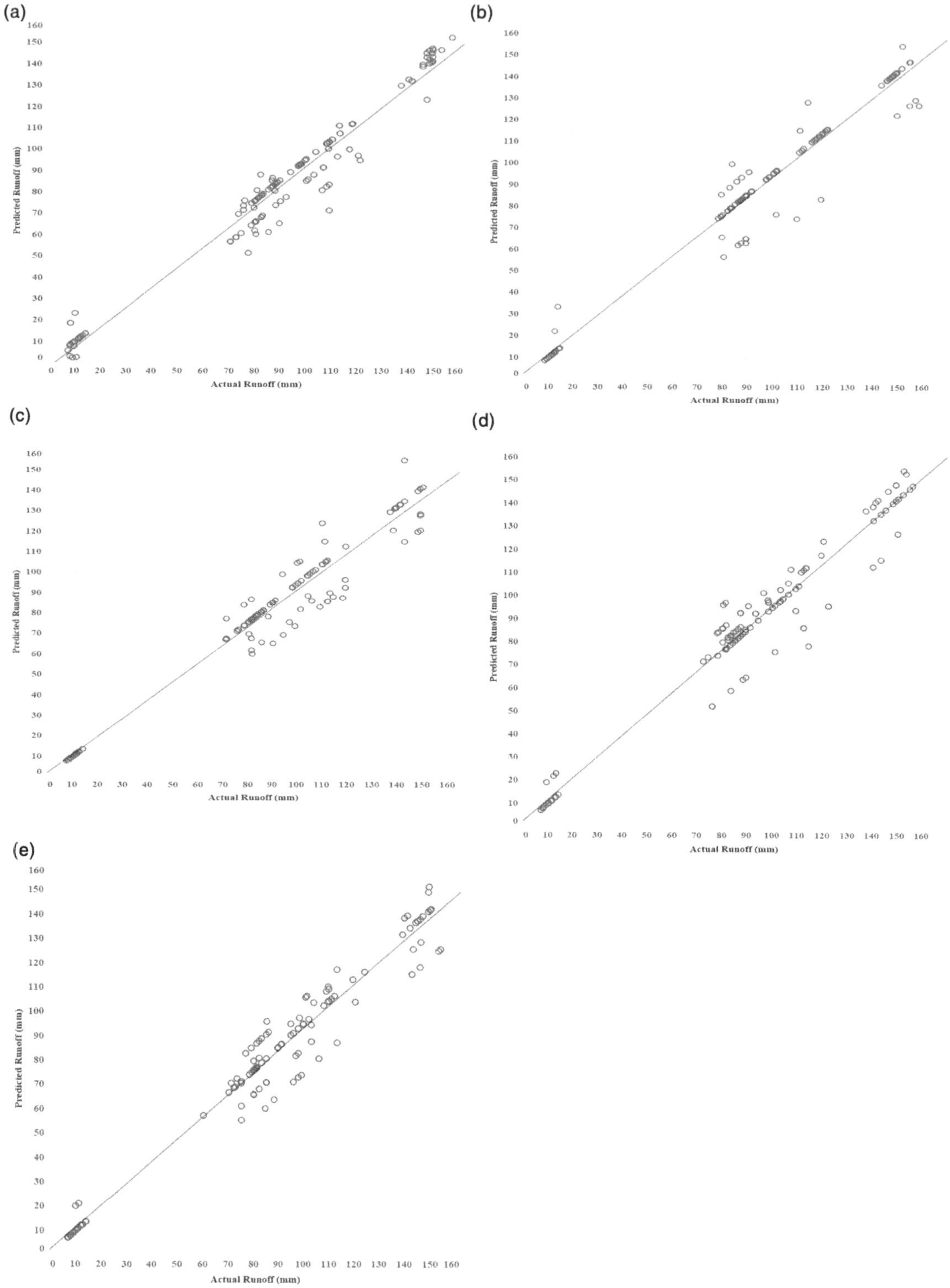

Figure 7 | Best fit lines by SVM-FFA model at (a) Balangir, (b) Loisinga, (c) Patnagarh, (d) Saintala and (e) Deogaon watersheds during training phase.

PSR-SVM-FFA models, respectively. Patnagarh, Saintala and Deogaon illustrate values for the SVM model of 0.9276, 0.9386 and 0.9459, respectively, during the testing phase. For the SVM-FFA model, the excellent values of WI are 0.953, 0.955 and 0.961 with respect to Patnagarh, Saintala and Deogaon. Similarly, the best values of the PSR-SVM-FFA model are 0.971, 0.972 and 0.982 for Patnagarh, Saintala and Deogaon, respectively.

doi: 10.2166/h2oj.2020.005

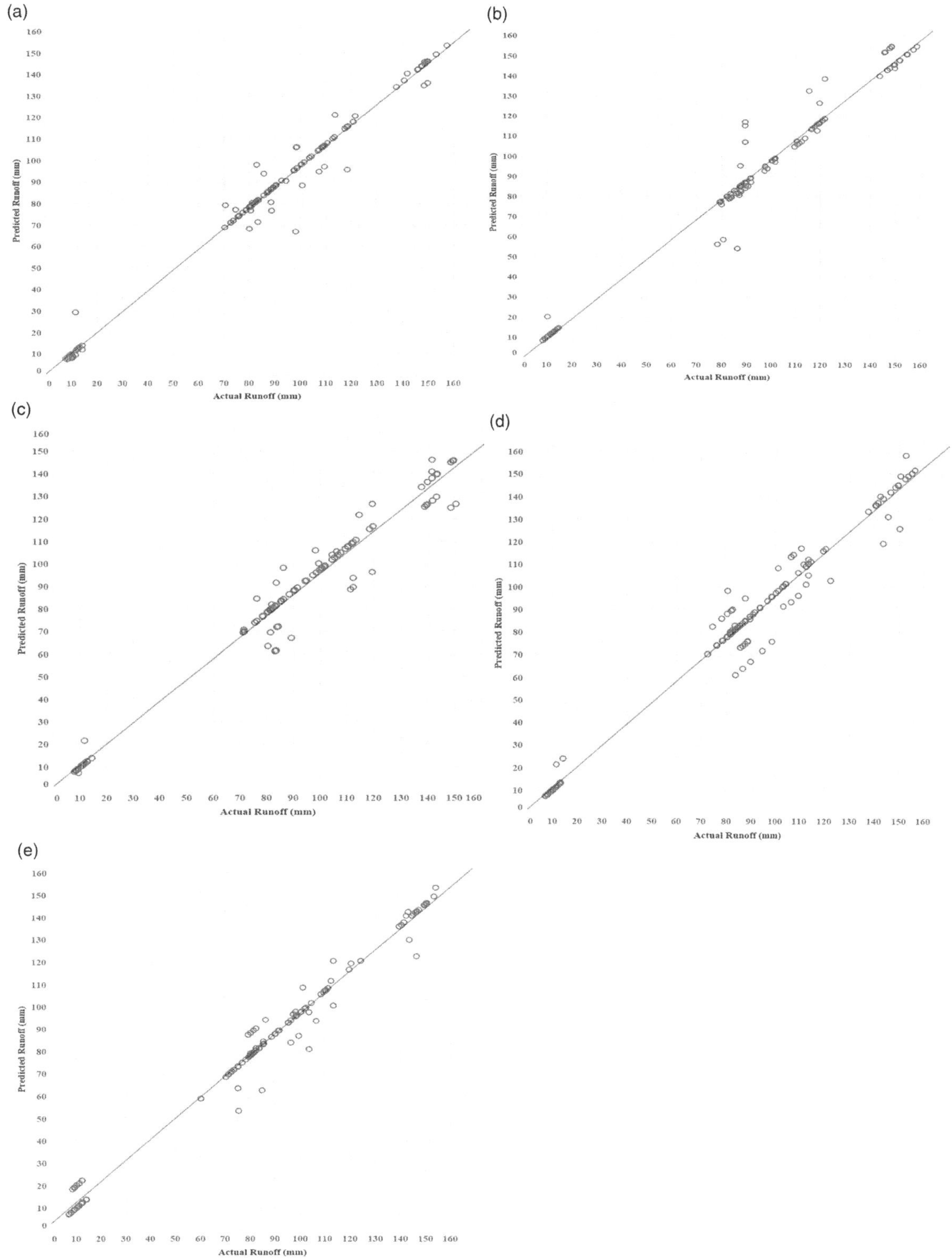

Figure 8 | Best fit lines by PSR-SVM-FFA model at (a) Balangir, (b) Loisinga, (c) Patnagarh, (d) Saintala and (e) Deogaon watersheds during training phase.

The linear scale plot of actual versus predicted monthly runoff for proposed model of projected area is shown in Figure 9. Results illustrate that estimated peak runoff is 141.208 mm, 142.313 mm and 149.272 mm for SVM, SVM-FFA and PSR-SVM-FFA against actual peak 157.22 mm for the Balangir watershed. The approximated peak runoffs are 126.4107 mm, 147.091 mm and 154.655 mm for SVM,

doi: 10.2166/h2oj.2020.005

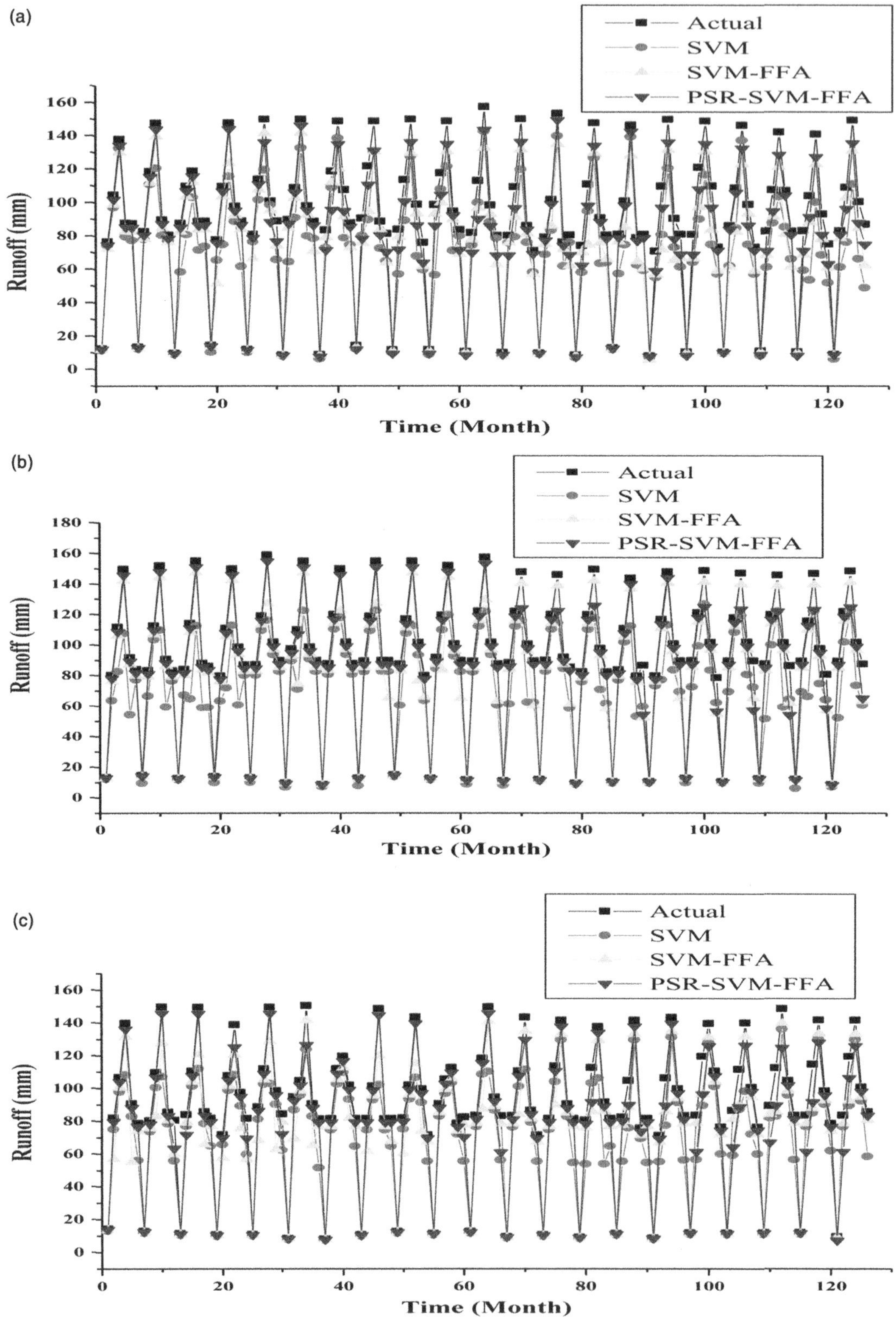

Figure 9 | Variation of measured runoff and predicted runoff at (a) Balangir, (b) Loisinga, (c) Patnagarh, (d) Saintala and (e) Deogaon watersheds during training phase. *(Continued.)*

SVM-FFA and PSR-SVM-FFA adjacent to the actual peak 158.8 mm for Loisinga division. For Patnagarh gauging station, tangible runoff is 150.43 mm aligned with predicted runoff of 136.032 mm, 141.329 mm and 145.744 mm for SVM, SVM-FFA and PSR-SVM-FFA, respectively. Correspondingly,

doi: 10.2166/h2oj.2020.005

(d)

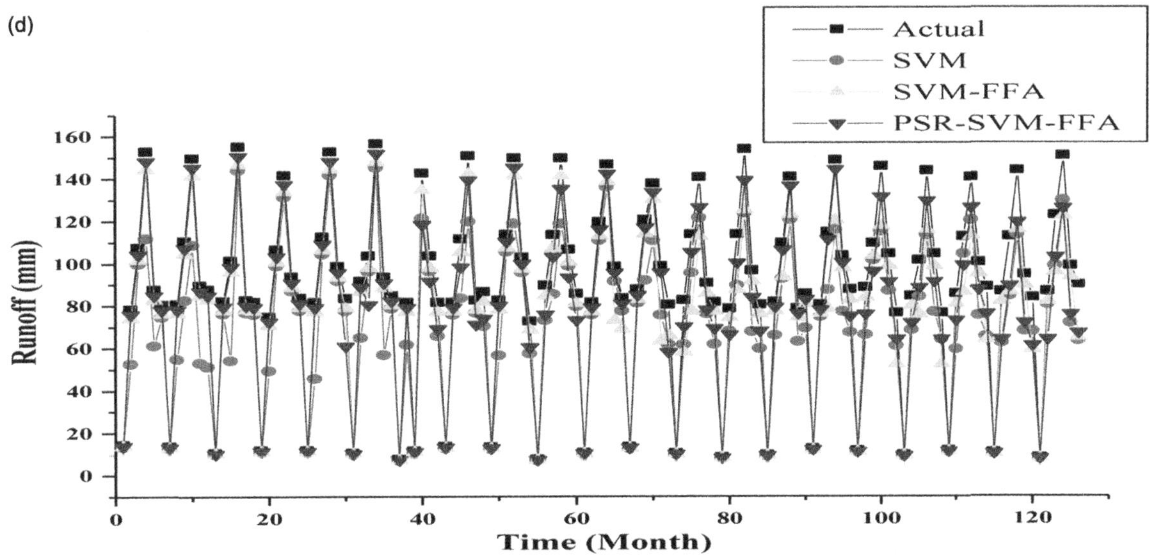

(e)

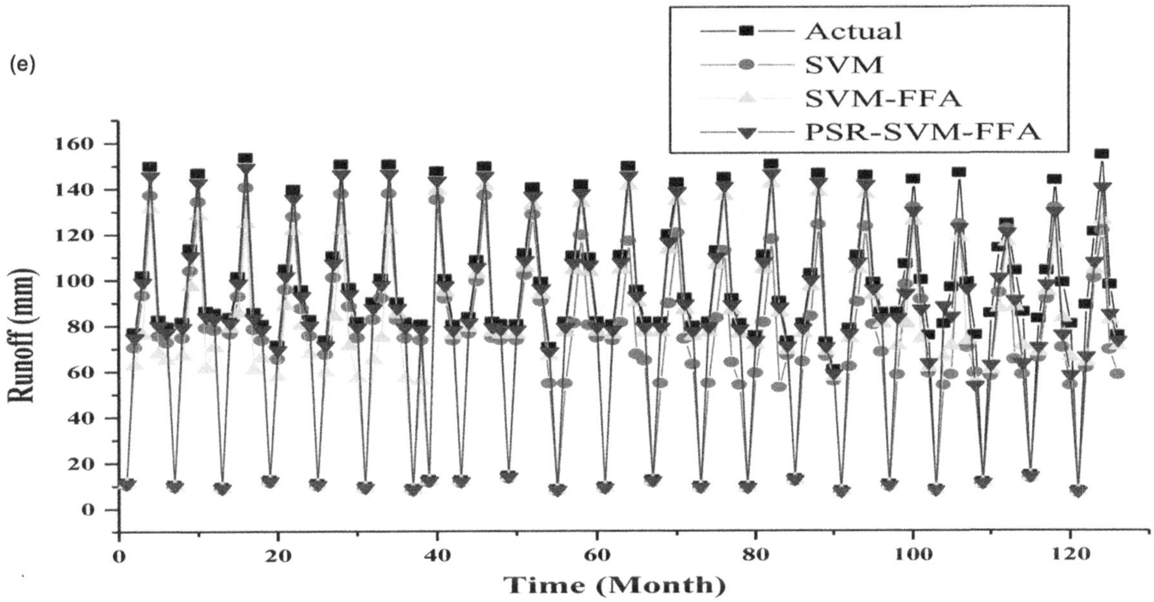

Figure 9 | Continued.

for Saintala watershed, the observed maximum runoff is 156.34 mm associated with estimated runoff of 151.728 mm, 147.522 mm and 144.911 mm for PSR-SVM-FFA, SVM-FFA and SVM, respectively. Similarly, evaluated peak runoffs are 149.322 mm, 141.591 mm and 140.376 mm, for PSR-SVM-FFA, SVM-FFA and SVM models with respect to genuine runoff of 154.23 mm for Deogaon watershed.

Comparison of best results

The SVM, SVM-FFA and PSR-SVM-FFA models are used to evaluate the performance of NSE, RMSE and WI indictors for five gauging watersheds. Assessments of performance indicators are specified in Table 6, which illustrates the efficiency of each model. Evaluating runoff is important and so the methods used here are significant for demonstrating runoff information. Therefore, calculation of RMSE, WI and NSE values are essential for predicting runoff. It is apparent that the PSR-SVM-FFA model is executed well, comparably to SVM-FFA and SVM. Accuracy of models is evaluated and assessed.

Assessment of each model is demonstrated in Figure 10 in terms of linear bar scheme.

Table 6 | Performance indices of different models for runoff forecasting

	SVM			SVM-FFA			PSR-SVM-FFA		
Stations	**NSE**	**RMSE**	**WI**	**NSE**	**RMSE**	**WI**	**NSE**	**RMSE**	**WI**
Balangir	0.653	73.92	0.942	0.791	70.72	0.971	0.874	42.27	0.985
Loisinga	0.601	70.21	0.925	0.783	68.49	0.961	0.908	37.96	0.967
Patnagarh	0.613	73.41	0.931	0.754	70.07	0.962	0.893	39.08	0.974
Saintala	0.658	69.73	0.947	0.793	67.88	0.966	0.869	36.11	0.976
Deogaon	0.648	74.05	0.953	0.779	70.11	0.974	0.897	40.35	0.988

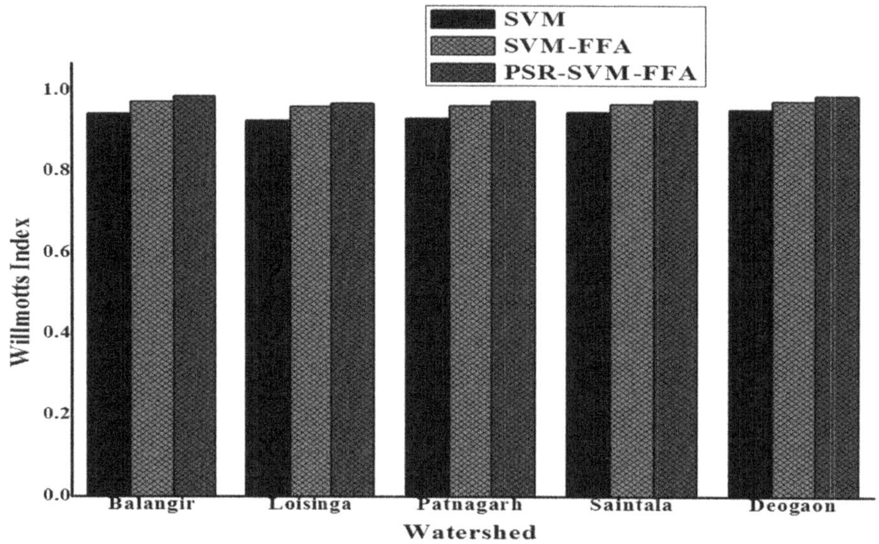

Figure 10 | Comparison of performance index for proposed watershed.

CONCLUSIONS

The efficiency of SVM, SVM-FFA and PSR-SVM-FFA approaches is explored for runoff prediction. Five gauging stations in Bolangir district, India are used for execution over a period of 28 years. The performance accuracy of PSR-SVM-FFA models is demonstrated in contrast to SVM-FFA and SVM models. It is appraised with various statistical indicators, WI, NSE and RMSE parameters, to reveal the projected PSR-SVM-FFA model is superior to SVM-FFA and SVM. It is observed that there is a significant improvement in the performance of the selected hybrid model for predicting the accuracy of each watershed. It is observed that the development of PSR coupled with the hybrid SVM-FFA model is superior to SVM models because of comprehensive withdrawal of input arrangements to produce targeted values. The PSR-SVM-FFA model is established to be a suitable model for runoff forecasting in arid watersheds. The research carried out over five gauge watersheds will help in the widespread good performance of non-gauging arid watersheds to predict runoff in similar conditions. Also, the present research findings can be used for irrigation and water resources engineering to design hydraulic structures regarding agriculture.

REFERENCES

Abd-Elazim, S. M. & Ali, E. S. 2018 Load frequency controller design of a two-area system composing of PV grid and thermal generator via firefly algorithm. *Neural Computing and Applications* **30** (2), 607–616.

doi: 10.2166/h2oj.2020.005

Al-Shammari, E. T., Mohammadi, K., Keivani, A., Ab Hamid, S. H., Akib, S., Shamshirband, S. & Petkovi, D. 2016 Prediction of daily dew point temperature using a model combining the support vector machine with firefly algorithm. *Journal of Irrigation and Drainage Engineering* **142** (5), 1–9.

Baydaroglu, O. & Kocak, K. 2014 SVR-based prediction of evaporation combined with chaotic approach. *Journal of Hydrology* **508**, 356–363.

Cha, S., Sohania, S. K., Kumara, D., Malikc, A., Chaharc, B. R., Nemac, A. K., Panigrahi, B. K. & Dhima, R. 2014 A support vector machine-firefly algorithm based forecasting model to determine malaria transmission. *Neuro Computing* **129**, 279–288.

Chen, W., Pourghasemi, H. R. & Naghibi, S. A. 2018 A comparative study of landslide susceptibility maps produced using support vector machine with different kernel functions and entropy data mining models in China. *Bulletin of Engineering Geology and the Environment* **77** (2), 647–664.

Cortes, C. & Vapnik, V. 1995 Support-vector networks. *Machine Learning* **20** (3), 273–297.

Das, G. (ed.) 2009 Runoff. In: *Hydrology and Soil Conservation Engineering Including Watershed*, 2nd edn (G. Das, ed.). PHI Learning, New Delhi, India, pp. 79–98.

Delafrouz, H., Ghaheri, A. & Ghorbani, M. A. 2017 A novel hybrid neural network based on phase space reconstruction technique for daily river-flow prediction. *Soft Computing* **22** (7), 2205–2215.

Dutta, S., Singh, M. & Kumar, A. 2018 Classification of non-motor cognitive task in EEG based brain-computer interface using phase space features in multivariate empirical mode decomposition domain. *Biomedical Signal Processing and Control* **39**, 378–389.

Fan, G. F., Peng, L. L. & Hong, W. C. 2018 Short term load forecasting based on phase space reconstruction algorithm and bi-square kernel regression model. *Applied Energy* **224**, 13–33.

Gandomia, A. H., Yang, X. & Alavic, A. H. 2011 Mixed variable structural optimization using Firefly Algorithm. *Computers and Structures* **89**, 2325–2336.

Ghose, D. K. & Samantaray, S. 2019 Estimating runoff using feed-forward neural networks in scarce rainfall region. In: *Smart Intelligent Computing and Application* (Satapathy, S. C., Bhatija, V. & Das, S., eds). Smart Innovation, system and technologies, Vol. 104. Springer, Singapore, pp. 53–64.

Gocic, M., Shamshirband, S., Razak, Z., Petkovik, C. S. & Trajkovic, S. 2016 Long-term precipitation analysis and estimation of precipitation concentration index using three support vector machine method. *Advances in Meteorology*. **2016**, 7912357, 1–11.

Hajiloo, R., Salarieh, H. & Alasty, A. 2018 Chaos control in delayed phase space constructed by the Takens embedding theory. *Communications in Nonlinear Science and Numerical Simulation* **54**, 453–465.

Ji, Y. & Sun, S. 2013 Multitask multiclass support vector machines: model and experiments. *Pattern Organisation* **46**, 914–924.

Lieu, Q. X., Do, D. T. & Lee, J. 2018 An adaptive hybrid evolutionary firefly algorithm for shape and size optimization of truss structures with frequency constraints. *Computers & Structures* **195**, 99–112.

Mehr, A. D., Nourani, V., Khosrowshahi, V. K. & Ghorbani, M. A. 2018 A hybrid support vector regression–firefly model for monthly rainfall forecasting. *International Journal of Environmental Science and Technology* **16**, 335–346.

Moghaddam, T. B., Soltani, M., Shahraki, H. S., Shamshirband, S., Noor, N. B. M. & Karim, M. R. 2016 The use of SVM-FFA in estimating fatigue life of polyethylene terephthalate modified asphalt mixtures. *Measurements* **90**, 526–533.

Raghavendra, S. N. & Deka, P. C. 2014 Support vector machine applications in the field of hydrology: a review. *Applied Soft Computing* **19**, 372–386.

Shamshirband, S., Mohammadi, K., Tong, C. W., Zamani, M., Motamedi, S. & Ch, S. 2015 A hybrid SVM-FFA method for prediction of monthly mean of global solar radiation. *Theoretical and Applied Climatology* **125** (1–2), 53–65.

Sivakumar, B., Berndtsson, R. & Persson, M. 2001 Monthly runoff prediction using phase space reconstruction. *Hydrological Sciences Journal* **46** (3), 377–387.

Sun, W. & Wang, Y. 2018 Short-term wind speed forecasting based on fast ensemble empirical mode decomposition, phase space reconstruction, sample entropy and improved back-propagation neural network. *Energy Conversion and Management* **157**, 1–12.

Tabari, H., Kisi, O., Ezani, A. & Talaee, P. H. 2012 SVM, ANFIS, regression, and climate based model for reference evapotranspiration modelling using limited climatic data in semiarid highland environment. *Journal of Hydrology* **444–445**, 78–89.

Tao, H., Sulaiman, S. O., Yaseen, Z. M., Asadi, H., Meshram, S. G. & Ghorbani, M. A. 2018 What is the potential of integrating phase space reconstruction with SVM-FFA data-intelligence model? Application of rainfall forecasting over regional scale. *Water Resources Management* **32**, 3935–3959.

Tayfur, G. 2014 *Soft Computing in Water Resources Engineering: Artificial Neural Networks, Fuzzy Logic and Genetic Algorithms*. WIT Press, Southampton, UK.

Thanh Noi, P. & Kappas, M. 2018 Comparison of random forest, k-nearest neighbor, and support vector machine classifiers for land cover classification using Sentinel-2 imagery. *Sensors* **18** (1), 18.

Vignesh, R., Jothiprakash, V. & Sivakumar, B. 2015 Stream flow variability and classification using false neighbour method. *Journal of Hydrology* **531**, 706–715.

Wang, H., Wang, W., Cui, Z., Zhou, X., Zhao, J. & Li, Y. 2018a A new dynamic firefly algorithm for demand estimation of water resources. *Information Sciences* **438**, 95–106.

Wang, H., Zheng, B., Yoon, S. W. & Ko, H. S. 2018b A support vector machine-based ensemble algorithm for breast cancer diagnosis. *European Journal of Operational Research* **267** (2), 687–699.

Wang, H., Wang, W., Cui, L., Sun, H., Zhao, J., Wang, Y. & Xue, Y. 2018c A hybrid multi-objective firefly algorithm for big data optimization. *Applied Soft Computing* **69**, 806–815.

Yang, X. S. 2009 Firefly algorithms for multimodal optimization. In: *International Symposium on Stochastic Algorithms*. Springer, Berlin, Heidelberg, Germany, pp. 169–178.

Zaini, N., Malek, M. A., Yusoff, M., Mardi, N. H. & Norhisham, S. 2018 Daily river flow forecasting with hybrid support vector machine-particle swarm optimization. *IOP Conference Series: Earth and Environmental Science* **140**, 012035.

First received 9 January 2020; accepted in revised form 12 March 2020. Available online 6 August 2020

IWA Publishing's authorised EU representative for General Product Safety Regulations is Diane D'Arras, 15 rue Duret, 75116 Paris, France, e-mail: safety@iwap.co.uk.

Printed and bound by CPI Group (UK) Ltd, Croydon, CR0 4YY
17/09/2026
02218811-0019